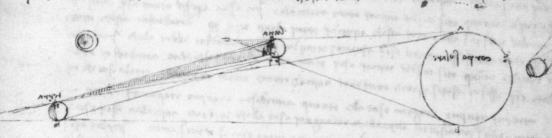

LEARNING FROM LEONARDO

LEARNING FROM
LEONARDO

Decoding the Notebooks of a Genius

Fritjof Capra

Berrett–Koehler Publishers, Inc.
San Francisco
a BK Currents book

Berrett-Koehler Publishers, Inc.
235 Montgomery Street, Suite 650
San Francisco, CA 94104-2916
Tel: (415) 288-0260   Fax: (415) 362-2512   www.bkconnection.com

Ordering Information

Quantity sales. Special discounts are available on quantity purchases by corporations, associations, and others. For details, contact the "Special Sales Department" at the Berrett-Koehler address above.

Individual sales. Berrett-Koehler publications are available through most bookstores. They can also be ordered directly from Berrett-Koehler: Tel: (800) 929-2929; Fax: (802) 864-7626; www.bkconnection.com

Orders for college textbook/course adoption use. Please contact Berrett-Koehler: Tel: (800) 929-2929; Fax: (802) 864-7626.

Orders by U.S. trade bookstores and wholesalers. Please contact Ingram Publisher Services, Tel: (800) 509-4887; Fax: (800) 838-1149; E-mail: customer.service@ingrampublisherservices.com; or visit www.ingrampublisherservices.com/Ordering for details about electronic ordering.

Berrett-Koehler and the BK logo are registered trademarks of Berrett-Koehler Publishers, Inc.

Printed in the United States of America

Berrett-Koehler books are printed on long-lasting acid-free paper. When it is available, we choose paper that has been manufactured by environmentally responsible processes. These may include using trees grown in sustainable forests, incorporating recycled paper, minimizing chlorine in bleaching, or recycling the energy produced at the paper mill.

PRODUCED BY WILSTED & TAYLOR PUBLISHING SERVICES
Copyediting Nancy Evans   *Design* Yvonne Tsang   *Index* Andrew Joron

LIBRARY OF CONGRESS CATALOGING-IN-PUBLICATION DATA
Capra, Fritjof, author.
  Learning from Leonardo : decoding the notebooks of a genius / Fritjof Capra. — First edition.
    pages cm. — (A BK currents book)
    ISBN 978-1-60994-989-1 (hardback)
  1. Leonardo, da Vinci, 1452–1519—Notebooks, sketchbooks, etc. 2. Discoveries in science.
3. Creative ability in science. 4. Science, Renaissance. I. Title.
  Q143.L5C368 2014
  509.2—dc23   2013032747

First Edition

18 17 16 15 14 13   10 9 8 7 6 5 4 3 2 1

JACKET IMAGES: Wing © Horacio Villalobos/epa/Corbis; Ms. B, folio 88v (detail).

ENDSHEETS: Codex Leicester, folios 35v and 2r (detail).

FRONTISPIECE: Profile of an old man (detail). Windsor Collection, *Landscapes, Plants, and Water Studies,* folio 48r.

FACING: Botanical specimen from "Star of Bethlehem," c. 1508 (detail, see plate 6).

PAGE VI: Studies of flexions of the spine in the movements of horses, cats, and dragons, c. 1508 (detail, see fig. 6-3).

To my brother Bernt
who has shared my fascination
with the genius of Leonardo da Vinci
from the very beginning

CONTENTS

PREFACE

In his classic *Lives of the Artists*, the Italian painter and architect Giorgio Vasari said of Leonardo da Vinci:

> His name became so famous that not only was he esteemed during his lifetime but his reputation endured and became even greater after his death.

Indeed, during the Renaissance Leonardo was renowned as an artist, engineer, and inventor throughout Italy, France, and other European countries. In the centuries after his death, his fame spread around the world, and it continues undiminished to this day.

I have been fascinated by the genius of Leonardo da Vinci for several decades and have spent the last ten years studying his scientific writings in facsimile editions of his famous Notebooks. My first book about him, *The Science of Leonardo*, published in 2007, is an introduction to his life and personality, his scientific method, and his synthesis of art and science. In this second book I go a step further, presenting an in-depth discussion of the main branches of Leonardo's scientific work from the perspective of twenty-first-century science—his fluid dynamics, geology, botany, mechanics, science of flight, and anatomy. Most of his astonishing discoveries and achievements in these fields are virtually unknown to the general public.

Leonardo da Vinci was what we would call, in today's scientific parlance, a systemic thinker. Understanding a phenomenon, for him, meant connecting it with other phenomena through a similarity of patterns. He usually worked on several projects in parallel, and when his understanding advanced in one area he would revise his ideas in related areas accordingly.

Thus, to appreciate the full extent of his genius, one needs to be aware of the evolution of his thinking in several parallel but interconnected disciplines. This has been my approach to absorbing and understanding Leonardo's scientific thought. Having explored and contributed to the systems view of life that has emerged in science in the last thirty years, and having written several books about it, I found it very natural to analyze and interpret Leonardo's science from that perspective. Indeed, I believe that the ever-present emphasis on relationships, patterns, qualities, and transformations in his writings, drawings, and paintings—the tell-tale sign of systemic thinking—was what initially attracted me to his work and kept me utterly fascinated for so many years.

What emerged from my explorations of all the branches of Leonardo's science and of his "demonstrations" (as he called them) in his drawings, paintings, and writings was the realization that, at the most fundamental level, Leonardo always sought to understand the nature of life. His science is a science of living forms, and his art served this persistent quest for life's inner secrets. In order to paint nature's living forms, Leonardo felt he needed a scientific understanding of their intrinsic nature and underlying principles; in order to analyze the results of his observations, he needed his artistic ability to depict them. I believe that this intersection of needs is the very essence of his synthesis of science and art.

Leonardo thought of himself not only as an artist and natural philosopher (as scientists were called in his time), but also as an inventor. In his view, an inventor was someone who created an artifact or work of art by assembling various elements into a new configuration that did not appear in nature. This definition comes very close to our modern notion of a designer, which did not exist in the Renaissance. Indeed, I have come to believe that the wide-ranging activities of Leonardo da Vinci, the archetypal Renaissance man, are best examined within the three categories of art, science, and design. In all three dimensions he uses living nature as his mentor and model. In fact, as I delved into the Notebooks, I discovered not only Leonardo the systemic thinker but also, to my great surprise, Leonardo the ecologist and ecodesigner.

The persistent endeavor to put life at the very center of his art, science, and design, and the recognition that all natural phenomena are fundamentally interconnected and interdependent, are important lessons we can learn from Leonardo today. Thus, Leonardo's synthesis is not only intellectually fascinating but also extremely relevant to our time, as I shall argue in the Coda of this book.

In previous decades, scholars of Leonardo's Notebooks tended to see them as disorganized and chaotic. My own sense, however, is that in Leonardo's mind, his science was not disorganized at all. In his manuscripts, we find numerous reminders to himself as to how he would eventually integrate the entire body of his research into a coherent whole. I have tried to follow these clues, arranging the material of this present book in a framework that I feel is consistent with Leonardo's thought. In fact, several of my chapter titles—"The Movements of Water," "The Elements of Mechanics," "The Human Figure"—are the ones Leonardo himself intended to use.

Leonardo's view of natural phenomena is based partly on traditional Aristotelian and medieval ideas and partly on his independent and meticulous observations of nature. The result is a unique science of living forms and their continual movements, changes, and transformations—a science that is radically different from that of Galileo, Descartes, and Newton.

A fundamental underlying idea is that nature as a whole is alive, and that the patterns and processes in the macrocosm of the Earth are similar to those in the microcosm of the human body. I have divided the contents of Leonardo's scientific work into these two basic categories: nature's forms and transformations in the macrocosm and in the microcosm. They constitute Parts I and II of the present book.

In the macrocosm, the main themes of Leonardo's science are the movements of water and air (chapter 1), the geological forms and transformations of the living Earth (chapter 2), and the botanical diversity and growth patterns of plants (chapter 3). In the microcosm, his main focus was on the human body—its beauty and proportions (chapter 4), the mechanics of its movements (chapter 6), and how it compared to other animal bodies in motion, in particular the flight of birds (chapter 7).

Unlike Descartes, Leonardo did not see the body as a machine, but he clearly recognized that the anatomies of animals and humans involve mechanical functions that can be appreciated only with an understanding of the basic principles of mechanics. Consequently, he reminded himself to "arrange it in such a way that the [chapter] on the elements of mechanics with its practice shall precede the demonstration of the movement and force of man and other animals." I have followed Leonardo's advice. My chapter on "The Elements of Mechanics" (chapter 5) precedes that on "The Body in Motion" (chapter 6).

As I have mentioned, Leonardo's ultimate goal—in his science as well as his art—was to understand the nature of life. This persistent quest culminated in his anatomies of the heart and blood vessels and in the embryological studies he undertook in his old age. Leonardo's explorations of the mystery of life in the human body (chapter 8) are the final highlight of my analysis of his science.

To follow Leonardo's meandering mind as he moves swiftly between interrelated phenomena—for example, from patterns of turbulence in water to similar patterns in the flow of air, the flight of birds, and on to the nature of sound and the design of musical instruments—is not easy within the linear constraints of written language. I have tried to facilitate this task by including in my text a network of cross-references, as well as copious references to Leonardo's manuscripts and to the works of the foremost Leonardo scholars. In addition, I have compiled a short chronology of Leonardo's life and work (see p. 326), which shows how he was constantly involved in several simultaneous projects.

In this and in my previous book, I discuss more than one hundred scientific discoveries made by Leonardo da Vinci during the fifteenth and sixteenth centuries. In the following pages, I present a timeline of his fifty or so most important discoveries, together with indications of the centuries when they were rediscovered by other scientists. This graphic summary is an impressive reminder of Leonardo's pioneering genius in so many scientific fields.

Leonardo did not publish any of his discoveries, nor do we have any records of written correspondence with the natural philosophers, mathematicians, engineers, doctors, and other intellectuals with whom he maintained regular contact. Although we can assume that he shared some of his insights and working methods in conversations with this circle, we have no evidence of any direct influence of his scientific achievements on subsequent generations of scientists.

Today, as we are developing a new systemic understanding of life with a strong emphasis on complexity, networks, and patterns of organization, we are witnessing the gradual emergence of a science of qualities that has some striking similarities with Leonardo's science of living forms. We cannot help but wonder how Western science might have developed had

Leonardo's Notebooks been studied by the founders of the Scientific Revolution in the seventeenth century.

From their correspondence it is evident that Galileo, Newton, and their contemporaries struggled with many of the same problems that Leonardo had recognized and often solved one or two centuries earlier. Moreover, they used similar metaphors and reasoned in similar ways, so they would have understood his Notebooks much better than we do today. If they had been aware of his discoveries, the development of science would doubtless have taken a very different path, and Leonardo da Vinci's influence on scientific thought might have been as profound as his impact on the history of art.

Fritjof Capra
Berkeley
February 2013

TIMELINE OF SCIENTIFIC DISCOVERIES

The following chart lists the most important scientific discoveries made by Leonardo da Vinci during the fifteenth and sixteenth centuries, together with the approximate dates when they were rediscovered by other scientists. It also includes references to the pages of this book (in parentheses) where the discoveries are discussed, as well as corresponding page references [in brackets] to my previous book, *The Science of Leonardo*.

Discovered by
Leonardo between
1485 and 1515

OPTICS

Wave nature of light
[p. 231]

Rayleigh scattering
(why the sky is blue)
[p. 233]

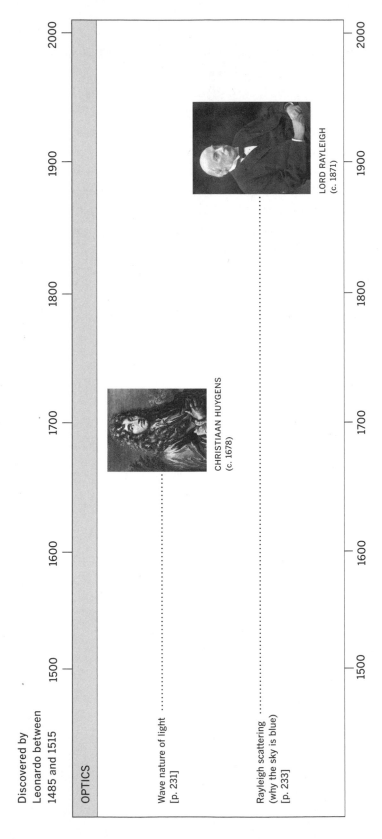

CHRISTIAAN HUYGENS
(c. 1678)

LORD RAYLEIGH
(c. 1871)

1500   1600   1700   1800   1900   2000

1500   1600   1700   1800   1900   2000

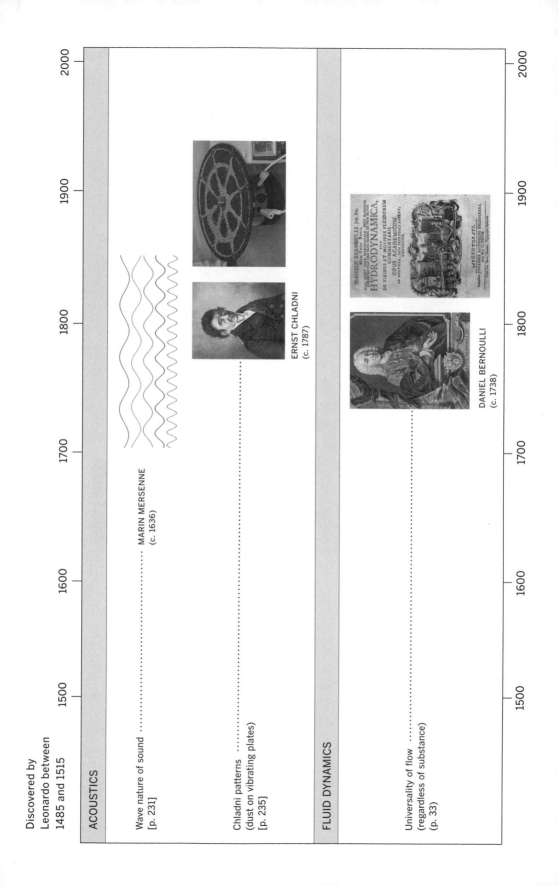

Discovered by
Leonardo between
1485 and 1515

ACOUSTICS

Wave nature of sound ·········
[p. 231]

Chladni patterns ·········
(dust on vibrating plates)
[p. 235]

MARIN MERSENNE
(c. 1636)

ERNST CHLADNI
(c. 1787)

FLUID DYNAMICS

Universality of flow ·········
(regardless of substance)
(p. 33)

DANIEL BERNOULLI
(c. 1738)

1500   1600   1700   1800   1900   2000

1500   1600   1700   1800   1900   2000

Discovered by
Leonardo between
1485 and 1515

FLUID DYNAMICS

Flow visualizations
(millet grains, dye, etc.)
(p. 35)

Viscosity
(p. 42)

Continuity principle
(conservation of mass)
(p. 45)

OSBORNE REYNOLDS
(c. 1883)

ISAAC NEWTON
(c. 1687)

BENEDETTO CASTELLI
(c. 1628)

1500 1600 1700 1800 1900 2000

1500 1600 1700 1800 1900 2000

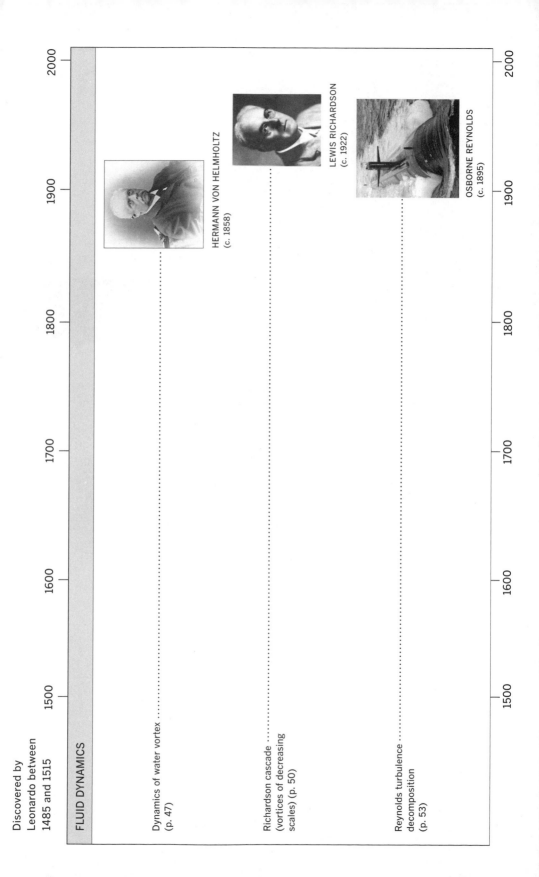

Discovered by
Leonardo between
1485 and 1515

FLUID DYNAMICS

Dynamics of water vortex
(p. 47)

HERMANN VON HELMHOLTZ
(c. 1858)

Richardson cascade
(vortices of decreasing
scales) (p. 50)

LEWIS RICHARDSON
(c. 1922)

Reynolds turbulence
decomposition
(p. 53)

OSBORNE REYNOLDS
(c. 1895)

1500 1600 1700 1800 1900 2000

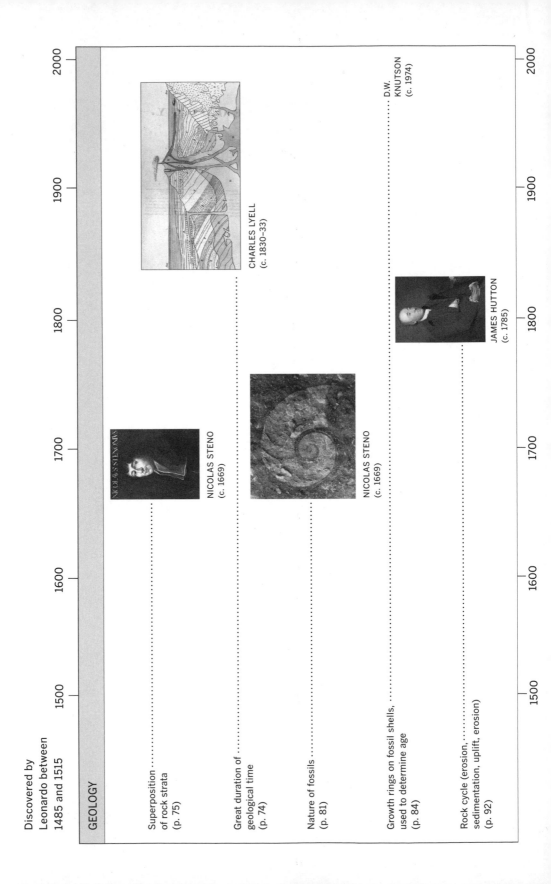

Discovered by
Leonardo between
1485 and 1515

GEOLOGY

Superposition
of rock strata
(p. 75)

NICOLAS STENO
(c. 1669)

Great duration of
geological time
(p. 74)

CHARLES LYELL
(c. 1830–33)

Nature of fossils
(p. 81)

NICOLAS STENO
(c. 1669)

Growth rings on fossil shells,
used to determine age
(p. 84)

D.W.
KNUTSON
(c. 1974)

Rock cycle (erosion,
sedimentation, uplift, erosion)
(p. 92)

JAMES HUTTON
(c. 1785)

1500   1600   1700   1800   1900   2000

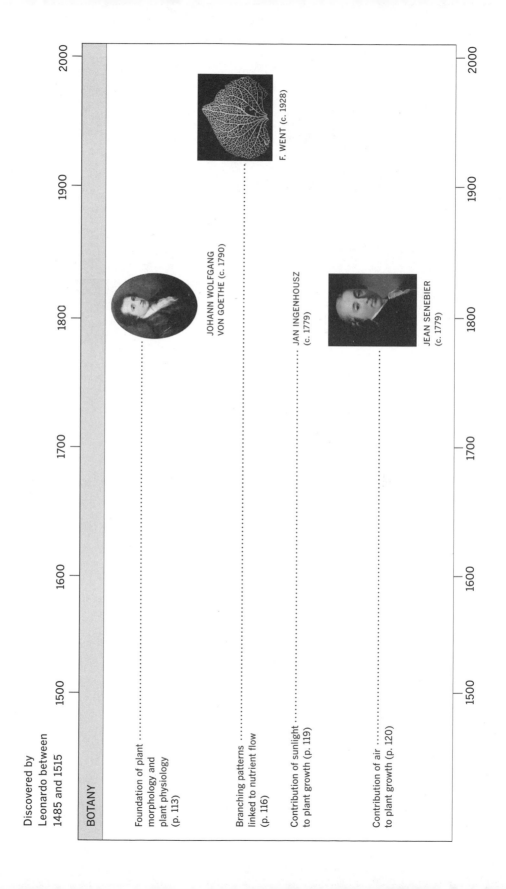

Discovered by
Leonardo between
1485 and 1515

BOTANY

1500   1600   1700   1800   1900   2000

Foundation of plant
morphology and
plant physiology
(p. 113)

JOHANN WOLFGANG
VON GOETHE (c. 1790)

Branching patterns
linked to nutrient flow
(p. 116)

F. WENT (c. 1928)

Contribution of sunlight
to plant growth (p. 119)

JAN INGENHOUSZ
(c. 1779)

Contribution of air
to plant growth (p. 120)

JEAN SENEBIER
(c. 1779)

BOTANY

2000 1900 1800 1700 1600 1500

Tropism (orientation of plants
in response to stimuli)
(p. 120)

CHARLES DARWIN
(c. 1880)

Regulation of plant growth
by hormones ("vital sap")
(p. 120)

P. BOYSEN-JENSEN
(c. 1913)

Migration of sap (auxins)
from light to dark side
of stem (p. 123)

F. WENT (c. 1928)

Annual growth rings,
linked to wet and dry years
(p. 121)

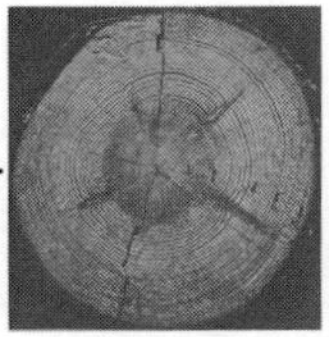

A.E. DOUGLAS
(c. 1937)

1500 1600 1700 1800 1900 2000

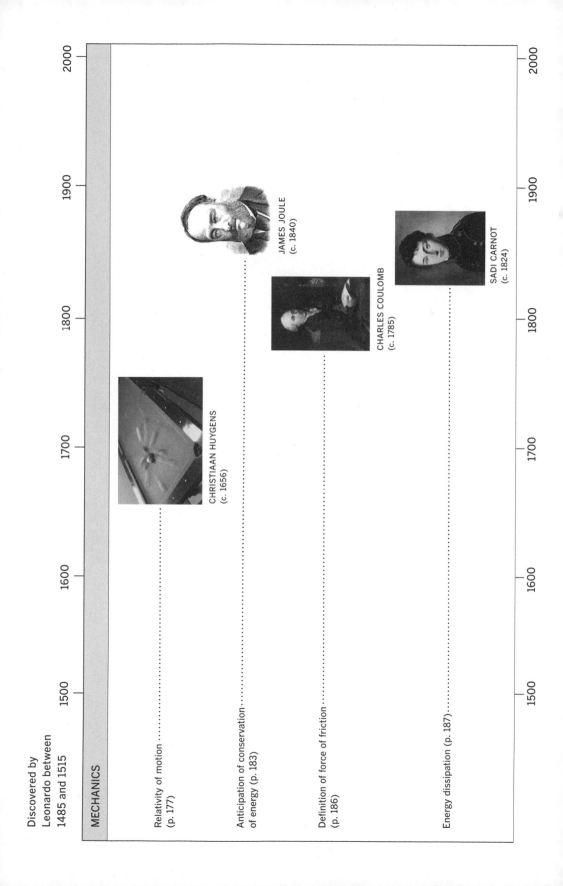

Discovered by
Leonardo between
1485 and 1515

MECHANICS

Relativity of motion
(p. 177)

CHRISTIAAN HUYGENS
(c. 1656)

Anticipation of conservation
of energy (p. 183)

JAMES JOULE
(c. 1840)

Definition of force of friction
(p. 186)

CHARLES COULOMB
(c. 1785)

Energy dissipation (p. 187)

SADI CARNOT
(c. 1824)

1500   1600   1700   1800   1900   2000

Discovered by
Leonardo between
1485 and 1515

1500 | 1600 | 1700 | 1800 | 1900 | 2000

MECHANICS

Anticipation of "arrow of time" (irreversible processes) (p. 188) ....

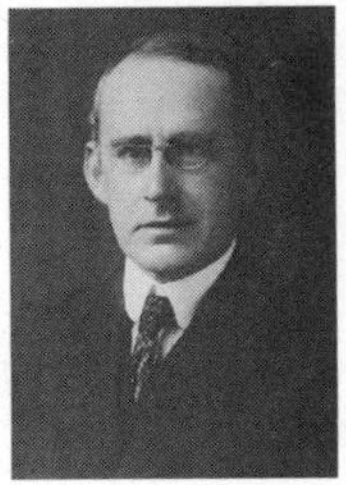

ARTHUR EDDINGTON
(c. 1928)

Parabolic nature of ballistic trajectories (p. 196) ....

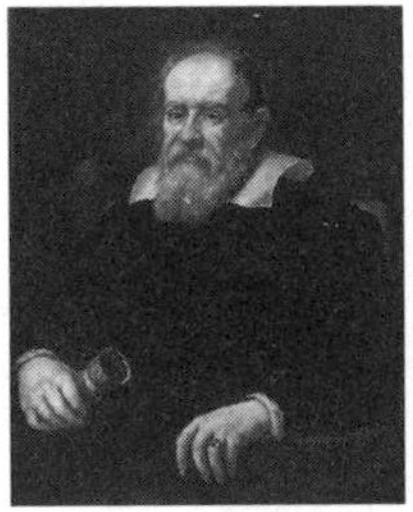

GALILEO GALILEI
(c. 1609)

Newton's third law of motion: action = reaction (p. 200) ....

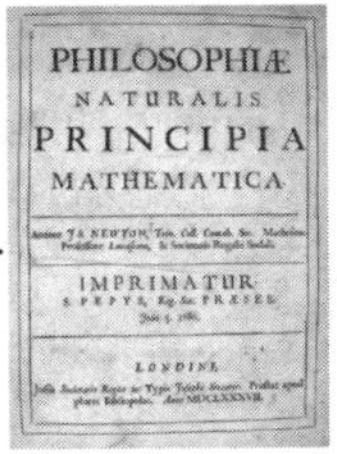

PHILOSOPHIÆ
NATURALIS
PRINCIPIA
MATHEMATICA

IMPRIMATUR

LONDINI

ISAAC NEWTON
(c. 1687)

1500 | 1600 | 1700 | 1800 | 1900 | 2000

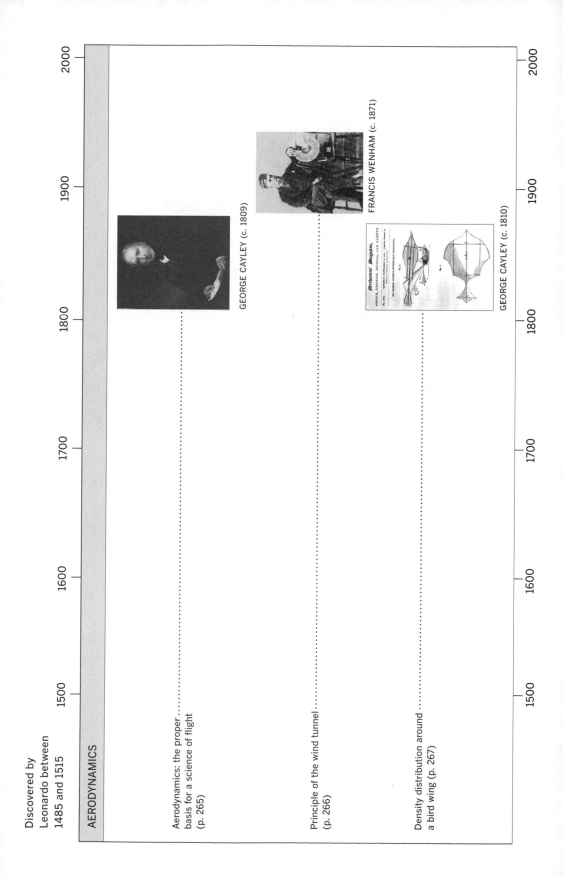

Discovered by
Leonardo between
1485 and 1515

AERODYNAMICS

Aerodynamics: the proper
basis for a science of flight
(p. 265)

GEORGE CAYLEY (c. 1809)

FRANCIS WENHAM (c. 1871)

Principle of the wind tunnel
(p. 266)

Density distribution around
a bird wing (p. 267)

GEORGE CAYLEY (c. 1810)

1500   1600   1700   1800   1900   2000

1500   1600   1700   1800   1900   2000

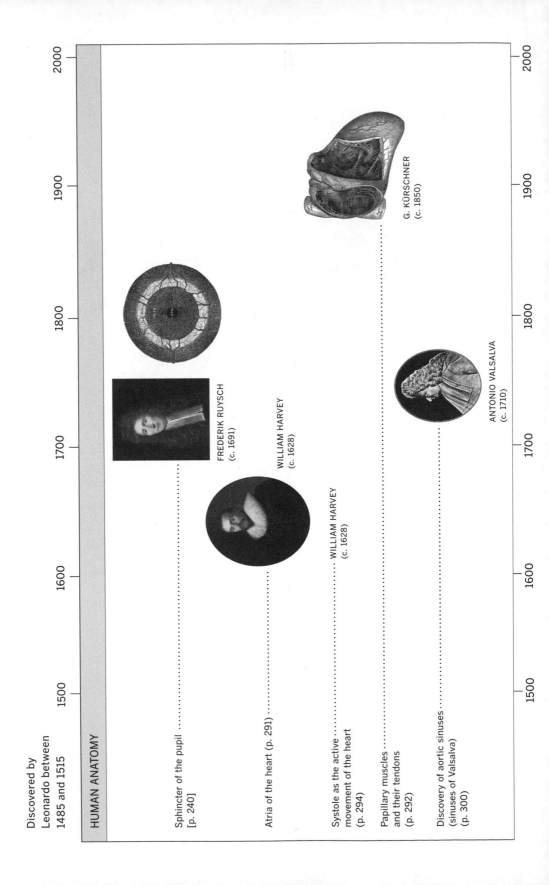

Discovered by
Leonardo between
1485 and 1515

HUMAN ANATOMY

2000   1900   1800   1700   1600   1500

G. KÜRSCHNER
(c. 1850)

FREDERIK RUYSCH
(c. 1691)

ANTONIO VALSALVA
(c. 1710)

WILLIAM HARVEY
(c. 1628)

WILLIAM HARVEY
(c. 1628)

Sphincter of the pupil ....
[p. 240]

Atria of the heart (p. 291) ....

Systole as the active ....
movement of the heart
(p. 294)

Papillary muscles ....
and their tendons
(p. 292)

Discovery of aortic sinuses ....
(sinuses of Valsalva)
(p. 300)

2000   1900   1800   1700   1600   1500

Discovered by
Leonardo between
1485 and 1515

HUMAN ANATOMY

	1500	1600	1700	1800	1900	2000

Gradual closure of aortic
valve by blood turbulence
(p. 301) .... C. LEE AND
L. TALBOT
(c. 1979)

Discovery and explanation
of arteriosclerosis (p. 307) .... JEAN LOBSTEIN
(c. 1833)

Definition of living organisms
as open systems (p. 311) .... LUDWIG VON
BERTALANFFY
(c. 1940)

Quantitative observations
of fetal growth (p. 316) .... CHARLES MINOT
(c. 1892)

Development of the embryo's
mental life (p. 318) .... HUMBERTO
MATURANA
(c. 1970)

ECOLOGY

	1500	1600	1700	1800	1900	2000

Anticipation of food chains
and food cycles (p. 282) .... CHARLES ELTON
(c. 1927)

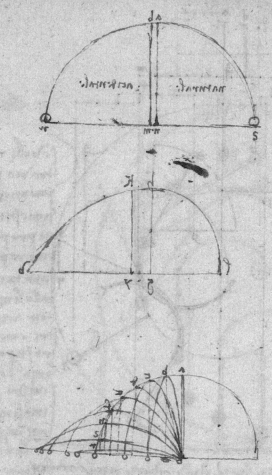

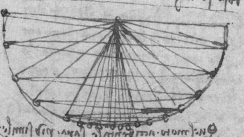

PROLOGUE

Leonardo's Genius

Before entering into the details of Leonardo's science, let us examine what is commonly referred to as his genius.

During Leonardo's time, the term "genius" did not have our modern meaning of a person endowed with extraordinary intellectual and creative powers.[1] The Latin word *genius* originated in Roman religion, where it denoted the spirit of the *gens*, the family. It was understood as a guardian spirit, first associated with individuals and then also with peoples and places. The extraordinary achievements of artists or scientists were attributed to their genius, or attendant spirit.

This meaning of genius was prevalent throughout the Middle Ages and the Renaissance. In the eighteenth century, the meaning of the word changed to its familiar modern meaning, denoting these individuals themselves, as in the phrase "Newton was a genius."

Regardless of the term used, the fact that certain individuals possess exceptional and inexplicable creative powers beyond the reach of ordinary mortals has been recognized throughout the ages. It has often been associated with divine inspiration, attributed first to poets and later on also to painters and other artists. In the Italian Renaissance, those individuals were given the epithet *divino*. Among the Renaissance masters, Leonardo as well as his younger contemporaries Raphael and Michelangelo were acclaimed as divine.

Since the development of modern psychology, neuroscience, and genetics, there has been a lively discussion about the origins, mental characteristics, and genetic makeup of geniuses. However, numerous studies of well-known historical figures have shown a bewildering diversity of hereditary, psychological, and cultural factors, defying all attempts to estab-

FACING Studies of the motion of a pendulum.
Codex Madrid I, folio 147r (detail, see fig. 5-14).

1

lish some common pattern.² While Mozart was a famous child prodigy, Einstein was a late bloomer. Newton attended a prestigious university, whereas Leonardo was essentially self-taught. Goethe's parents were well educated and of high social standing, but Shakespeare's seem to have been relatively undistinguished; and the list goes on.

However, psychologists have been able to identify a set of mental attributes that, in addition to exceptional talent in a particular field, seem to be distinctive signs of genius.³ All these were characteristic of Leonardo to a very high degree. Identifying these signs of genius in the mind and working methods of Leonardo da Vinci is an exercise that can inspire our own lives, both as individuals and as a society.

Relentless Curiosity and Intellectual Fearlessness

The first distinctive characteristic of a genius is an intense curiosity and great enthusiasm for discovery and understanding. This was indeed an outstanding quality of Leonardo, whom art historian Kenneth Clark called "the most relentlessly curious man in history."⁴ Throughout Leonardo's life, this boundless curiosity was his main driving force. Wherever he looked, there were new discoveries to be made, and for forty years he explored almost the entire range of natural phenomena known in his time, as well as many others previously unknown.

This curiosity was matched by incredible mental energy, so much so that following the trains of thought in Leonardo's Notebooks can be quite exhausting.⁵ As I did so over the years, I was struck again and again by the fact that he never seemed to have the slightest hesitation about entering into new fields of knowledge. In the chapter on geology (chapter 2), this is illustrated in some detail with Leonardo's extensive research on fossils. I offer it here as an example of his intellectual fearlessness.

Marine fossils represented an enigma to Leonardo that natural philosophers had debated intensely since antiquity. If fossil shells were remnants of marine organisms, how did they end up in sedimentary strata that lie in the high mountains? Leonardo studied a wide variety of fossils with the utmost care, precisely described their specific sites, and reconstructed the process of fossilization in remarkable detail. He also studied the classical texts and then set out to refute the theories current in his time, the most popular being that the fossil shells had been carried to the mountains from the sea by the biblical flood.

Based on highly sophisticated observations, Leonardo presented several brilliant arguments that invalidated this and other theories involving supernatural forces and showed convincingly that the fossils found in mountain rocks had been formed in the oceans where these creatures had lived in the distant past. Having done so, however, he still had to show how those layers of marine sediments ended up in the high mountains. In other words, he needed to posit a theory of how mountains were formed during extremely long periods of geological time.

Leonardo did not hesitate to take on this formidable challenge. Again he studied the principal classical and medieval texts, this time on the formation of the Earth, and he used some of their key ideas to formulate his own tectonic theory—an elaborate blend of Aristotelian and medieval ideas, combined with his own observations and with astonishing conceptions that are not unlike those of our modern plate tectonics.

In all these endeavors, Leonardo attempted to explain the phenomena he investigated in terms of natural processes. He scoffed at any belief in supernatural forces, repeatedly referred to nature (instead of God) as the source of all creation, and held a firm belief that nature's creations could be understood rationally, while also acknowledging the limitations of the human mind.

Intense Concentration and Attention to Detail

Another striking sign of genius is an extraordinary capacity for intense concentration over long periods of time. Isaac Newton apparently was able to hold a mathematical problem in his mind for weeks until it surrendered to his mental powers. When asked how he made his remarkable discoveries, Newton is reported to have replied, "I keep the subject constantly before me and wait until the first dawnings open little by little into the full light."[6] Leonardo seems to have worked in a very similar way, most of the time not on just one but on several problems simultaneously.

Leonardo combined his powers of concentration with tremendous patience. He might let weeks pass between putting successive layers of paint on an oil painting, and would rework and refine his panels for years, reflecting on every detail of their conception, engaging with himself in what he called a "mental discourse" (*discorso mentale*). He showed the same patience and attention to detail in his scientific observations and experiments.

Holistic Memory

Closely associated with the power of intense concentration that is characteristic of geniuses seems to be their exceptional holistic memory—an ability to memorize large amounts of information in the form of a coherent whole, a single gestalt. Goethe is said to have entertained his fellow passengers on long coach journeys by reciting his novels to them, word for word, before committing them to paper. Mozart, as a child, wrote out a note-perfect score of a complex choral composition after hearing it only once. Leonardo would follow people with striking facial features for hours, memorize their appearance, and then draw them, reportedly with complete accuracy, when he was back in his studio.

We have a vivid testimony of Leonardo's exceptional powers of concentration, his great patience, and his holistic memory from a contemporary writer, Matteo Bandello, who described how, as a boy, he watched the artist paint *The Last Supper*. He would see the master arrive early in the morning, climb up onto the scaffolding, and immediately start to work:

> He sometimes stayed there from dawn to sundown, never putting down his brush, forgetting to eat and drink, painting without pause. He would also sometimes remain two, three, or four days without touching his brush, although he spent several hours a day standing in front of the work, arms folded, examining and criticizing the figures to himself. I also saw him, driven by some sudden urge, at midday, when the sun was at its height, leaving the Corte Vecchia, where he was working on his marvelous clay horse, to come straight to Santa Maria delle Grazie, without seeking shade, and clamber up onto the scaffolding, pick up a brush, put in one or two strokes, and then go away again.[7]

The mental attributes discussed so far—relentless curiosity, intellectual fearlessness, a capacity for intense concentration, attention to detail, and holistic memory—are characteristics of genius that seem to be timeless, independent of historical and cultural contexts. In addition, Leonardo displayed signs of genius that can only be appreciated within the historical context of the Middle Ages and the Renaissance. Two of these in particular are defining characteristics of his scientific thought: his empirical method and his systemic thinking.

Leonardo's Empirical Method

In the mid-fifteenth century, when the young Leonardo received his training as a painter, sculptor, and engineer in Florence, science in the modern sense, as a systematic empirical method for gaining knowledge about the natural world, did not exist. The worldview of natural philosophy, as it was then called, had been handed down from Aristotle and other philosophers of antiquity and then fused with Christian doctrine by the Scholastic theologians who presented it as the officially authorized creed. The religious authorities condemned scientific experiments as subversive, seeing any attack on Aristotle's science as an attack on the Church. Leonardo da Vinci broke with this tradition:

> First I shall do some experiments before I proceed farther, because my intention is to cite experience first and then with reasoning show why such experience is bound to operate in such a way. And this is the true rule by which those who speculate about the effects of nature must proceed.[8]

One hundred years before Galileo Galilei and Francis Bacon, Leonardo single-handedly developed a new empirical approach to science, involving the systematic observation of nature, logical reasoning, and some mathematical formulations—the main characteristics of what is known today as the scientific method.[9] In the intellectual history of Europe, Galileo, born 112 years after Leonardo, is usually credited with being the first to develop this kind of rigorous empirical approach and is often hailed as the father of modern science. There can be no doubt that this honor would have been bestowed on Leonardo da Vinci had he published his scientific writings during his lifetime, or had his Notebooks been widely studied soon after his death.

The empirical approach came naturally to Leonardo. He was gifted with exceptional powers of observation, which were complemented by great drawing skills. He was able to draw the complex swirls of turbulent water or the swift movements of a bird in flight with a precision that would not be reached again until the invention of serial photography.

What turned Leonardo from an artist with exceptional gifts of observation into a scientist was his recognition that his observations, in order to be scientific, needed to be carried out in an organized, methodical fashion.

Scientific experiments are performed repeatedly and in varying circum-stances so as to eliminate accidental factors and technical flaws as much as possible. This is exactly what Leonardo did. He never tired of repeating his experiments and observations again and again, with fierce attention to the minutest detail, and he would often systematically vary his parameters to test the consistency of his results.

The systematic approach and careful attention to detail that Leonardo applied to his observations and experiments are characteristic of his en-tire method of scientific investigation. He would usually start from com-monly accepted concepts and explanations, often summarizing what he had gathered from the classical texts before proceeding to verify it with his own observations. After testing the traditional ideas repeatedly with careful observations and experiments, Leonardo would adhere to tradi-tion if he found no contradictory evidence; but if his observations told him otherwise he would not hesitate to formulate his own alternative explana-tions.

As I have mentioned, Leonardo generally worked on several problems simultaneously and paid special attention to similarities of patterns in dif-ferent areas of investigation. When he made progress in one area, he was always aware of the analogies and interconnecting patterns to phenom-ena in other areas, and would revise his theoretical ideas accordingly. This method led him to tackle many problems not just once but several times during different periods of his life, modifying his theories in successive steps as his scientific thought evolved over his lifetime.

Leonardo's practice of repeatedly reassessing his theoretical ideas in various areas meant that he never saw any of his explanations as final. Even though he believed in the certainty of scientific knowledge (as did most philosophers and scientists for the next three hundred years), his successive theoretical formulations in many fields are quite similar to the tentative theoretical models that are characteristic of modern science. For example (as discussed in chapter 8), he proposed several different mod-els for the functioning of the heart and its role in maintaining the flow of blood before he concluded that the heart is a muscle pumping blood through the arteries.

Leonardo also used simplified models—or approximations, as we would say today—to analyze the essential features of complex phenom-ena. For instance, he represented the flow of water through a channel of varying cross section by using a model of rows of men marching through a street of varying width (see chapter 1). This technique of using simplified theoretical models to understand complex phenomena put him centuries ahead of his time.

Like modern scientists, Leonardo was always ready to revise his models when he felt that new observations or insights required him to do so. In his art as in his science, he always seemed to be more interested in the process of exploration than in the completed work or final results. Thus many of his paintings and all of his science remained unfinished works in progress.

This is a general characteristic of the modern scientific method. Although scientists publish their work in various stages of completion in papers, monographs, and textbooks, science as a whole is always a work in progress. Old models and theories continue to be replaced by new ones, which are judged superior but are nevertheless limited and approximate, destined to be replaced in their turn.

Since the Scientific Revolution in the seventeenth century, this progress in science has been a collective enterprise. Scientists continually exchange letters, papers, and books and discuss their theories at various meetings and conferences. With Leonardo, the situation was quite different. He worked alone and in secrecy, did not publish any of his findings, and only rarely dated his notes. Having pioneered the scientific method in solitude, he did not see science as a collective, collaborative enterprise. Leonardo's secrecy about his scientific work is the one significant respect in which he was not a scientist in the modern sense.

Systemic Thinking

Throughout the history of Western science, there has been a basic conceptual tension between the parts and the whole. The emphasis on the parts has been called mechanistic, reductionist, or atomistic; the emphasis on the whole holistic, organismic, or ecological. In twentieth-century science, the holistic perspective has become known as "systemic" and the way of thinking it implies as "systemic thinking."

At the dawn of Western philosophy and science, Pythagoras distinguished "number," or pattern, from substance, or matter, viewing it as something that limits matter and gives it shape. Ever since the days of early Greek philosophy, there has been this tension between substance and pattern, between matter and form. The study of matter begins with the question, "What is it made of?" This leads to the notion of fundamental elements, building blocks that can measured and quantified. The study of form asks, "What is the pattern?" And that leads to the notions of order, organization, relationships. Instead of quantity, it involves quality; instead of measuring, it involves mapping.

These two very different lines of investigation have been in competition with one another throughout our scientific and philosophical tradition. The study of matter was championed by Democritus, Galileo, Descartes, and Newton; the study of form by Pythagoras, Aristotle, Kant, and Goethe. Leonardo clearly followed the tradition of Pythagoras and Aristotle in developing his science of living forms, their patterns of organization, and their processes of growth and transformation. Indeed, systemic thinking lies at the very core of his approach to scientific knowledge.

Leonardo's science is a science of natural forms, of qualities, quite different from the mechanistic science that would emerge two hundred years later. Leonardo's forms are living forms, continually shaped and transformed by underlying processes. Throughout his life he studied, drew, and painted the rocks and sediments of the Earth, shaped by water; the growth of plants, shaped by their metabolism; and the anatomy of the animal (and human) body in motion.

Nature as a whole was alive for Leonardo. He saw the patterns and processes in the microcosm as being similar to those in the macrocosm. At the most fundamental level, as already mentioned, Leonardo always sought to understand the nature of life. This has often escaped earlier commentators, because until recently the nature of life was defined by biologists only in terms of cells and molecules, to which Leonardo, living two centuries before the invention of the microscope, had no access. But today, a new understanding of life is emerging at the forefront of science—an understanding in terms of metabolic processes and their patterns of organization. And those are precisely the phenomena Leonardo explored throughout his life.

Leonardo's studies of the living forms of nature began with their appearance to the painter's eye and then proceeded to detailed investigations of their intrinsic nature. His science is a science of qualities. He preferred to *depict* the forms of nature rather than *describe* their shapes, and he analyzed them in terms of their proportions rather than measured quantities.

Another important aspect of systems science is its inherently dynamic nature. Since the earliest days of biology, scientists and philosophers have recognized that living form is more than shape, more than a static configuration of components in a whole. There is a continual flow of matter through a living system, while its form is maintained; there is growth and decay, regeneration and development. Hence, the understanding of living structure is inextricably linked to the understanding of metabolic and developmental processes.

This was very much Leonardo's approach. His science is utterly dynamic. He portrays nature's forms—in mountains, rivers, plants, and the human body—in ceaseless movement and transformation. He studies the multiple ways in which rocks and mountains are shaped by turbulent flows of water, and how the organic forms of plants, animals, and the human body are shaped by their metabolism. The world Leonardo portrays, both in his art and in his science, is a world in development and flux, in which all configurations and forms are merely stages in a continual process of transformation.

Inspiration for Our Time

Here, then, are the principal signs of Leonardo's genius: his relentless curiosity, intellectual fearlessness, capacity for intense concentration, attention to detail, holistic memory, commitment to the empirical method, and pervasive systemic thinking. Most of us will not be able to develop these characteristics of genius to anywhere near Leonardo's degree. But we can all be inspired by his specific ways of work—as a scientist, artist, and designer—and learn valuable lessons from his method.

The great challenge of our time is to build and nurture sustainable communities—communities designed in such a way that their ways of life, businesses, economy, physical structures, and technologies respect, honor, and cooperate with nature's inherent ability to sustain life. The first step in this endeavor, naturally, must be to understand how nature sustains life. It turns out that this involves a new ecological understanding of life, also known as "ecological literacy," as well as the ability to think systemically—in terms of relationships, patterns, and context.

Indeed, such a new understanding of life has emerged over the last thirty years.[10] Contemporary science no longer sees the universe as a machine composed of elementary building blocks. We have discovered that the material world, ultimately, is a network of inseparable patterns of relationships; the planet as a whole is a living, self-regulating system. The view of the human body as a machine and of the mind as a separate entity is being replaced by one that sees not only the brain but also the immune system, the bodily tissues, and even each cell as a living, cognitive system. Evolution is seen not as a competitive struggle for existence, but rather as a cooperative dance in which creativity and the constant emergence of novelty are the driving forces. With the new emphasis on complexity, networks, and patterns of organization, a new science of qualities is slowly emerging.

This new science is being formulated in a language quite different from Leonardo's. As we shall see throughout this book, however, the underlying conception of the living world as being fundamentally interconnected, highly complex, creative, and imbued with cognitive intelligence is quite similar to Leonardo's vision. This is the main reason, in my view, why the science and art of this great genius of the Renaissance can be a tremendous inspiration for our time.

The new systemic understanding of life that has been developed at the forefront of science comprises biological, cognitive, social, and ecological dimensions. It applies to all living systems—individual organisms, social systems, and ecosystems. Hence, it is relevant to virtually all professions and endeavors, besides being fascinating in itself. Our intellectual curiosity to find out more about it may encounter demanding obstacles at first, but in the end will be richly rewarded.

At the core of the new understanding of life is a shift of metaphors from seeing the world as a machine to understanding it as a network. Exploring this shift without prejudice, driven by intellectual curiosity, will be beneficial in many ways. Individually, it will help us to better deal with our health, seeing our organism as a network of components with both physical and cognitive/emotional dimensions. As a society, the exploration of networks will help us to build a sustainable future, grounded in the awareness of ecological networks and the interconnectedness of our major problems. Such exploration will also help us manage our organizations, which are social networks of increasing complexity.

We may not be able to match Leonardo's capacity for intense concentration and attention to detail over long periods of time, but we will be more successful in dealing with the challenges of the frenetic pace of our Industrial Age if we give ourselves adequate time to reflect on a problem, keeping in focus both the problem and its various ramifications. Creating extended periods of time for reflection in order to carefully think through our solutions before applying them is what environmental educator David

Orr calls "slow knowledge"—the equivalent of Leonardo's *discorso mentale*.[11] In our human organizations, the challenge will be to create these periods of reflection for the benefit of all members and the organization as a whole.

Very few people have the capacity for what I have called "holistic memory"—the ability to memorize large amounts of information in the form of a coherent whole. But we all can train ourselves to improve our associative memory, to remember relationships and connections, which is crucial for systemic thinking. Today, with information at our fingertips in our laptops and smart phones, what is important is to know how things are interconnected rather than to remember individual facts exactly. As the great playwright and statesman Václav Havel put it: "Education is the ability to perceive the hidden connections between phenomena."[12]

Leonardo developed his empirical method single-handedly, in a cultural vacuum. Today, the scientific method is practiced worldwide, but it is still ignored or even rejected by many individuals and institutions outside of science. This is true, for example, of many conservative politicians in the United States, who are often ignorant or in denial of the scientific facts about climate change, or even about evolution. We will all be much better off, as individuals and as a society, if we respect the empirically based and carefully honed insights of scientists and act accordingly.

As I have mentioned, Leonardo was always respectful of the classical Greek and Latin texts and familiarized himself with them as much as possible, accumulating a considerable personal library and often borrowing manuscripts from other scholars. He would usually start his investigations from commonly accepted concepts and explanations, but then always proceeded to examine the classics critically, never afraid of correcting them in the light of his own observations.

As we develop our ability to think systemically, together with our creativity and intuition, we need to be aware of the constant interplay between tradition and innovation. We need new ideas for many of our systemic problems, but we also need to be educated—that is, familiar with tradition—to even formulate our questions and to avoid reinventing the wheel. Leonardo was a master of acknowledging tradition before examining it critically in the light of his empirical method. His method can be a great inspiration for us when we try to manage the pervasive tensions between tradition and innovation.

For most of us, intellectual fearlessness can mean learning how to trust our intuition and creativity, which may lead to novel ideas or solutions. If we have the courage to explore these new ideas without fear of rejection or ridicule, we will often be highly rewarded.

The spontaneous emergence of novelty in social networks, often referred to simply as "emergence," has been recognized as a key characteristic of life. Complexity theory has revealed the underlying dynamics of emergence: a network of communications involving multiple feedback loops, open to disturbances from the environment; then a critical point of instability; and finally a breakthrough to a new order, or new idea.[13] Trusting our collective intuition and creativity creates an environment conducive to that emergence of novelty.

This is the basis of a new understanding of leadership that is now being explored by organizational theorists and business executives.[14] The traditional idea of a leader is that of a person who is able to hold a vision, to articulate it clearly, and to communicate it with passion and charisma. This is still important, but another kind of leadership facilitates the emergence of novelty by creating conditions rather than giving directions and by using the power of authority to empower others.

Leaders need to recognize and understand the different stages of this fundamental life process. Emergence requires an active network of communications. Moreover, the emergence of novelty is a property of open systems, which means that the organization needs to be open to new ideas and new knowledge. Facilitating emergence, therefore, means first of all building up and nurturing networks of communications and then creating openness—a learning culture in which continual questioning is encouraged and innovation is rewarded. In the end, leaders need to be able to recognize the emergent novelty, articulate it, and incorporate it into the organization's design. Not all emergent solutions will be viable, however, and hence a culture that fosters emergence must include the freedom to make mistakes. In such a culture, experimentation is encouraged and learning from failures is valued as much as success. Leonardo da Vinci's relentless curiosity and intellectual fearlessness can thus be highly inspiring to a new generation of business and community leaders.

When we look at the state of the world today, it is clear that the major problems of our time—energy, the environment, climate change, food security, and financial security—cannot be understood in isolation. They are systemic problems, which means that they are all interconnected and interdependent; they require systemic thinking to be solved. We need to learn how to take into account the interdependence of our problems, and we often need to work on several of them simultaneously to solve any one of them.

This was exactly Leonardo's method, and this—together with his deep respect for nature—is perhaps his greatest legacy to us. I shall return to this legacy in the Coda of this book; but first let us explore Leonardo da Vinci's marvelous world of living forms and transformations.

PART I

Form and Transformation in the Macrocosm

1

The Movements of Water

Among the four classical elements, water held by far the greatest fascination for Leonardo. Throughout his life, he studied its movements and flows, drew and analyzed its waves and vortices, and speculated about its role as the fundamental "vehicle of nature" (*vetturale della natura*) in the macrocosm of the living Earth and the microcosm of the human body.[1]

Leonardo's notes and drawings about his observations and ideas on the movement of water fill several hundred pages in his Notebooks. They include elaborate conceptual schemes and portions of treatises in the Codex Leicester and in Manuscripts F and H, as well as countless drawings and notes scattered throughout the Codex Atlanticus, the Codex Arundel, the Windsor Collection, the Codices Madrid, and Manuscripts A, E, G, I, K, and L.[2] The sheer bulk of Leonardo's writings on water duly impressed his contemporaries and succeeding generations of historians. In fact, water was the only subject, apart from painting, of which an extensive compilation of handwritten transcriptions from the Notebooks was made. This collection of notes, transcribed in the seventeenth century and comprising 230 folios, was published in 1828 in Bologna under the title *Della natura, peso, e moto dell'acque* (On the Nature, Weight, and Movement of Water).[3]

Carrier and Matrix of Life

Leonardo was fascinated by the nature and movements of water for several reasons. I believe that, ultimately, they all have to do with his persistent quest to understand the nature of life, which informed both his science and his art. Leonardo's science is a science of living, organic forms,

PRECEDING A stream running through a rocky ravine, c. 1483 (detail, see fig. 2-5).
FACING "Water falling upon water," c. 1508–9 (see fig. 1-13).

and he clearly recognized that all organic forms are sustained and nourished by water:

> It is the expansion and humor of all living bodies.
> Without it nothing retains its original form.[4]

The term "humor" is used here in its medieval sense of a nourishing bodily fluid. In another Notebook, Leonardo wrote: "[Water] moves the humors of all kinds of living bodies."[5] Being a painter, he had ample experience with water as a solvent and accurately described this chemical property: "It has nothing of itself, but moves and takes everything, as is clearly shown when distilled."[6]

Leonardo's view of the essential role of water in biological life is fully borne out by modern science. Today we know not only that all living organisms need water for transporting nutrients to their tissues but also that life on Earth began in water. The first living cells originated in the primeval oceans more than three billion years ago, and ever since that time all the cells that compose living organisms have continued to flourish and evolve in watery environments. Leonardo was completely correct in viewing water as the carrier and matrix of life.

One of the fundamental principles of Leonardo's science is the similarity of patterns and processes in the macro- and microcosm. Accordingly, he compared the "water veins" of the Earth to the blood vessels of the human body (see p. 26).[7] As blood nourishes the tissues of the body, so water nourishes the Earth's vegetation with its "life-giving moisture."[8] And as water expands when it vaporizes in the heat of the sun and "becomes mingled with the air," so blood by its warmth spreads into the periphery of the body.[9] Indeed, we shall see that Leonardo described in great detail how blood carries nutrients to the bodily tissues and that he developed an ingenious, though incorrect, theory of how body heat is generated by the turbulent flows of blood in the chambers of the heart (see p. 296).

In his paintings, Leonardo represented water as the carrier of life not only in the scientific sense but also symbolically, in the religious sense. According to the Christian theology that shaped the culture in which he lived, the faithful receive a new spiritual life in the sacrament of baptism, and water is the medium that conveys this sacrament. In the words of the

Bible, baptism is rebirth of water and spirit (John 3:5). Several of Leonardo's paintings contain variations on this fundamental religious theme, often integrating the religious symbolism with his scientific understanding of the life-giving quality of water.

This integration is already apparent in the very first record we have of Leonardo as a painter, when he was still an apprentice in the workshop of Andrea del Verrocchio in Florence. Around 1473, when Leonardo was twenty-one, Verrocchio let the youth paint one of two angels and parts of the background in his picture of the *Baptism of Christ* (plate 2).[10] Leonardo painted a wide, romantic stretch of hills and pinnacles of rocks of the kind that would form the backgrounds in many of his later paintings, and to that he added a long watercourse, flowing from a pool in the far distance all the way to the foreground, where it forms small waves rippling around the legs of Christ. While these ripples in the foreground represent the life-giving water of the sacrament, the watercourse in the background, cutting through arid rocks and flowing into a fertile valley, portrays water as the carrier of biological life in the macrocosm of the Earth.

This theme is expanded and elaborated in several of Leonardo's later paintings, in particular, in three of his masterpieces—the *Virgin of the Rocks* (plate 8), the *Mona Lisa* (plate 11), and the *Madonna and Child with Saint Anne* (plate 7). In the *Virgin of the Rocks*, Leonardo depicts a prophetic meeting of the infant Christ with the infant Saint John long before the Baptism. According to a fourteenth-century legend, this meeting took place during the Holy Family's flight into Egypt, where they lived in the wilderness after their escape from Herod's massacre of the innocents. Leonardo has placed the scene in front of a rocky grotto and turned it into a complex meditation on the destiny of Christ, expressed through the gestures and relative positions of the four protagonists, as well as in the intricate symbolism of the surrounding rocks and vegetation.[11] An angel conspicuously points to the Baptist, directing our attention to his spiritual dialogue with Christ, while Mary tenderly protects the children with her outstretched arms.

As in Verrocchio's *Baptism*, a mountain stream emerges in the far distance from the misty atmosphere surrounding pinnacles of rocks and breaks through the rocky landscape, flowing all the way to the foreground of the painting where it runs through a small pool—an allusion to the Baptism. However, the rocks are rendered here in much more detail and with astonishing geological accuracy (see pp. 77ff.), and the luxuriant vegetation in the grotto's moist environment is clear testimony to the generative powers of water, presented by the artist in a subtle synthesis of scientific knowledge and religious symbolism (see pp. 102ff.).

The *Mona Lisa* is Leonardo's deepest meditation on the mystery of the origin of life—the theme that was foremost in his mind during his old age. The central theme of the artist's most famous painting is life's procreative power, both in the female body and in the body of the living Earth. Essential to this power is the fundamental role of water as the life-giving element (see pp. 318ff.).

The theme of the origin of life is taken up again in the *Saint Anne*, which Leonardo painted around the same time as the *Mona Lisa*. Here the artist returned once more to exploring the mystery of life within a religious context. The painting shows Mary, her mother Saint Anne, and the Christ child together with a lamb in a highly original composition. Its theological message can be viewed as a continuation of Leonardo's long meditation on the destiny of Christ, which began with the *Virgin of the Rocks*.[12]

Once more, the familiar mountain lakes and jagged rocks rise high into the background, although they are less imposing than those behind the *Mona Lisa*. In both paintings, the central theme is the mystery of the origin of life in the human body and in the body of the Earth. In the *Saint Anne*, this is rendered even more complex by the presence of three generations and by the myth of the virgin birth. There is a double mystery here: the immaculate conception of Mary by Saint Anne and that of Christ by Mary. To emphasize the analogy between human nature and the Earth, Leonardo has mirrored the three generations in the painting's foreground by three tiers of mountain lakes, interlinked by small waterfalls, in the background.

What these four paintings—the *Baptism*, the *Virgin of the Rocks*, the *Mona Lisa*, and the *Saint Anne*—have in common is Leonardo's extended reflection on water as the life-giving element in the macrocosm of the Earth and the microcosm of human existence. Drawing on his scientific understanding, his artistic genius, and his great familiarity with religious symbolism, Leonardo expressed this meditation in a series of masterpieces that have become enduring icons of European art.

Nature's Fluid Forms

Another reason Leonardo was so fascinated by water is that he associated it with the fluid and dynamic nature of organic forms. Ever since antiquity, philosophers and scientists had recognized that biological form is more than shape, more than a static configuration of components in a whole. There is a continual flux of matter through a living organism, while its form is maintained; there is growth and decay, regeneration and development. This dynamic conception of living nature is one of the main themes in Leonardo's science and art.[13] He portrayed nature's forms—in mountains, rivers, plants, and the human body—in ceaseless movement and transformation. And, knowing that all organic forms are sustained by water, he sensed a deep connection between their fluidity and the fluidity of water.

As Leonardo observed the flow of the life-giving element, he marveled at its endless versatility and adaptability. "Running water has within itself an infinite number of movements," he noted in Manuscript G, "sometimes swift, sometimes slow, and sometimes turning to the right and sometimes to the left, now upwards and now downwards, turning over and back on itself, now in one direction and now in another, obeying all the forces that move it."[14] In the Codex Atlanticus he wrote: "Thus, joined to itself, water turns in a continual revolution. Rushing this and that way, up and down, it never rests, neither in its course nor in its nature. It owns nothing but seizes everything, taking on as many different characters as the places it crosses."[15]

In addition, Leonardo carefully studied the actions of water in the erosion of rocks and river banks, its transformations into solid and gaseous forms (known in science today as phase transitions), and its properties as a chemical solvent. He never divided these diverse properties into separate categories but saw them all as different aspects of the fundamental role of water in nourishing and sustaining life:

> Without any rest, it is ever removing and consuming whatever borders upon it. So at times it is turbulent and goes raging in fury, at times clear and tranquil it meanders playfully with gentle course among the fresh pastures. At times it falls from the sky in rain, snow, or hail. At times it forms great clouds out of fine mist. At times it moves of itself, at times by the force of others. At times it

increases the things that are born with its life-giving moisture. At times it shows itself either fetid or full of pleasant odors. Without it nothing can exist among us.[16]

For Leonardo, the fluid and ever-changing forms of water were extreme manifestations of the fluidity that he saw as a fundamental characteristic of all the forms of nature. He also noticed, however, that certain flows of water can produce forms that are surprisingly stable: eddies, vortices, and other forms of turbulence known to scientists today as coherent structures (see p. 55). He observed and sketched a great variety of these relatively stable turbulent structures, and I believe that his lifelong fascination with them came from his deep intuition that, somehow, they embodied an essential characteristic of living, organic forms.

Today, from our modern perspective of complexity theory and the theory of living systems, we can say that Leonardo's intuition was absolutely correct. The fundamental characteristic of a water vortex—for example, the whirlpool that is formed as water drains from a bathtub—is that it combines stability and change. The vortex has water continuously flowing through it, and yet its characteristic shape, the well-known spirals and narrowing funnel, remains remarkably stable. This coexistence of stability and change is also characteristic of all living systems, as complexity and systems theorists recognized in the twentieth century.[17]

The process of metabolism, the hallmark of biological life, involves a continual flow of energy and matter through a living organism—the intake and digestion of nutrients and the excretion of waste products—while its form is maintained. Thus, metaphorically, one could visualize a living organism as a whirlpool, even though the metabolic processes at work are not mechanical but chemical.

Leonardo never used the analogy between the dynamic of a water vortex and that of biological metabolism, at least not in the Notebooks that have come down to us. However, he was well aware of the nature of metabolic processes. Indeed, we shall see that his detailed description of tissue metabolism in connection with the flow of blood in the human body must be seen as one of his most astonishing scientific insights (see p. 312). Thus, it seems not too far-fetched to assume that he was so fascinated by whirlpools and vortices because he intuitively recognized them as symbols of life—stable and yet continually changing.

A Source of Power

Leonardo saw water not only as the life-giving element but also as the principal force shaping the Earth's surface and as a major source of power, which could be harnessed by human ingenuity. In his time, three hundred years before the Industrial Revolution, the windmill, the water wheel, and the muscles of beasts provided the only power to drive human technologies, and among those Leonardo thought that water had the greatest potential. At the age of fifty, when he was famous as a painter throughout Europe and known as one of Italy's leading military and hydraulic engineers, he dreamed of a grand scheme for a kind of "industrial" canal along the river Arno between Florence and Pisa.[18] He imagined that such a waterway would provide irrigation for the surrounding fields as well as energy for numerous mills that could produce silk and paper, drive potters' wheels, saw wood, forge iron, burnish arms, and sharpen metal.[19] Leonardo's ambitious project was never realized, but it was a prophetic vision. Centuries later, the powers of steam and hydroelectricity would indeed transform human civilization.

As an engineer, Leonardo was also well aware of the destructive power of water. In the plains of northern Italy, at the foot of the Alps, an elaborate system of canals had been built for irrigation and for commercial navigation, and one of the main challenges faced by hydraulic engineers was how to protect these canals from the flooding of their tributaries (see p. 32). This flooding happened periodically during heavy autumn rains and after a sudden spring melting of the Alpine snows. Leonardo paid great attention to these inundations, which could be very violent. He had witnessed a catastrophic flooding of the Arno in his native Tuscany at the age of fourteen. This childhood experience must have left a deep impression on him and perhaps was the cause of his morbid fascination with floods, which he considered the most frightening of all cataclysmic events.[20] "How can I find words to describe these abominable and frightening evils, against which there is no human defense?" he wrote in the Codex Atlanticus. "With swollen waves rising up, it devastates high mountains, destroys the strongest embankments, and tears out deeply rooted trees. And with voracious waves, laden with the mud of plowed fields, it carries off the fruits of the hard work of the miserable and tired tillers of the soil, leaving the valleys bare and naked with the poverty it leaves in its wake."[21]

As a hydraulic engineer, Leonardo invented special machines for digging canals, improved the existing systems of locks, drained marshes, and modified the flows of rivers to prevent damage to properties along their banks. As an architect, he designed elaborate landscape gardens with splendid fountains, running water for cooling wine, sprinkler systems for refreshing guests during the hot summers, and automatic musical instruments played by water mills.[22]

He decided early on that his reputation and skills in hydraulic engineering and landscape design would be grounded in a thorough understanding of the flow of water. In his science and his art, Leonardo never tired of observing, analyzing, drawing, painting, and studying how water moves through the air, the blood vessels of the human body, the vascular tissues of plants, and the seas and rivers of the living Earth.

The Water Cycle

Since Leonardo's science was based on repeated observations of natural phenomena combined with meticulous analysis,[23] it is not surprising that he had an accurate understanding of the evaporation and condensation of water and was able to describe it clearly. "Readily it rises up as vapors and mists," he wrote in Manuscript A, "and, converted into clouds, it falls back as rain because the minute parts of the cloud fasten together and form drops."[24] A slightly more detailed description can be found in the Codex Arundel:

> At times it is bathed in the hot element and, dissolving into vapor, becomes mingled with the air; and drawn upward by the heat, it rises until, having found the cold region, it is pressed closer together by its contrary nature, and the minute particles become attached together.[25]

He was also well aware of the fact that water continually cycles through the earth and atmosphere: "We may conclude that the water goes from the rivers to the sea, and from the sea to the rivers, thus constantly circulating and returning."[26] Taken together, these statements seem to indicate that Leonardo had a clear understanding of the essential phases of the water cycle—how water in the oceans, heated by the sun, evaporates into the air; how it rises into the atmosphere until cooler temperatures cause it to condense into clouds; how minute particles in the clouds coalesce into larger drops that precipitate as rain or snow; and how this precipitation eventually flows into rivers that carry it back into the oceans.

In actual fact, however, Leonardo's views of the water cycle were far from clear. He considered several different explanations, struggled for many years because none of them satisfied his critical mind, and arrived at the correct view only in his old age, in his early sixties. How are we to understand that? What prevented a man of his genius from understanding a natural process that seems so evident to us today?

The answer to the puzzle provides a fascinating example of the tremendous power of the philosophical framework known today as a scientific paradigm—the constellation of concepts, values, and perceptions that form the intellectual context of all scientific investigations.[27] One of the foundations of the medieval worldview was the conviction that nature as a whole was alive, and that the patterns and processes in the macrocosm were similar to those in the microcosm. This analogy between macro- and microcosm, and in particular between the Earth and the human body, goes back to Plato and had the authority of common knowledge in the Middle Ages and the Renaissance.[28] Leonardo fully embraced it as one of the guiding principles of his science and discussed it repeatedly (see pp. 65ff.). Whenever he explored the forms of nature in the macrocosm, he also looked for similarities of patterns and processes in the human body, and so it was natural for him to compare the "water veins" of the Earth to the blood vessels of the body.

Our modern systemic conception of life fully validates Leonardo's method of exploring similarities between patterns and processes in different living systems, and his view of the Earth as being alive has reappeared in today's science, where it is known as Gaia theory.[29] However, Leonardo ran into difficulties with his comparisons between the living Earth and the living human body because he extended them beyond the similarity of patterns to comparisons of forces and material structures. One of the important insights of modern systems and complexity theories has been that, even though patterns of relationships between the components and processes of two different living systems may be similar, the processes themselves and the forces and structures involved in them may be quite different.[30] It took Leonardo the better part of his life to realize this, but he clearly did so in his old age.

Since the total amount of water on Earth is finite, Leonardo argued, the water carried into the sea by the rivers must somehow cycle back to their sources, "thus constantly circulating and returning."[31] Since he conceived of water as a "humor" that nourishes the Earth just as the blood nourishes the human body, he imagined that there must be water veins inside the body of the living Earth corresponding to the blood vessels in the bodies of animals and humans:

> The body of the Earth, like the bodies of animals, is interwoven with a network of veins which are all joined together, and are formed for the nutrition and vivification of that Earth and of its creatures. They originate in the depths of the sea, and there after many revolutions they have to return through the rivers, created high up by the bursting of these veins.[32]

This was the traditional view, put forward by philosophers from Aristotle to the Renaissance: inside the living Earth, there is a system of water veins, in which the water circulates like the blood in a living body, until the veins eventually break in the high mountains. There, the water emerges from mountain springs, is collected by the rivers, and flows back into the sea. Leonardo realized, of course, that rivers are also fed by rainwater and melting snow. But for many years he maintained that their principal sources were the internal veins of the Earth. Even though he encountered many logical inconsistencies, he was unwilling for the longest time to abandon the powerful analogy between the circulation of water in the Earth and that of blood in the human body. "The water that rises within the mountains," he wrote in his early forties, "is the blood that keeps these mountains alive."[33]

Leonardo's scientific mind was not content with the beautiful metaphorical description of water as "the blood that keeps the mountains alive." He needed to explain how the water actually rises up to the mountain springs through internal channels. It was clear to him that some forces counteracting gravity had to be at work:

> The water which sees the air through the broken veins of the high mountain summits is suddenly abandoned by the power which brought it there, and when it escapes from these forces that elevated it to the summit, it freely resumes its natural course.[34]

But what exactly were these forces? To find an answer, Leonardo used a method that was characteristic of all his investigations. Understanding a phenomenon, for him, always meant connecting it with other phenomena through a similarity of patterns. In this case, he identified two similar phenomena—how the blood in the human body rises to the head and how the sap in a plant rises up from its roots—and he assumed that the same forces were acting in all three examples:

> The same cause which moves the humors in all kinds of living bodies against the natural course of gravity also propels the water through the veins of the Earth, wherein it is enclosed, and distributes it through small passages. As the blood rises from below and pours out through the broken veins of the forehead, and as the water rises from the lower part of the vine to the branches that are cut, so from the lowest depth of the sea the water rises to the summits of the mountains where, finding the veins broken, it pours down and returns to the low-lying sea.[35]

Having established this similarity of patterns, Leonardo then set out to identify the common forces underlying them. Over the years, he tried and then rejected several explanations.[36] At first, he thought that the water was drawn up inside the mountains as steam by the heat of the sun, and he suggested that this process was similar to blood rising to a man's head when it is hot:

> When the sun warms a man's head, the blood increases and rises so much with other humors that, by pressuring the veins, it often causes headaches.[37]

"The heat of fire and sun by day," Leonardo argued, "have the power to extract the moisture from the low places of the mountains and draw it up high in the same way as it draws the clouds and extracts their moisture from the bed of the sea."[38]

Subsequently, however, he discovered two reasons why this explanation could not work. He noted that on the highest mountain tops, closest to the heating sun, the water remains cold and is often icy. Moreover, with this mechanism the greatest amount of water should be drawn up in summer when the sun is hottest, but mountain rivers are often lowest at this time.

In a second explanation, Leonardo suggested that the water might be drawn up in a process of distillation, fueled by the Earth's internal heat. He was aware of the presence of fire within the Earth from observations of hot springs and volcanoes, and he had also experimented with several types of distillation apparatus.[39] Perhaps, he suggested, the interior fires of the Earth boil water in special caverns until it rises as vapor to the roofs of those caverns, "where, coming upon the cold, it suddenly changes back into water, as one sees happen in a retort, and goes falling down again and forming the beginnings of rivers."[40] Again, Leonardo found an argument against his own explanation. Such extensive distillation, he realized, would keep the roofs of internal caverns wet from the rising steam, but he remembered from his explorations of mountain caves that they were often bone dry.

A third proposal was based on the observation that water rises in a vacuum within an enclosed space. Leonardo was quite familiar with this phenomenon. One of his early inventions, when he was still working in Verrocchio's *bottega*, had been a method of creating a vacuum to raise water by means of a fire burning in a closed bucket.[41] Now he hypothesized that the internal fires might rarefy the air in the Earth's caverns and thus raise the water to the top. However, he soon realized that this would not work, because additional air would enter the cavern through the openings of the mountain springs and would thus stop the siphoning action of the vacuum.

On a folio of the Codex Leicester, Leonardo summarized both the distillation model and the siphoning model together with their counterarguments.[42] He illustrated the discussion with a drawing showing the cross section of a mountain with interior veins running from the sea all the way up to the top where they connect with two large caverns. Right below, we see clear sketches of the two processes of distillation and siphoning (fig. 1-1). An accompanying folio contains numerous drawings illustrating experiments with various siphons.[43]

As yet another alternative, Leonardo suggested that the water might be drawn up inside the mountains by some process similar to the action of a sponge, but that vague idea did not satisfy him either. "If you should say that the Earth's action is like that of a sponge," he countered, "the answer is that, even if the water rises to the top of that sponge by itself, it cannot then pour down any part of itself from this top, unless it is squeezed

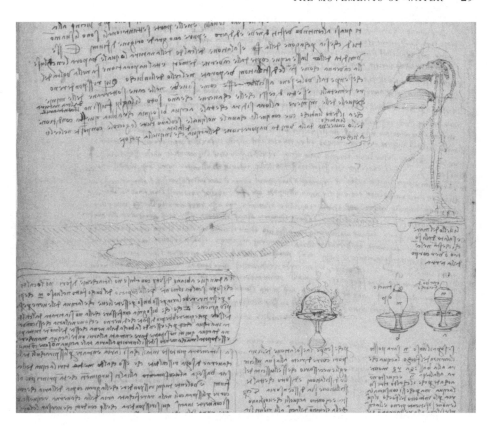

FIG. 1-1. Models of water circulation by distillation and
by siphoning action. Codex Leicester, folio 3v (detail).

by something else, whereas with the summits of the mountains one sees
the opposite, for there the water always flows away by itself without being
squeezed by anything."[44]

After many years of considering various explanations and finding
counterarguments to all of them, Leonardo finally realized that his analogy between the blood vessels of the human body and the water veins
of the Earth was too narrow; that in the water cycle, the water does not
circulate inside the mountains but rises as vapor through the air, drawn
up by the heat of the sun, and then falls as rain on the mountaintops. On
a folio of the Windsor Collection, written after 1510 when he was around
sixty, Leonardo stated unequivocally that "the origin of the sea is contrary
to the origin of the blood," because the rivers "are caused entirely by the
aqueous vapors raised up into the air."[45]

Around the same time, in a note in Manuscript G about water as the carrier of minerals, Leonardo stated quite casually, as a matter of fact, that the rivers are produced by clouds:

> The saltiness of the sea is due to the numerous water veins, which in penetrating the earth find the salt mines, and dissolving parts of these carry them away with them to the ocean and to the other seas from whence the clouds, originators of the rivers,* never raise them up.[46]

From our contemporary perspective, Leonardo's long intellectual struggle to understand the water cycle is extremely interesting. His successive theoretical formulations are quite similar to the theoretical models that are characteristic of modern science.[47] Like scientists today, he continually tested his models and was ready to replace them when he found that they contradicted some empirical evidence. Moreover, as he progressed, he kept in mind the analogies and interconnecting patterns to phenomena in other areas, and revised his theories about those other phenomena accordingly. Thus, as he modified his explanations of the water cycle, he also modified similar models of the functioning of the heart and the flow of blood in the human body (see pp. 284–85).

In the end, Leonardo came to realize that, although water and blood both carry nutrients to living systems (as we would say today) and both cycle continually, the pathways of the two cycles and the forces driving them are quite different. During the years of 1510–15, when he finally reached a clear understanding of the water cycle, he also came to the conclusion that the blood in the human body is moved by the pumping action of the heart (see pp. 290ff.). That Leonardo was able, in his old age, to abandon the narrow analogy between the circulation of blood in the human body and the circulation of water in the body of the Earth, which had been firmly established in medieval philosophical thought, is an impressive testimony to his intellectual integrity, his perseverance, and the power of his scientific method.

When he was in his early sixties and reached his full understanding of the water cycle and the movement of blood in the human body,

---

* The phrase "the clouds, originators of the rivers" seems to be no more than a passing comment here. However, as Augusto Marinoni pointed out in a footnote to his transcription of Manuscript G, folio 48v, Leonardo highlighted this passage with a special mark of his pen to signal its importance.

Leonardo also produced his most sophisticated writings in botany, in which he described the transport of "vital sap" through the vascular tissues of plants (see pp. 120ff.). It would be fascinating to know how his insights into the circulation of water and blood affected his ideas of how water rises through the plant tissues from the roots to the top. Today we know that this is a consequence of the evaporation of water from the leaves and of its intermolecular forces—the "cohesion in itself," as Leonardo called it (see p. 41). Unfortunately, we are not likely to ever know Leonardo's last thoughts on these matters since the manuscript that may have contained his definitive treatise on botany has been lost.[48]

From Hydraulic Engineering to the Scientific Study of Flow

The majority of Leonardo's extensive collections of notes and drawings on the flow of water were concerned with problems of hydraulic engineering and with the phenomenon of flow itself. In the Renaissance, the latter was a subject unique to Leonardo. The movement of rigid bodies had been studied since antiquity. In contrast, although hydraulic engineers had produced magnificent works—from the great aqueducts and luxurious thermal baths of the Romans to the ingenious navigation locks of the early fifteenth century—it had not occurred to any of them to wonder how flowing water could be described mathematically. Nor did they attempt to explore the fundamental laws of fluid flow, the subject of our modern discipline of fluid dynamics. Leonardo did both. His investigations, drawings, and attempted mathematical descriptions of flow patterns in water and air must be ranked among his most original scientific contributions, leading him to discoveries that would reappear only centuries later.

When he was first employed as painter and "ducal engineer" at the Sforza court in Milan in 1490, Leonardo had already spent eight years in the capital of Lombardy, which was a vibrant trading center of tremendous wealth and a major seat of political power in northern Italy. During those years, he had not only painted the *Virgin of the Rocks* and a highly original portrait of the mistress of Ludovico Sforza, but had also undertaken an extensive program of self-education during which he systematically studied the principal fields of knowledge of the time.[49]

From his first years in Lombardy, Leonardo was fascinated by the engineering problems involved in the region's elaborate system of canals. During the previous three centuries, hydraulic engineering in northern Italy had reached a level of considerable sophistication.[50] The wealth of the Lombard region was dependent on the control of water and on land reclamation from marshes. Hydraulic engineering was needed to reduce damage from the periodic flooding of Alpine rivers, to supply the cities with water, to keep ports working, for irrigation, and for commercial navigation. The great canals of Lombardy, wide enough to let two large barges pass, interconnected the principal rivers of the area and featured a series of sophisticated locks for overcoming differences of water levels.

As ducal engineer at the Sforza court, Leonardo was probably in charge of all hydraulic works and thus became thoroughly familiar with the existing technologies and the problems that needed to be solved.[51] Indeed, the Codex Leicester contains a vast number of observations on practical hydraulic problems in rivers and canals. Before Leonardo, such knowledge had been transmitted mostly orally, and the approach of the Lombard engineers was purely empirical: all their practices and rules were based on the success or failure of previous similar works. This did not satisfy Leonardo's scientific mind. He needed to know the reasons behind the empirical rules, and so he embarked on his lifelong studies of the laws of fluid flow, beginning with the basic dynamics of the flow of rivers and proceeding to complex patterns of turbulent flow.

Even in the midst of his theoretical studies, Leonardo always kept their practical applications in mind. For example, during a discussion of the flow of water around immersed obstacles, he noted: "The science of these objects is of great usefulness, for it teaches how to bend rivers and avoid the ruins of the places struck by them."[52] In Manuscript F, written during the same period as the Codex Leicester, we find the following admonition: "When you put together the science of the movements of water, remember to put beneath each proposition its applications, so that such science may not be without its uses."[53]

In Leonardo's time, the scientific study of flow phenomena, now known as fluid dynamics, was entirely new. It was a field of study he himself created single-handedly. However, in view of his dynamic conception of the world and his practice of portraying nature's forms in his drawings and paintings as being in ceaseless movement and transformation, such a study must have seemed completely natural to him. Indeed, flow was one of the dominant themes in his science and art. In the words of hydraulic engineer and Leonardo scholar Enzo Macagno, "To Leonardo, if not everything, almost everything was flowing or could be in one state of flow or another."[54]

In early Greek philosophy, the idea that everything in the world is in a process of constant change was expressed in the famous saying by Heraclitus of Ephesus, "Everything flows." There is no evidence that Leonardo was familiar with the philosophy of Heraclitus, but in an intriguing double portrait by the famous architect Donato Bramante, who was a close friend of Leonardo, Bramante represented his friend as Heraclitus and himself as Democritus.[55]

Since he saw movement and transformation as fundamental characteristics of all natural forms, Leonardo assumed that the basic properties of flow were the same for all fluids, and he found this confirmed by his observations. He emphasized especially the similarity between flows of water and air. "In all cases of motion, there is great conformity between water and air," he noted in Manuscript A,[56] and in the Codex Atlanticus: "The movement of water within water acts like that of air within air."[57] However, Leonardo was well aware that air differs from water in being "infinitely compressible,"* whereas water is incompressible.[58]

As far as flows of liquids were concerned, Leonardo experimented not only with water but also investigated the flows of blood, wine, oil, and even those of grains like sand and seeds.[59] His experiments with granular materials are especially remarkable. He realized that he could learn something about the flow of water by observing a similar but somewhat simpler phenomenon—the flow of grains in which the individual flowing particles are actually visible. This method of using simplified models to analyze the essential features of complex phenomena is an outstanding characteristic of our modern scientific method.[60] The fact that Leonardo used it repeatedly is truly remarkable. In his view of flow as a universal phenomenon of gases, liquids, and granular materials, and his attempts to use the latter as models of the former, Leonardo's thought showed a level of scientific abstraction that was centuries ahead of his time.

Modern Fluid Dynamics

In modern science, the field of fluid dynamics (also referred to as "fluid mechanics," "hydrodynamics," and "hydraulics"†) is notoriously difficult because of the pervasive appearance of turbulent flows that have so far

---

* While water, too, is compressible, the pressures required to obtain significant volume changes are so large that it can be assumed to be incompressible for all practical purposes.

† Strictly speaking, fluid mechanics is the study of fluids (gases and liquids), both static and in motion; fluid dynamics the study of fluids in motion; hydrodynamics the study of water in motion; while hydraulics (more of an engineering term) refers to the study of problems that have to do with water, both static and in motion.

eluded a comprehensive mathematical analysis.[61] In an oft-quoted phrase, physicist and Nobel laureate Richard Feynman called turbulence "the last unsolved problem of classical physics."

Turbulent flows are composed of eddies, also known as vortices, in a broad range of sizes, continually forming and breaking down—those swirling and randomly moving patches of fluid that so fascinated Leonardo. The problem in fluid dynamics is that turbulence is not the exception but the rule. At low velocities, the fluid's internal friction, or viscosity, prevents turbulence from occurring, but as the flow velocity increases turbulence sets in, first at very small scales and eventually spreading throughout the entire fluid.

In the nineteenth century, the mechanical engineer Osborne Reynolds discovered that the onset of turbulence can be characterized in terms of a single parameter, now known as the Reynolds number, which is proportional to the flow velocity, the fluid's viscosity, and the dimensions of the physical object containing the flow. At low values of the Reynolds number, the flow remains smooth, or laminar, and at a certain critical value it becomes turbulent. This discovery marked a major advance in fluid dynamics. Since different values of the Reynolds number correspond to different types of turbulence, this parameter allows scientists and engineers to compactly characterize turbulent flows.

In spite of this advance, however, scientists have so far not been able to formulate a comprehensive theory of turbulence. The mathematical difficulties arise from the fact that the basic equations of motion governing fluid flow, known as the Navier-Stokes equations, are nonlinear and notoriously difficult to solve. This nonlinearity is the mathematical equivalent of the chaotic nature of turbulence. Any turbulent flow contains a very large number of interrelated variables. The value of any one of these at a particular point depends on the flow at many other points, so that solutions must be obtained at many points simultaneously. This is made even more complicated by the fact that turbulent flows display a broad spectrum of scales. The size of the largest eddies may be over a thousand times that of the smallest, which makes their simultaneous mathematical description exceedingly difficult.

The use of powerful computers to simulate and analyze turbulent flows has recently made it possible to find some solutions for special cases. However, even with the new concepts and techniques of complexity theory, or nonlinear dynamics, scientists and mathematicians have had only limited success.[62] They are able to simulate the onset of turbulence and some simple flow patterns—for example, slow, two-dimensional flows around an obstacle—but full turbulent flows are still frustrating all efforts of comprehensive analysis.

Thus, fluid dynamics today consists of a multitude of theories, each valid only for special cases and each invariably invoking some heuristic hypotheses based on experimental observations.[63] In view of this patchwork of theories and of the tremendous difficulties faced by today's engineers, physicists, and mathematicians in their attempts to solve the enigmas of turbulent flows, Leonardo's achievements in this field are truly impressive. At a time long before the development of sophisticated mathematical techniques and powerful computers, Leonardo was able to gain many remarkable insights into the nature of fluid flow.

Leonardo's voluminous notes on the movements of water remained hidden for several centuries after his death, and hence had no influence on the development of science and engineering. The first theoretical analyses of fluids were undertaken only in the eighteenth century when the great mathematician Leonhard Euler applied the Newtonian laws of motion to an idealized "perfect" fluid (that is, a fluid without viscosity), and the physicist and mathematician Daniel Bernoulli discovered some of the basic energy relations exhibited by liquids. Unlike Leonardo two centuries earlier, however, the hydraulic engineers of that time were not interested in theory, and the theorists did not compare their models with observations of the actual flow of fluids.[64] Real progress in fluid dynamics had to wait until the nineteenth century when Claude-Louis Navier and George Stokes generalized Newton's equations for the description of the flow of viscous fluids and Reynolds discovered the parameter that now bears his name. It was only then that physicists and mathematicians rediscovered many of the theoretical ideas about fluid motion that Leonardo had clearly formulated more than three hundred years earlier.

Rivers and Tides, Waves and Flows

In his lifelong studies of "the movements of water," Leonardo observed the flows of rivers and tides, drew beautiful and accurate maps of entire watersheds, and investigated currents in lakes and seas, flows over weirs and waterfalls, the movement of waves, as well as flows through pipes, nozzles, and orifices. A thorough analysis of all his observations, drawings, and theoretical ideas would fill an entire book, and indeed such a book ought to be written. In this chapter I shall concentrate on Leonardo's discussions of the main characteristics of fluid flow and of turbulence.

Leonardo was well aware of the difference between flow and wave motion. As I discussed at some length in my previous book,[65] he recognized from precise observations of circular ripples in a pond that the water particles do not move along with the wave but merely move up and down as the wave passes by. What is transported along the wave is the disturbance causing the wave phenomenon—the "tremor," as Leonardo called it—but not any material particles. To make it easier for the eye to follow the precise movements of the water particles, Leonardo threw small pieces of straw into the pond, and he also compared their motion to waves in a wheat field:

> The wave flees from the place of its creation without the water changing its position, like the waves which the course of the wind makes in wheat fields in May, when one sees the waves running over the fields without the ears of wheat changing their place.[66]

The comparison between water waves and waves in a wheat field was quite natural for Leonardo, because he saw wave motion as a universal form of propagation of physical effects in all four elements—earth, water, air, and fire (or light).[67] Moreover, he masterfully portrayed the effects of "waves of emotion" in his paintings. This is especially apparent in the ebb and flow of movements in his most grandiose work, *The Last Supper*.[68]

As several art historians have pointed out, the internal dynamics of the painting can be perceived as a wave movement, emanating from the figure of Christ in the center, spreading outward in both directions, and then being reflected at the end of the table and the edge of the fresco to return once more to the center.[69] A few years before painting the *Last Supper*, Leonardo had carefully observed interpenetrating waves generated by pebbles thrown into the still water of a pond.[70] Now he portrayed a similar effect in the realm of human emotions. The words of Christ, "One of you will betray me," are dropped into the solemn silence of the assembled company, where they generate powerful waves of emotion that propagate and interpenetrate through all the figures in the composition.

During his extensive travels in central Italy, Leonardo often spent time at the seashore, both on the Adriatic and the Ligurian seas.[71] During these visits, he invariably observed and analyzed the nature and types of waves, the breaking of surf upon the shore and the generation of mist, the deposits of debris on the beaches, the ebb and flow of the tides, the impacts of waves on rocky coastlines during sea storms, and other phenomena associated with waves. However, by far the largest part of his work on the movements of water was concerned with the nature of fluid flow.

Leonardo not only observed smooth and turbulent flows of water in rivers and canals and over weirs and waterfalls but also carried out flow experiments in controlled laboratory settings. For example, he would fill vessels of different shapes with water, disturb their surfaces with his hand, and observe the effects. In a simple experiment, he produced a rotational vortex. "When the hand is turned in circular movement in a vase half-filled with water," he noted, "it generates an artificial eddy that will expose the bottom of this vase to the air."[72] With the same method, he also generated more complex forms of turbulence: "The hand drawn frequently back and forth across the vase produces strange movements and surfaces of different heights."[73]

In a series of more sophisticated experiments, Leonardo built special flow channels for observing fine details of water movements. Here is how he described his experimental setup:

> Make one side of the channel of glass and the remainder of wood; and let the water that strikes there have millet or fragments of papyrus mixed in it, so that you can better see the course of the water from their movements. And when you have made the experiment of these rebounds, fill the bed with sand mixed with small gravel; then smooth this bed and make the water rebound upon it; and watch where it rises and where it settles down.[74]

These careful and detailed experiments are typical of the empirical method Leonardo used in all his scientific investigations. In his studies of fluid flow, he designed and tested several experimental methods that are still routinely employed in our modern laboratories five hundred years later. Foremost among them were his techniques of flow visualization. Even though he had exceptional powers of observation, he must have realized how difficult it is to accurately perceive the streamlines of turbulent flows. He repeatedly described how he added millet seeds and other types of grains to the water to make it easier for the eye to follow these complex movements:

> This experiment you will make with a square glass vessel, keeping your eye at about the center of one of these walls; and in the boiling water with slow movement you may drop a few grains of millet, because by means of the movement of these grains you can quickly know the movement of the water that carries them with it. And from this experiment you will be able to proceed to investigate many beautiful movements.[75]

To analyze turbulent flows in rivers, Leonardo observed the movements of leaves or pieces of wood, and he also used sawdust for making complex streamlines visible:

> If you throw sawdust down into a running stream, you will be able to observe where the water, turned upside down after striking against the banks, throws this sawdust back toward the center of the stream, and also the revolutions of the water, and where other water either joins it or separates from it, and many other things.[76]

In other experiments, he used dye to stain one of two coalescing water currents to find out how exactly they merge. These methods of flow visualization were rediscovered in the nineteenth century and have become standard practice in modern fluid dynamics to observe turbulent flows.[77] Leonardo applied similar methods for visualizing the streamlines in flows of air, observing the movement of clouds, snowflakes, smoke, dust, leaves, and other things carried by the currents of the wind. "The air moves like a river," he explained, "and drags the clouds with it just as running water drags all the things that float upon it."[78]

On a folio of the Codex Atlanticus, we find a particularly clear and elegant description of flow visualization with millet seeds (fig. 1-2). Leonardo has sketched a glass vessel filled with water and with millet seeds dropping in from above, and he explains that the purpose of the experiment is to find out how exactly the water empties out through a central hole at the bottom. He notes that the seeds drift in from the sides, as the pathlines in the drawing indicate. In another similar experiment, he uses

black seeds at the center and white seeds at
the periphery in order to see which parts of
the water pass through the hole first.[79] It is
noteworthy that Isaac Newton drew a diagram similar to Leonardo's in his famous
Principia (Proposition XXXVI, Problem
VIII) but got the pathlines wrong because
he had no means to visualize them.[80]

The Fundamentals of Flow

After many years of studying and analyzing the movements of water, Leonardo's
sharp observations and methodical experiments gave him a full understanding of the
main characteristics of fluid flow. He recognized the two principal forces operating
in flowing water—the force of gravity and

FIG. I-2. Flow visualization with
millet seeds. Codex Atlanticus,
folio 219r (detail).

the fluid's internal friction, or viscosity—and he correctly described many
phenomena generated by their interplay. He also realized that water is
incompressible and that, even though it assumes an infinite number of
shapes, its volume is always conserved. In addition, he recognized that
water changes its properties when materials are suspended or dissolved in
it and also when its temperature changes.

In a branch of science that did not even exist before him, Leonardo's
deep insights into the nature of fluid flow must be ranked as a momentous
achievement. That he also drew some turbulent structures erroneously
and imagined some flow phenomena that do not occur in reality does
not diminish his accomplishment, especially in view of the fact that
even today scientists and mathematicians encounter considerable difficulties in their attempts to predict and model the complex details of turbulent flows.

Leonardo was the first to study the dynamics of the smooth flow of a
river methodically. To do so, he watched corks float down a straight stretch
at various distances from the bank and counted "beats of time" to find out
how long it took the corks to pass through 100 *braccia* (approximately 200
feet). He also designed surveying instruments for measuring differences in
levels in order to calculate "how much a river falls per mile."[81] In addition,

he used specially designed floats, suspended at varying depths of a flowing river, to determine the water's relative speeds at different levels.[82]

Leonardo's analysis of these systematic experiments begins with the recognition that the flow of water is caused by the force of gravity. "One cannot describe the process of the movement of water unless one first defines what gravity is," he declares,[83] and then proceeds to explain the origin of gravity according to the Aristotelian four-element theory.[84] According to this view, commonly held in the Middle Ages and the Renaissance, the elements, when left to themselves, settle into concentric spheres with the Earth at the center, surrounded successively by the spheres of water, air, and fire (or light). However, the four elements do not always remain in their assigned realms but are constantly disturbed and pushed into neighboring spheres, whereupon they naturally try to return to their proper places. This is why, according to Aristotle, rain falls downward through the air, while air drifts upward in water, and the flames of fire rise up into the air.

Leonardo's discussion of gravity in connection with the flow of water clearly follows the Aristotelian theory. When the water is raised out of its sphere, he explains, it acquires weight and is pulled back by gravity, and this is true for all four elements:

> All the elements, though they are without weight in their own sphere, possess weight outside their sphere, that is, when moved toward the sky, but not when moved toward the center of the Earth.[85]

For most of his life, Leonardo held on to the Aristotelian concept of gravity as arising from the natural tendencies, or goals, inherent in all matter. But he often struggled when trying to explain various phenomena of mechanics within that traditional teleological framework (see p. 173), and in his mid-fifties, while he was working on organizing his notes on water in the Codex Leicester, he also expressed some doubts about the Aristotelian view:

> The water that moves through the river is either called, or chased, or it moves by itself. If it is called, or rather commanded, who commands it? If it is chased, who chases it? If it moves of itself, it shows itself to be endowed with reason, but bodies that continuously change shape cannot possibly have reason, for in such bodies there is no judgment.[86]

However, in his old age Leonardo was much more interested in the growth patterns and metabolic processes in plants, the movements of the human body, and the mystery of the origin of life, than in formulating a theory of mechanics that would go beyond the Aristotelian framework. He never revisited his doubts about the nature of gravity, and it took another two hundred years until the Aristotelian teleological view of forces was replaced by a radically different concept in the physics of Galileo and Newton (see p. 179).

At any rate, for Leonardo's analysis of fluid flow the Aristotelian view of gravity was sufficient, as it did not contradict any of his observations. When he was in his late forties, he designed a highly ingenious experiment to measure the acceleration due to gravity by using an inclined plane, as Galileo would do more than a hundred years later (see p. 193). Leonardo reasoned that, "although the motion is oblique, it observes in each of its degrees an increase in motion and in velocity in arithmetic progression."[87] In other words, he asserted that the velocity of a freely falling body is a linear function of time.

Realizing that, in cases of negligible friction, a ball falling freely through the air obeys the same mathematical law as one rolling down an inclined plane, Leonardo then applied the same reasoning to the flow of rivers. "If the rivers were straight with equal breadth, depth, and slant," he argued, "you would find that with each degree of movement they would acquire degrees of speed."[88]

It was clear to Leonardo, however, that the analogy between a ball rolling down an inclined plane and water flowing through an inclined riverbed was an idealization that neglected the effects of friction. As an engineer, he paid special attention to friction in his designs of machines and, in fact, was the first to systematically analyze frictional forces (see pp. 185–86). Accordingly, he was keenly aware of the internal friction of fluids, known as viscosity, and dedicated numerous pages in his Notebooks to analyzing its effects on fluid flow.

In modern physics textbooks, viscosity is defined as the internal friction, or "stickiness," between molecules in a fluid. Except for the concept of molecules, this was exactly how Leonardo described it. "Water has always cohesion in itself," he wrote in the Codex Leicester, "and this is the more potent as the water is more sticky."[89] The viscosity of flowing water generates friction between the liquid and its solid container, and also between adjacent layers flowing at different velocities. Leonardo clearly understood the effects of these frictional forces, known in modern fluid dynamics as shear stress. In particular, he recognized that the friction between a solid and a liquid is much larger than that between different layers of the fluid. "The bottom offers more resistance; this is why [the water] moves more on the surface than at the bottom," he noted in Manuscript H,[90] and in Manuscript I he recorded that "rivers, when straight, flow with much greater impetus in the center of their breadth than they do at their sides."[91]

Leonardo's understanding of the dynamics of the flow of rivers far surpassed that of his contemporaries. Until the second half of the eighteenth century, it was commonly believed that the velocity of water in a river increased from its surface to the bottom. The decrease of the flow velocity near solid walls, such as those of a riverbed, which Leonardo described clearly and accurately in several of his Notebooks, was rediscovered only in the nineteenth century.[92]

The third fundamental characteristic of fluid flow identified by Leonardo (in addition to the effects of gravity and viscosity), was the conservation of mass. He realized that the mass of any portion of flowing water is always conserved, and that, since water is incompressible, this portion will have the same volume no matter what shape it assumes. Leonardo was so impressed with this conclusion that he tried to develop a special type of geometry that would allow him to describe with mathematical precision the continual movements and transformations of water, in which its mass and volume are always conserved.[93] He called this geometry of transformations "geometry done with motion" and worked on it intensely during the last twelve years of his life. Today, we can see that Leonardo's mathematical transformations were early forms of topology, one of the most important fields in modern mathematics, which was fully developed by Henri Poincaré at the beginning of the twentieth century, some five hundred years after Leonardo's death.[94]

Leonardo's explorations of topological transformations never reached a stage where they could actually be used to model fluid flows. However, even though his mathematical techniques were not sophisticated enough for this ambitious task, he seems to have been on the right track. Instead of his topological transformations of visible geometric forms, physicists today use a mathematical method known as tensor calculus to describe continuous transformations of infinitely small volumes of fluids under the influence of gravity and shear stresses. The conservation of mass,

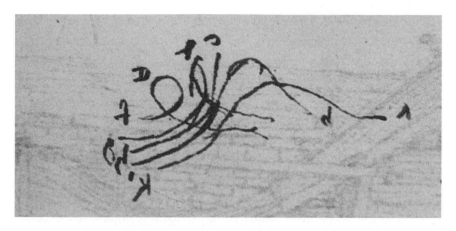

FIG. 1-3. Streamlines of a breaking wave. Ms. F, folio 20r (detail).

expressed in the so-called continuity equation, is a cornerstone of the modern mathematical theory, as it was for Leonardo.

Realizing that the mathematics of his time was inappropriate for describing the ceaseless movements and transformations of flowing water, and that his own geometry of transformations was too rudimentary for modeling complex flow phenomena, Leonardo chose a third option. Instead of mathematics, he used his exceptional facility of drawing to document his observations in pictures that can be strikingly beautiful while at the same time playing the role of mathematical diagrams. His numerous drawings of turbulent flow patterns are not realistic representations of single instances of observation, but are syntheses of many observations in the form of theoretical models. They range from simple wave patterns, in which the individual streamlines are labeled with letters (fig. 1-3), to highly complex combinations of swirling eddies and bubbles, as in the drawing of "water falling upon water" in the Windsor Collection (see fig. 1-13).

Measuring the Flow of Water

During the late fifteenth century, when Leonardo was employed as ducal engineer at the Sforza court in Milan and had to oversee numerous projects of hydraulic engineering as part of his duties (see p. 32), the measurement of the flow of water was an important technical and administrative issue. Agriculture in the expansive Lombard plain depended crucially on effective irrigation through the region's large network of rivers and canals, and all administrative and legal questions connected with water rights, concessions, and payments required accurate measurements of rates of flow under varying conditions.

However, the unit of measurement for the rate of flow (the volume flowing through a given cross section per unit of time) that was commonly used in Lombardy was based on an erroneous assumption, which made a fair and equitable distribution of water virtually impossible. Known as the *oncia milanese* ("Milanese ounce") this measure had been in use since the twelfth century. It was defined as the quantity of water discharged by a rectangular orifice of a certain dimension (equivalent to approximately 6" × 8"). While it is evident to us that the actual rate of flow also depends on the velocity of the water discharged by the orifice, which in turn depends on the level of water upstream, rates of flow were measured for hundreds of years simply in terms of cross-section areas without taking velocity into account.[95]

Leonardo's discussions of measuring and calculating quantities of flow, by contrast, exhibit a conceptual clarity that is truly impressive. It is a clarity derived from his thorough understanding of the fundamental characteristics of fluid flow and from his considerable powers of visualization. He realized that the rate of flow through a given cross section must be defined as the volume flowing through that area per unit of time and that, since the flow velocity is defined as the distance traveled by the water per unit of time, the rate of flow equals the product of cross-section area times flow velocity. The fact that the rate of flow is proportional not only to the area but also to the flow velocity is clearly stated in several Notebooks. In Manuscript F, for example, Leonardo writes:

> The quantities of ounces of water delivered through an orifice will be greater or lesser according to the higher or lower velocity of the water. Doubling the velocity will double the water in the same time; and in the same way, tripling the velocity will triple the quantity of water in the same time; and so it would go on to infinity.[96]

In fact, Leonardo was not the first to understand the concept of the rate of flow. The Greek engineer Heron of Alexandria had done so already in the first century A.D. In his *Dioptra*, a book on land surveying, Heron discussed the measurement of the flow of water from springs, and in this context he correctly defined the rate of flow and also stated the precise re-lationship between rate of flow, area, and velocity. However, Heron's work was subsequently forgotten and remained unnoticed for almost fifteen hundred years, until his insights were rediscovered by Leonardo at the end of the fifteenth century.

When he studied the flow of rivers, Leonardo correctly conceived of the rate of flow as the volume flowing through a particular cross section in a "beat of time," and since he knew that volume is conserved in all flows, he concluded that under steady flow conditions, "the river transports in every section of its length in the same time the same quantity of water."[97] He also realized that this implies that in steady flows, cross-sectional areas and flow velocities are inversely proportional:

> A river of uniform depth will have a more rapid flow at the narrower width than at the greater, to the extent that the greater width sur-passes the narrower.[98]

This is a clear and succinct statement of an important principle of fluid dynamics, known today as the continuity principle. It is the mathemati-cal expression of the fact that in a steadily flowing river, the flow needs to speed up in narrower passages to accommodate the same quantity of wa-ter. To illustrate this flow dynamic, Leonardo used the ingenious analogy of a regiment marching through a street of varying width while keeping its compact formation. A sketch in Manuscript A schematically shows a street with men marching through three sections with widths widening progres-sively in the ratios 1 to 4 to 8. In the accompanying text, Leonardo explains:

> Suppose that the men fill these avenues with their bodies, and that they must march in a continuous manner. When the men in the widest place take one step, . . . those in the intermediate width will take 2, and those in the third place, with a width one fourth of the second, will take 8 steps during that same time. And you will find this proportion in all movements that pass through places of vary-ing widths.[99]

In representing the flow of water in a river of varying cross section by a regiment of men marching through a street of varying width, Leonardo used a simplified, or approximate, model to analyze the essential features of the phenomenon he studied. In subsequent centuries, this technique became a central feature of the modern scientific method.[100] Most advances in science today are presented first in terms of such approximate models and are cast into more comprehensive and elaborate theories only much later. That Leonardo already used this modern scientific approach in the fifteenth century is quite extraordinary.

Leonardo's precise formulations of the rate of flow and the principle of continuity remained unpublished until the nineteenth century and had no influence on hydraulic engineering in his time, so far as we know. Rates of flow continued to be measured in terms of the flawed unit of the *oncia milanese* for another three hundred years, and the continuity principle, although known to hydraulic engineers at least qualitatively, was not formulated precisely again until Benedetto Castelli, a pupil of Galileo, did so in the seventeenth century. Thus, until the middle of the nineteenth century, Castelli rather than Leonardo da Vinci was considered the founder of the Italian school of hydraulics.[101]

The Water Vortex

At the center of Leonardo's investigations of turbulent flows lies the water vortex, or whirlpool. Throughout the Notebooks, there are countless drawings of eddies and whirlpools of all sizes and types.* These often very beautiful drawings are testimony to Leonardo's endless fascination with the ever-changing yet stable nature of this fundamental type of turbulence. As I have suggested earlier, this fascination likely came from a deep and correct intuition that the dynamics of vortices, which combine stability and change, embody an essential characteristic of all living forms (see p. 22).

Leonardo described correctly how eddies are formed when flowing water encounters an obstacle, such as stagnant water. "The current of moving water seeks to maintain its course according to the force that caused it," he wrote in Manuscript A, "and when it finds an opposing obstacle, it completes the length of the course it began by circular and whirling

---

* Eddies are small whirlpools; "vortex" is the scientific term for any fluid in whirling or rotary motion. Leonardo uses many terms for water vortices, including *retroso* (counter-flow), *revertigine* (reverting flow), *volta revertiginosa* (reverting turn), *moto cocleare* (spiral movement), and *voragine* (hollow).

FIG. I-4. Formation of counter-rotating vortices by
water flowing into a tank. Ms. A, folio 6or (detail).

movements."[102] On the verso of the same folio, Leonardo illustrated this
process with a sketch of water flowing into a tank (fig. 1-4). His observation that the circular flow will run its course until the energy is dissipated,
just as it would if it moved in a straight line, is completely correct. The
drawing itself is partly incorrect, however, as in reality only one pair of
counter-rotating vortices is formed.[103]

Leonardo was the first to study and understand the detailed motions
of water vortices, often drawing them accurately even in complex situations, for example in wakes behind boats and other obstacles in turbulent
water. He correctly distinguished between flat circular eddies in which
the water essentially rotates as a solid body, and spiral vortices (such as the
whirlpool in a bathtub) that form a hollow space, or funnel, at their center.* "The spiral or whirling motion of every liquid," he noted, "is so much
swifter as it is nearer to the center of its revolution. This fact, which we
point out, is worth noting, since the circular motion of a wheel is so much
slower as it is nearer to the center of the revolving object."[104]

In fact, circular eddies are not exactly flat but dip slightly in their
center. As the rotational velocity increases, this dip becomes more
pronounced and the circular eddy turns into a spiral vortex. Leonardo

---

* In modern fluid dynamics, circular whirlpools are known as rotational, and spiral whirlpools as
irrotational vortices.

observed accurately that the hollow space at the center of such a vortex will deepen with increasing rotational velocity:

> The eddy with the deeper hollow will be the one produced in water of swifter movement; and that eddy will have a smaller hollow if it is produced in deeper water which has not the same movement but is slower.[105]

Based on these observations, Leonardo identified three types of eddies: those level with the surface, those with raised centers (for example, in air turbulence), and those with depressed centers (for example, the whirlpool in a bathtub). He described the three types as "leveling the bottom," "filling up the bottom," and "hollowing out the bottom," respectively.[106] Leonardo's detailed studies of vortices in turbulent water and air were not taken up again for another 350 years, when the physicist Hermann von Helmholtz developed a mathematical analysis of vortex motion in the mid-nineteenth century.

For many years, Leonardo observed whirlpools and vortices in the currents of rivers and lakes, around piers and jetties, at the confluence of rivers, in the basins of waterfalls, behind objects of various shapes immersed in flowing water, and in laboratory tanks. He repeatedly attempted to classify different types of vortices according to their shapes—their topology, as we would say today. He made long lists of these types of turbulence. Here is a typical example from Manuscript F:

> Of the eddies on the surface and of those created at various depths of the water; of those that take up the whole of that depth, and of the moving and the stable ones; of the long and the round ones; of those that change their movement periodically and those which divide; of those which convert into the eddies they merge with; and of those that are mixed with falling and reflecting water and make it spin around.

> What kind of eddies turn light objects on the surface without submerging them? What kind are those that submerge them and make them spin around at the bottom, and then leave them there? What kind are those that detach things from the bottom and throw them

up to the surface of the water? Which are the slanting eddies, which are the upright ones, and which the level ones?[107]

If this list sounds confusing, it will be worth pointing out that even today, five hundred years after Leonardo, there is still no commonly accepted classification scheme for different types of eddies. As fluid dynamicist Ugo Piomelli observes wryly: "If you ask 200 turbulence experts, you will get between 190 and 200 different answers. And, what is more interesting, you will even find 10 or 20 different definitions of what an eddy is."[108] In other words, Leonardo's seemingly confusing classification schemes reflect, more than anything else, the great dynamic complexity of those swirling, ever-changing vortices.

Turbulent Flows

Looking through page after page of Leonardo's depictions and descriptions of turbulent flows, one is astonished to see how many subtle and complex features of turbulence he identified that are confirmed in modern fluid dynamics. For example, as noted earlier, turbulent flows often display a broad spectrum of scales with eddies of many different sizes forming continually (see p. 34). Leonardo was well aware of this fact:

> The great revolutions of eddies are rare in the currents of the rivers, and the small eddies are almost innumerable; and large objects are turned around only by large eddies and not by small ones, whereas the small objects are made to spin both by small eddies and by large ones.[109]

Leonardo recognizes here not only a whole spectrum of eddies of different sizes, but also a corresponding range of angular momenta, illustrated with objects of different sizes carried by the eddies.*

The Windsor Collection contains two beautiful drawings of a jet of water falling into a pool (figs. 1-5 and 1-13). In figure 1-5, Leonardo depicts the formation of a broad range of turbulent structures of various sizes. The region of turbulence is divided into two parts: a "frozen" regime showing different types of stable vortices, and a regime drawn in lighter chalk, in

---

* In physics, the angular momentum of a rotating body is related to the force necessary to overcome its rotational inertia.

FIG. 1-5. Study of turbulence generated by a jet of water
falling into a pond, c. 1509–11. Windsor Collection,
Landscapes, Plants, and Water Studies, folio 44 (detail).

which the turbulent kinetic energy cascades down through vortices of
smaller and smaller scales until it dissipates in the water's viscosity. (The
small sketch above the main drawing summarizes the principal movement
of falling and reflected water, while the sketch on the left is a visual reminder of the connection between circular waves and wave fronts.)

The entire drawing is an elaborate diagram of Leonardo's understanding of turbulence generated by the jet of water, which is shown to be correct in its essential aspects in modern flow visualizations. Its most remarkable feature is the energy cascade through vortices of decreasing scales and
subsequent energy dissipation, known as a Richardson cascade in modern
fluid dynamics, which Leonardo qualitatively anticipated five hundred
years before it was formulated by the physicist Lewis Richardson.[110]

A large number of Leonardo's studies of turbulence are concerned with
the flow of water around immersed objects, known in modern fluid dynamics as bluff body flows. In the small Notebook known as Manuscript
H, for example, which Leonardo carried with him during his observations of flowing water in nature, we find a series of sketches of bluff body
flows around objects of various shapes (fig. 1-6). As in many other similar
drawings, Leonardo here correctly shows how the flow pattern is attached
to the object upstream but generates separated vortices (so-called vortical
shedding) downstream.

When looking at these sketches, it is important to realize that they are not realistic, instantaneous pictures of actual flow lines. Leonardo probably observed a series of fluctuating flow patterns, visualizing them with the help of a sequence of millet seeds, identified certain types of turbulence, and then recorded his observations in summary diagrams. Hence, the sketches we see in the Notebooks are superpositions of observed instantaneous structures, resulting in pictures of average flow patterns. They are Leonardo's conceptual models of turbulent flows.

For example, the vortex pairs in the top two sketches in figure 1-6 would not appear simultaneously but alternately in rapidly changing, fluctuating flows. What the sketches show are the average flow patterns. The same technique is used by scientists today with the help of sophisticated time-lapse photography. Centuries earlier, Leonardo relied on simple methods of flow visualization and on his acute powers of observation to achieve remarkably similar results.

The Codex Leicester is full of sketches of bluff body flows, neatly arranged along the right-hand margin of the text. Figure 1-7, for example, shows water flowing around a column. Again, Leonardo has drawn a symmetrical picture recording the

FIG. 1-6. Flow patterns around immersed objects, c. 1493–94. Ms. H, folio 64r.

FIG. 1-7. Bluff body flow around a column. Codex Leicester, folio 22r (detail).

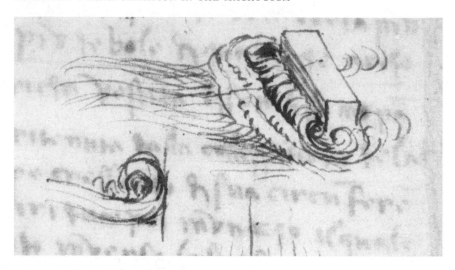

FIG. 1-8. Horseshoe vortex.
Codex Leicester, folio 25v (detail).

average flow pattern. The actual flow would consist of staggered fluctuating vortices. Similarly, the counter-rotating pairs of vortices in figure 1-4 should be interpreted as symmetrical averages of the actual flow pattern.

In another remarkable drawing, also from the Codex Leicester, Leonardo correctly depicts what is now called a horseshoe vortex around a prism-shaped obstacle (fig. 1-8). The basic shapes of the head and the two legs of the horseshoe are clearly shown. However, the direction of rotation of the head is incorrect. It should be clockwise, not counterclockwise as drawn by Leonardo. (Since the water on the surface flows faster than that below, the top layer will turn down and then rise up again when the flow encounters the obstacle.)

In addition to the relatively simple sketches in the Codex Leicester, Leonardo produced several elaborate drawings of highly complex patterns of turbulence, generated by placing various obstacles into flowing water. Figure 1-9, from the Windsor Collection, shows the turbulent flows around a rectangular plank inserted at two different angles. (Additional variations are suggested in the small sketches to the right of the main drawing.) The upper drawing clearly shows a pair of counter-rotating vortices (fluctuating in the actual flow) at the head of a stream of random wake. The essential details of this complex pattern of turbulence are completely accurate—a testimony to Leonardo's powers of observation and conceptual clarity.

FIG. 1-9. Turbulent wakes behind a rectangular
plank, c. 1509–11. Windsor Collection, *Landscapes,
Plants, and Water Studies*, folio 42v (detail).

As noted earlier, the origins of modern fluid dynamics go back to the
nineteenth century, when Osborne Reynolds identified a parameter that
allowed scientists and engineers to neatly characterize different types of
turbulence (see p. 34). At the same time, Reynolds also discovered that
the mathematical description of turbulent flows can be decomposed into a
main flow (averaged over time) and a fluctuating part. Leonardo also recognized the difference between these two components of turbulent flows,
albeit only in a qualitative way. On a famous sheet of the Windsor Collection, he compared the turbulent flow of water to the growth of curly
hair. He drew three turbulent wakes behind rectangular obstacles next
to a drawing of a curly lock of hair that looks virtually identical (fig. 1-10).
Below the drawings, he noted: "Observe the motion of the surface of the
water, which resembles that of hair, which has two motions, one of which
is caused by the weight of the hair, the other by the direction of the curls."
And then he continued:

> Thus, the water has its eddying motions, one part of which is due to
> the impetus of the principal current, the other to the incident and
> reflected motion.[111]

With this statement, Leonardo qualitatively anticipated what is now
known as the Reynolds turbulence decomposition nearly four hundred
years before Osborne Reynolds formulated it mathematically.[112]

In all his observations of turbulent flows, Leonardo clearly recognized

FIG. 1-10. Turbulent wakes and strands of curly hair, c. 1510–13. Windsor Collection, *Landscapes, Plants, and Water Studies*, folio 48r.

that, in spite of incessant changes and fluctuations, turbulence also produces surprisingly stable forms. With this recognition he anticipated another important concept of modern fluid dynamics, where these organized forms of turbulence—whirlpools and vortices at all scales—are known as coherent structures. Leonardo drew them repeatedly, and he commented on their stability in several Notebooks. In Manuscript I, for example, we find the following rather precise observation:

> Eddies, with various revolving movements, proceed to consume the initial impetus; and they do not remain in the same positions, but after they have been generated, turning thus, they are borne by the impetus of the water in the same shape, so that they come to make two movements: one they make within themselves by their revolution; the other as they follow the course of the water that carries them along.[113]

Like so many of Leonardo's acute observations on turbulence, this one, too, is fully borne out by modern flow visualizations.

Water Falling upon Water

A type of turbulent pattern that Leonardo found particularly intriguing is that caused by a jet of water falling into a calm pool. Manuscripts A, F, and I, as well as the drawings in the Windsor Collection, contain many studies of waterfalls and jets impinging on still water. In several of them, the jet enters the area above the pool through a rectangular orifice, an "industrial" feature that is reminiscent of Leonardo's frequent work on locks and canals.

In some of these studies, Leonardo draws turbulent structures of a special kind, which have been called "corkscrew waves" by art historians (fig. 1-11). These helical structures do indeed occur in special types of turbulence. I believe, however, that what we are seeing in Leonardo's sketches are not records of actual observations, but rather a particular theoretical model that he uses to explain the vortices generated in waterfalls. Indeed, in the Codex Atlanticus, there is an elaborate drawing of a rope coiled first into a helix and then into a double helix (fig. 1-12), which is reminiscent of the helically coiled vortices in the waterfall drawings. This strongly suggests that Leonardo used such a rope as a model for his torus- (or "doughnut"-) shaped vortices.

Leonardo's most elaborate studies of water falling into a pool are the two drawings in the Windsor Collection mentioned earlier (see p. 49). His celebrated drawing of "water falling upon water," in particular, is an impressive synthesis of numerous previous studies, which integrates various types of turbulence into a compact and beautiful picture (fig. 1-13). It is obvious that this is not a realistic snapshot of a jet of water falling into a

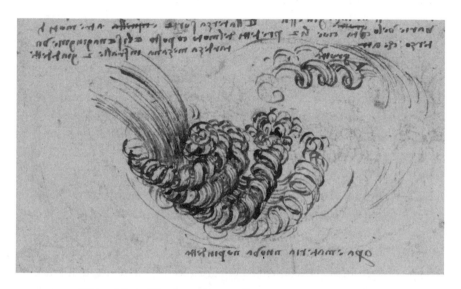

FIG. 1-11. Waterfall and "corkscrew waves."
Windsor Collection, *Landscapes, Plants, and Water
Studies,* folio 45r (detail).

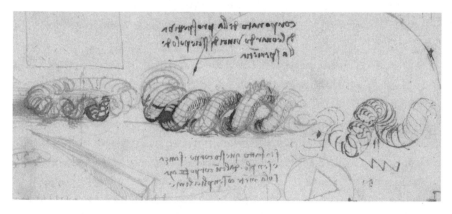

FIG. 1-12. Rope coiled into a helix and a double helix.
Codex Atlanticus, folio 520r (detail).

pond, but an elaborate diagram of Leonardo's analysis of the complex set of turbulences caused by the jet.

The accompanying description begins with the statement that there are four types of turbulence involved:

> The movements of the falling water after it has entered the pool are three in kind, and to these a fourth is added, which is that of the air submerging itself with the water.[114]

Leonardo then continues with a detailed description of these movements, which goes over several paragraphs and is rather confusing. The four movements he has identified seem to be the movement of water and trapped air vertically down into the pool; the return of water and air bubbles up to the surface; the bursting of air bubbles and their water falling

FIG. 1-13. "Water falling upon water," c. 1508–9. Windsor Collection, *Landscapes, Plants, and Water Studies*, folio 42r (detail).

back into the pool; and the swirling movement of water on the surface in the form of large eddies.

Leonardo's analysis is a mixture of accurate observations and erroneous assumptions. He correctly observes that air is drawn into the water (a process known today as "entrainment"), and that the water bubbles rise to the surface where they burst into clusters of rosettes. However, he erroneously assumes that after the bubbles burst, their water not only falls back onto the surface but penetrates the pool down to the bottom. In the drawing, this strange assumption is illustrated with a series of oval, downward slanting eddies. The fourth type of turbulence, finally—the large eddies generated on the periphery around the jet's impact—is again an illustration of a correct observation. Modern flow visualizations indeed show that the eddies become progressively larger as they move away from the zone of impingement.

In a striking contrast between his written and visual analysis, Leonardo's verbal analysis is tortuous and clumsy, while his visual analysis is compact and beautiful. Both are partly incorrect, but Leonardo was able to take a very chaotic situation and simplify it in terms of a drawing that serves as a concise mathematical diagram. This is exactly what chaos theorists do today. They bring order into chaos in terms of visual lines and shapes, albeit in an abstract mathematical space.[115]

Turbulence in Water and Air

As noted earlier, Leonardo studied not only the flow of water but also the flows of blood, wine, oil, and even flows of various grains. He recognized that the basic properties of flow are the same for all fluids, and he emphasized especially the similarities between flows of water and air (see p. 33).

The currents and turbulences of air were of special interest to him because an understanding of them was crucial to the pursuit of one of his great passions—the science of flight and the design of flying machines (see pp. 250ff.). "In order to give the true science of the movement of birds in the air," he declared, "it is necessary first to give the science of the winds."[116] To develop such a science of the winds, Leonardo observed and analyzed the swirling movements of air as meticulously and persistently as he studied the turbulences in flowing water.[117]

In addition, Leonardo frequently studied turbulences in mixtures of water and air. Throughout his life, he was fascinated by natural catastrophes, and he produced numerous drawings of floods and tempests full of swirling currents of air, mixed with rain and pieces of uprooted trees, sand, and rocks.[118] These studies culminated in a series of a dozen extraordinary drawings in somber black chalk known as the "deluge drawings," which are now part of the Windsor Collection and are accompanied by power-ful narratives of apocalyptic visions. Leonardo created them in Rome at the age of sixty-two when he was lonely, depressed, and given to morbid thoughts.[119]

The deluge drawings are violent and disturbing. They show trees be-ing uprooted and torn to pieces, rocks crumbling in the torrential rains, houses and entire towns being demolished by raging storms. The air is dark and gloomy. The landscape can just barely be made out. There are no human victims to be seen; they are rendered irrelevant by the huge scale of the deluge. The overwhelming impression is one of despair, of human frailty and futility in the face of nature's cataclysmic forces.

The first drawing in the series (fig. 1-14) shows the beginning of the deluge. In the upper right corner, clouds gather ominously and the first

FIG. 1-14. The beginning of the deluge, c. 1514.
Windsor Collection, *Landscapes, Plants, and Water Studies*, folio 57.

FIG. I-15. *Cyclone overcoming a town*, c. 1514. Windsor Collection, *Landscapes, Plants, and Water Studies*, folio 62.

swirling vortices of rain-soaked air have formed. In the landscape below, trees are bent down to the ground and water in a lake is stirred up into huge waves by the violent storm. In a further drawing (fig. I-15), we see the deluge in full force. The sky is filled with with giant menacing vortices; torrential rain is falling; a town is being devastated by the cyclone; rocks are whirled through the air by powerful eddying currents.

Figure I-16 shows the same raging vortices of water, air, and crumbling rocks, but here Leonardo has formalized his representation of the cataclysm to such an extent that the emotional impact is diminished and the drawing is much closer to a mathematical diagram. In the accompanying narrative, accordingly, highly emotional passages are interspersed with detached, analytical ones that include precise descriptions of cascades and currents and detailed instructions on how to paint optical effects generated by storm clouds and falling rain.

These drawings are superb examples of Leonardo's synthesis of art and science. He drew them from memory in his old age, not to present the results of observations or experiments, but to picture a mythical situation—the great deluge, the ultimate turbulence. The drawings carry a powerful emotional message, and at the same time they are astonishingly accurate in their renderings of cascades and currents in water and air.

Looking through the entire set of drawings, one can recognize many of the shapes Leonardo explored in his lifelong observations of turbulent flows. There are two-dimensional eddies (for example, in the upper left corner of fig. 1-16); three-dimensional spirals (top of fig. 1-15); torus-shaped vortices (upper part of fig. 1-14); pairs of counter-rotating eddies (upper part of fig. 1-16); and "corkscrew waves," created by the impact of rain on the ground (right side of fig. 1-16). An archetypal, idealized vortex is shown

FIG. 1-16. Formalized deluge study, c. 1514. Windsor Collection, *Landscapes, Plants, and Water Studies*, folio 59.

again and again in different scales and from different perspectives. Taken together, these deluge drawings are nothing less than a visual catalogue of cascades and vortices. They are Leonardo's final summing-up of his knowledge of turbulence.

The Spiral Form

In most of the deluge drawings, as well as in the study of "water falling upon water" (see fig. 1-13), Leonardo drew large eddies with small spirals at their ends. These forms are inaccurate. In reality, the turbulences would be circular or elliptical, but the differences can be very subtle and difficult to detect without sophisticated experimental techniques.

I believe that Leonardo liked to draw these eddies with spirals at their ends because he saw the spiral as an archetypal form of turbulent flow and, more generally, as a symbol of life. I have argued earlier that he intuitively recognized the dynamic of the spiral vortex—stable and yet continually changing—as symbolic of all living forms (see p. 22), and I believe that this can also be said of the spiral form in general. In his wide-ranging observations of natural forms, Leonardo could not fail to notice the spiral growth patterns of marine shells and plants. He paid special attention to the ways in which leaves or branches spiral around a central axis in many species of plants and trees, and he drew these spiral growth patterns with complete botanical accuracy (see p. 106). Sometimes, Leonardo also drew exaggerated forms of spiraling foliage, the most famous being that of the "Star of Bethlehem" (plate 6), which bears a striking resemblance to a water vortex.

Leonardo's fascination with spiral movements can also be seen in many of his paintings, especially in the portraits. With his frequent use of spiral body configurations (for example, in the *Saint John the Baptist* and the *Leda*), Leonardo created the form of the serpentine figure that was used extensively by Michelangelo and became one of the fundamental forms of elegance in the High Renaissance. In the words of art historian Daniel Arasse: "Leonardo used the serpentine form to bring life and movement into his figures, thus inventing the classic 'gracefulness' of representation."[120]

For Leonardo, the spiral form was the archetypal code for the ever-changing yet stable nature of living forms. He saw it in the growth patterns of plants and animals, in curling locks, in human movements and gestures, and above all in the swirling vortices of water and air. The movement of water is the grand unifying theme in Leonardo's science of living forms. Water is the life-giving element flowing through the veins of the Earth and the blood vessels of the human body. It nourishes and sustains all living bodies. Its forms, like theirs, are fluid and always varying. It is a major source of power and for eons has shaped the surface of the living Earth, gradually turning arid rocks into fertile soil. With its infinite variety of form and movement—as rivers and tides, clouds and rain, cascades and currents, eddies and whirlpools—water flows through Leonardo's art and interlinks the main fields of his science.

2

The Living Earth

From his early youth, Leonardo was fascinated by pinnacles of rocks, carved out by water and eventually turning into gravel and fertile soil, and he came to see water as the chief agent in the formation of the Earth's surface. As a young boy, he explored the rocky outcroppings, waterfalls, and caves in the Tuscan countryside around Vinci. Later on, they became the defining elements of his personal mythical landscape— the fantastic rock formations that would forever appear in the shadowy backgrounds of his paintings.

Leonardo's keen awareness of the continual interaction of water and rocks motivated him to undertake extensive studies in geology. During his travels in central and northern Italy as a military and hydraulic engineer, he studied the erosion of rocks, deposits of gravel and sand, and strata of sedimentation and produced many detailed maps of the regions he visited. His geological observations, like those in all the other branches of his science, display an astonishing accuracy. On the basis of these extensive and methodical analyses, he formulated a series of geological principles with a clarity that would not be achieved again until the twentieth century— principles that are still taught in geology courses today.[1]

The Bones, Flesh, and Blood of the Earth

Leonardo's expositions of geology, and especially his ideas about the formation of the Earth, sound so modern because, unlike most of his contemporaries, he was fully aware of the long duration of geological time, and, like geologists today, he viewed the Earth as a dynamic and continually changing entity. The philosophical basis of this conception was the idea—central to Leonardo's science—that the Earth as a whole is alive and that the patterns and processes in the microcosm, the human body, are similar to those in the macrocosm, the body of the living Earth.

As in many of his scientific investigations, Leonardo begins by summarizing this ancient doctrine. The earliest of his summaries is found in Manuscript A, written around 1490–92, during his first period in Milan:

> Man has been called by the ancients a lesser world, and certainly the term is well applied; because, since man is composed of earth, water, air, and fire, this body of the Earth is similar. If man has within himself bones as support and framework for the flesh, the world has the rocks as support of the earth. As man has within himself the pool of blood where the lungs increase and decrease with the breathing, so the body of the Earth has its ocean tide, which also increases and decreases every six hours with the breathing of the world. As from the said pool of blood originate the veins that spread their branches through the human body, so the ocean fills the body of the Earth with infinite veins of water.[2]

After this summary of the views of the ancients, Leonardo immediately notes that there is also an important difference between the body of the Earth and the human body: the Earth has no tendons. He argues that there is no need for tendons in the Earth, because the Earth (according to Aristotelian cosmology) does not move:

> In the body of the Earth, the tendons are lacking, and these are not there because tendons are made for the purpose of movement. And as the world is perpetually stable, no movement takes place; and since no movement takes place, the tendons are not necessary. But in all other things [man and the Earth] are very similar.[3]

Eighteen years later, Leonardo restates the idea of the living Earth in a more elaborate way. Whereas in his earlier statement, the analogy between the human body and the body of the Earth was based on the Aristotelian notion that both are composed of the four elements, he now goes beyond the level of a mere analogy. He justifies the conception of the Earth being alive by observing that the processes of growth and renewal, which are common to all life, are pervasive on Earth:

Feathers grow on birds and change every year; hairs grow on animals and every year they change, except in some parts, like the hairs of the beards of lions, cats, and the like. Grass grows in the fields and leaves on the trees, and every year they largely renew themselves.[4]

He concludes from this observation that the Earth must be endowed with a vital force, or soul,* which generates forms and processes that are similar to those in the human body:

We may therefore say that the Earth has a vital force of growth, and that its flesh is the soil; its bones are the successive strata of the rocks which form the mountains; its cartilage is the porous rock, its blood the veins of the waters. The lake of blood that lies around the heart is the ocean. Its breathing is the increase and decrease of the blood in the pulses, just as in the Earth it is the ebb and flow of the sea.[5]

As noted earlier, late in his life Leonardo abandoned the analogy between the blood vessels of the human body and the water veins of the Earth as being too restricted; he realized that the pathways of the water cycle in the macrocosm and the blood cycle in the microcosm, as well as the forces driving them, are quite different (see p. 29). However, he always maintained the basic conception of the living Earth. Whenever he explored the forms of nature in the macrocosm, he also looked for similarities of patterns and processes in the human body. In doing so, he went beyond the general analogies that were common in his time and drew parallels between very sophisticated observations in both realms. For example, as I shall discuss in subsequent chapters, he used his knowledge of turbulent flows of water to understand subtle details of the movement of blood in the heart and aorta (see pp. 300ff.). He saw the "vital sap" of plants as their essential life fluid and observed that sap nourishes the plant tissues as blood nourishes the tissues of the human body (see pp. 117–18). He took these observations as compelling testimonies to the unity of life at all scales of nature.

* Leonardo uses the Aristotelian term *anima vegetativa* (the "vegetative soul") to describe the vital force that was traditionally viewed as the source of all life (see Capra, *The Science of Leonardo*, pp. 147–48).

Leonardo's conception of the Earth as being alive, manifesting patterns and processes common to all living systems, was a forerunner of the modern Gaia theory, which views our planet as a living, self-organizing, and self-regulating system.[6] In fact, the originator of the contemporary theory, atmospheric chemist James Lovelock, even used analogies in one of his books that are somewhat similar to Leonardo's, except that Lovelock compared the Earth to a giant redwood tree rather than to the human body.[7]

Leonardo's views of rocks, soil, and water as the bones, flesh, and blood of the living Earth was the philosophical and perhaps even spiritual foundation of his lifelong fascination with dramatic rock formations, shaped by the action of water. He not only analyzed in his scientific writings how the body of the Earth is continually transformed by the interplay of erosion and sedimentation but also pictured these processes in many of his paintings and drawings with astonishing accuracy and persuasive power. "No artist, before or since," writes art historian Martin Kemp, "has quite equaled the suggestive magic with which [Leonardo] insinuates his vision of the inner and outer oneness of created forms into drawings and paintings."[8]

All of Leonardo's science is utterly dynamic. He portrays nature's forms in ceaseless movement and transformation, recognizing that living forms are continually shaped and transformed by underlying processes. This dynamic conception of nature is evident in his studies of anatomy, botany, and fluid dynamics, and it is perhaps most striking in his geology. In Leonardo's view, the Earth, being a living body, is continually shaped and transformed over long periods of time. This is why his geological concepts sound so modern to us.

In the introductory chapter of a geology textbook published in 1995, we find the following passage.

The Earth is a dynamic planet that has continuously changed during its 4.6-billion-year existence. . . . We can easily visualize how mountains and hills are worn down by erosion and how landscapes are changed by the forces of wind, water, and ice. Volcanic eruptions and earthquakes reveal an active interior, and folded and fractured rocks indicate the tremendous power of the Earth's internal forces.[9]

This passage, except for a couple of modern terms, would not look out of place in Leonardo's Notebooks.

The Sculpting of Mountains and Valleys

Like other hydraulic engineers in the Renaissance, Leonardo was very familiar with the erosion of river banks, especially during the periodic flooding of Alpine rivers in the Lombard region (see p. 32). His Notebooks, notably the Codex Leicester, contain numerous suggestions for dealing with these practical problems. Unlike his fellow engineers, however, Leonardo went beyond empirical rules, integrating his acute observations into a far-reaching theory of how water, over long periods of time, has sculpted mountains and valleys on the surface of the Earth.

To begin with, Leonardo explains that erosion results from the friction between water and earth: "The actual rivers have a clouded flow because of the earth that rises in them as a result of the friction of their waters on the bed and the banks."[10] Then he describes in great detail how this process of erosion creates entire valleys over time:

> Although it is almost level, many rivulets will originate in the lowest parts of the surface, and these will begin to hollow out and form receptacles for other surrounding waters. In this way, every part of their course will become wider and deeper, their waters steadily increasing, until all this water will drain away. And these concavities will become the courses of the torrents that receive water from the rains, and thus the banks of these rivers will continue to erode until the spaces in between become steep hills.[11]

From these observations, Leonardo concludes that all valleys are created by flowing water: "Each valley has been created by its river, and there is the same proportion between valleys as between rivers."[12] Remarkably, he states that this assertion can be verified by correlating layers of rocks on the two sides of a valley. He argues that the occurrence of the same sequence of superimposed rock strata on opposite slopes of a valley is the clearest proof of the fact that valleys were created by water cutting into high mountains:

> The rivers have all sawn through and divided the members of the great Alps one from another; and this is manifest from the order of the layered rocks in which, from the summit of the mountain down to the river, one sees the strata on one side of the river corresponding with those on the other.[13]

With this argument, Leonardo was two hundred years ahead of his time. The superposition of rock strata would not be recognized and studied in similar detail until the second half of the seventeenth century.*

In his careful studies of erosion, Leonardo gave detailed descriptions of the transport and deposition of rock fragments by rivers and streams. In particular, he noted the sequence of what a river deposits as it flows from the high mountains down to the river's mouth:

> The river that flows out from the mountains deposits a great quantity of large stones in its gravelly bed, and these stones still retain some of their angles and sides; and as it proceeds on its course it carries with it smaller stones with angles more worn away, and so the large stones become smaller; and farther down it deposits coarse and then minute gravel, . . . and after this follows sand, at first coarse and then fine, and then coarse and fine mud . . . and then the mud becomes so fine that it seems almost like water . . . and this is the white earth that is used for making jugs.[14]

Today, this sequence of processes—known as rounding, sorting, and sedimentation—is recognized as a basic principle of geology.† "I learned this rule," remembered evolutionary biologist Stephen Jay Gould, "as principle number one on day number one in my college course in beginning geology."[15]

Like modern geologists, Leonardo viewed the erosion, transformation, and sedimentation of rock fragments as one single process through which water continually shapes the surface of the Earth. "Water wears away the mountains and fills up the valleys," he wrote in the Codex Atlanticus. "If it could, it would like to reduce the Earth to a perfect sphere."[16] He carried out detailed studies of the shapes and dynamics of rivers, the hollowing out and silting of their beds, and the formation and development of meanders. In addition, he produced many beautiful and accurate maps of watersheds and lakes.

---

* Stratigraphy, the branch of geology concerned with the layering, or stratification, of rocks, was pioneered in the seventeenth century by the Danish anatomist and geologist Nicolas Steno (Niels Stensen); see Gohau, *A History of Geology*, pp. 61ff.

† Sedimentary particles are classified in modern geology according to their sizes and are known, in the order of decreasing size, as gravel, sand, silt, and clay.

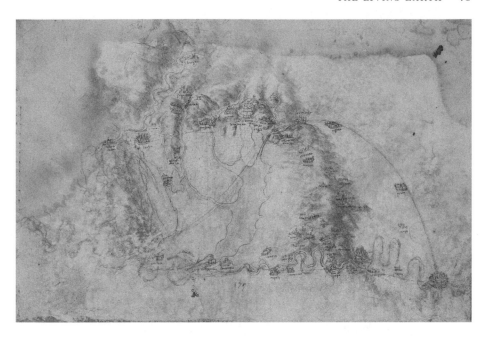

FIG. 2-1. Topographical map of the region northwest of Florence, showing Lucca, Pistoia, and Prato in the north and the Arno valley in the south, as well as the trajectory of the canal envisaged by Leonardo, c. 1503–4. Windsor Collection, *Drawings and Miscellaneous Papers*, RL 12685r.

Leonardo's Maps

Even the casual viewer cannot fail to notice the surprisingly modern look of Leonardo's geographical maps. He used cartographic techniques that surpassed anything attempted by medieval and Renaissance mapmakers. His maps often show distances based on elaborate odometer readings, involving ingenious instruments he himself had designed;[17] elevations are indicated by washes of different colors and shades, as in our modern atlases; and in some maps he uses a special relief technique to create "aerial views" of the depicted landscapes (figs. 2-1 and 2-2). In these beautiful and very detailed topographical renderings, the rivers, valleys, and settlements are pictured in such a realistic manner that one can have the eerie feeling of looking at the landscape from an airplane.

In most of his maps, Leonardo focused specifically on networks of rivers and lakes, often because he contemplated some hydraulic project in that region. For example, several detailed maps of the Arno basin (for example, fig. 2-1) were produced in connection with his plan of building a

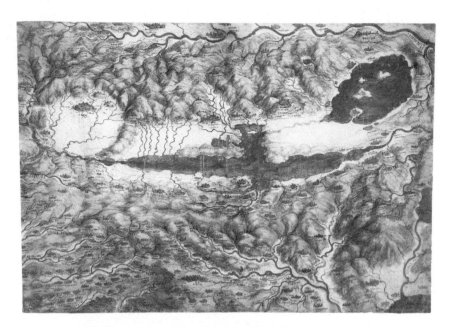

FIG. 2-2. Bird's-eye view of the Chiana Valley, showing the
cities Arezzo, Perugia, Cortona, and Siena, c. 1502.
Windsor Collection, *Drawings and Miscellaneous Papers*, RL 12278r.

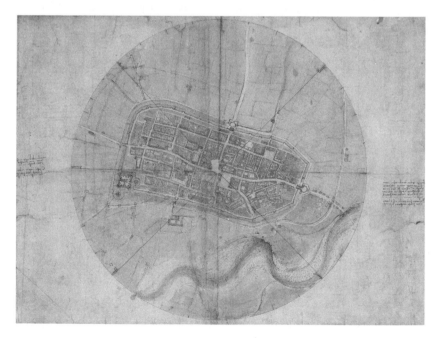

FIG. 2-3. Map of Imola, 1502.
Windsor Collection, *Drawings and Miscellaneous Papers*, RL 12284.

giant canal to bypass the river Arno.[18] Similarly, the beautiful large map
of the Chiana Valley (fig. 2-2) may have served to illustrate the idea to use
Lake Trasimene, pictured prominently in the map, as a source of water for
the Arno during the dry summer months.

A curious feature of this map is the fact that Leonardo has severely
distorted the relative scales of its center and periphery in order to fit the
outlying parts into the given format, while correctly picturing the continuity of the terrain and its intricate waterways. I have argued that this may
be viewed as an example of the "topological" thinking that is also evident
in many of Leonardo's architectural studies and mathematical diagrams.[19]

Because they served as supporting documents for military strategies
or projects of hydraulic engineering, Leonardo's maps often had to be
very precise and accurate, but many are also works of art that depict human settlements and built structures within their natural environment.
In these drawings, the watercourses and geological features embody the
vitality and continual movement of the Earth's living forms. This strong
sense of the living Earth is conveyed even in some of Leonardo's maps of
cities. The most famous is a very detailed map of Imola, a small town in
Emilia-Romagna where Leonardo spent several months during the winter
of 1502 as military engineer in the service of Cesare Borgia.[20] He designed
new fortifications for the town's citadel and drew a highly original circular map (fig. 2-3), in which the main streets, town walls, and strategic
routes are pictured with great precision, along with various bearings and
distances to other cities.

What makes the map of Imola so remarkable and so beautiful is Leonardo's artistic choice of placing the precise "aerial" view of the fortified
town within a natural environment of fields, paths, and various dwellings
near the town, including in particular the sweeping meanders of the river
Santerno. In the words of art historian Daniel Arasse,

> This map, a real synthesis of science and art, reflects the feeling that
> Leonardo has for the natural world, and for the way, simultane-
> ously glorious and fragile, human activity expresses its rationality
> in it. While the survey of the housing of Imola impresses by the
> elegance of its geometry, the striking curves made by the flow of the
> river Santerno reveal, like the vein-like tributaries of the rivers and
> streams of central Italy, the feeling for the "life of the earth," this
> organism, this living macrocosm.[21]

Geological Time

Leonardo carried out careful and detailed studies not only of erosion but also of the corresponding process of sedimentation—the deposition of suspended rock fragments at the mouths of rivers and in the ocean. He recognized that sedimentation often occurs in successive layers:

> The stratified rocks of the mountains are all in layers of mud deposited one above another by the floods of the rivers. . . . The different thicknesses of the strata of the rocks are created by the different floods of the rivers, that is the greater and the lesser floods.[22]

In particular, Leonardo discerned a specific sedimentary structure known to modern geologists as "graded bedding," in which the grain size of the sediments decreases upward within a single layer, or "bed," and he correctly explained its formation:

> Each layer is composed of heavier and lighter parts; the lowest being the heaviest. And the reason for this is that these layers are formed by the sediments from the waters discharged into the sea by the current of the rivers that flow into it. The heaviest part of this sediment was the part that was discharged first in the sequence.[23]

In his drawings and paintings, Leonardo pictured many geological structures with extraordinary accuracy. In his masterwork *Madonna and Child with Saint Anne*, a beautiful example of graded bedding is shown at the feet of Saint Anne (plate 7). The decrease of grain size from the bottom of the bed to the top is clearly visible.

Leonardo's recognition of temporal sequences in the strata of soil and rock and of the way in which valleys are created by flowing water led him to a momentous conclusion: that the forms of the Earth are the result of slow processes taking place over long epochs of what we now call geological time. "Since things are far more ancient than letters,"* he wrote in the Codex Leicester,

> it is not to be wondered at if in our days no record exists of how these seas extended over so many countries. . . . But for us it is sufficient to have the testimony of things born in the salt waters and found again in the high mountains, sometimes far away from the seas.[24]

---

* This passage appears in Leonardo's elaborate analysis of fossils; see pp. 81ff.

FIG. 2-4. Horizontal outcrop of rock, c. 1510–13.
Windsor Collection, *Landscapes, Plants, and Water Studies*, folio 53r.

With this view, Leonardo was centuries ahead of his time. Geologists became aware of the great duration of geological time only in the early nineteenth century with the work of Charles Lyell, who is often considered the father of modern geology.[25] Leonardo fully realized the importance of the understanding of geological time for scientific knowledge as a whole. "Knowledge of the past and of the surface of the Earth is ornament and food for human minds," he wrote in the Codex Atlanticus.[26]

Leonardo was also the first to introduce the notion of folds of rock strata (*piegamenti delle falde delle pietre*).[27] His ideas of how rocks are formed over enormously long periods of time in layers of sedimentation and subsequently shaped and folded by powerful geological forces come close to an evolutionary perspective. Leonardo arrived at this perspective three hundred years before Charles Darwin, who also found inspiration for evolutionary thought in geology, in particular in the works of Lyell.[28]

The Windsor Collection contains a drawing of a horizontal outcrop of rock (fig. 2-4) that illustrates Leonardo's conception of evolutionary geological processes in dramatic fashion. It shows horizontal strata of sedimentary rock that have been severely eroded and warped. "This drawing

FIG. 2-5. A stream running through a rocky ravine, c. 1483.
Windsor Collection, *Landscapes, Plants, and Water Studies*, folio 3r.

clearly manifests the shaping hand of time," writes Kemp, "and bears wit-
ness to the great forces that have molded the strata into their contorted
configurations."[29]

One geological formation that fascinated Leonardo throughout his life
was the pinnacle of jagged rocks that appears so often in the backgrounds
of his paintings. These rocky towers are remnants of long erosional processes that have stripped away softer layers of rock and exposed harder,
more weather-resistant rock underneath. These processes of weathering
and erosion are shown clearly in the very first of Leonardo's drawings of
rock formations, a picture of a rocky ravine with water birds produced
around 1483, shortly after his arrival in Milan (fig. 2-5).

The dominant feature of the drawing is a cliff that has lost almost all
its soil and vegetation except for a few twisted trees that cling precariously
to its rim. The entire rock has been split into large vertical blocks, a cluster
of pinnacles, and boulders of ever-smaller sizes by the persistent action of
weathering and water through the ages. As in the previous drawing, there
is a palpable sense of continuous movement and transformation.

The Virgin of the Rocks

Leonardo's most detailed and sophisticated depiction of rock formations
is to be found in his early masterpiece, the *Virgin of the Rocks*, now in the
Louvre (plate 8). The painting, created between 1483 and 1486, caused a
sensation in Milan's artistic and intellectual circles and marked the beginning of Leonardo's great fame as a painter.[30] The work was unprecedented on many levels. It was revolutionary in its rendering of light and
dark (*chiaroscuro*). Its low tones of olive green and gray were in stark contrast to the bright colors of Italian fifteenth-century art, and its composition represented a complex and controversial meditation on the destiny
of Christ.[31] In addition, the *Virgin of the Rocks* is testimony to Leonardo's
powers of scientific observation and profound knowledge, unmatched in
his time, of geological formations and plant growth.

Geologist Ann Pizzorusso, who has carried out a detailed study of the
painting's geological features, has called it "a geological tour de force" because of "the subtlety with which Leonardo represents a complicated geological formation."[32] As Pizzorusso shows in her article, the different rocks
in the grotto, which were unnamed when Leonardo studied and painted
them, are rendered with such accuracy that they can be readily identified
and described with proper technical terms by a modern geologist.

A key distinction made in geology today is that between sedimentary and igneous rocks. Sedimentary rocks form at or near the Earth's surface by sedimentation and subsequent consolidation of rock fragments and dissolved minerals. Igneous rocks, by contrast, are formed within the Earth's crust when liquid magma cools and crystallizes. Since molten rock is less dense than solid rock, it tends to move upward toward the surface. It may end up being ejected as lava by volcanoes, or it may intrude into layers of sedimentary rock below the surface and crystallize there into igneous rock. The fluid pressure of the intruding magma can be so great that it actually lifts the overlying rock layers.[33]

From his extensive geological studies, Leonardo was familiar with the processes of sedimentation and compaction of fragments into sedimentary rock, but it is doubtful that he understood the origin of igneous rock. Although igneous rock can be seen after erosion has exposed it at the Earth's surface, its formation within the crust can be studied only indirectly and presents a great challenge to geologists even today. Yet Leonardo was able, based on meticulous observations, to draw numerous fine details of the texture of rocks to such an extent that modern geologists can easily recognize telltale signs of the geological processes that formed.

In the *Virgin of the Rocks*, Leonardo pictured a complex geological formation that resulted from the intrusion of an extremely hard igneous rock, known as diabase, into soft layers of sandstone, one of the most common sedimentary rocks.* Both the sandstone and the diabase are weathered, and Leonardo has pictured accurately how the surfaces of the two types of rock have weathered differently in accordance with their respective hardness. "The result," explains Pizzorusso, "is an accurate portrayal of weathered sandstone intersected by diabase, which is much more resistant and therefore retains its structural characteristics."[34] Pizzorusso illustrates her analysis of the geology in the *Virgin of the Rocks* with a sketch of the painting's principal features, to which she has added the technical terms for the rock formations shown (fig. 2-6).

At the top of the grotto, we see rounded ("spherically weathered") mounds of sandstone, which have decomposed sufficiently to allow roots to take hold in them and plants to grow. Above the Virgin's head, jutting out in vertical relief, is the igneous diabase, which had intruded into the layers of sandstone and lifted the top layer in the distant geological past, thus

---

* Among European geologists, diabase is more commonly known as dolerite, and commercially it is also called "black granite."

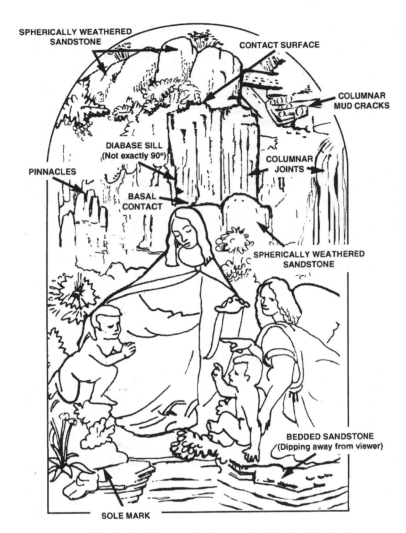

FIG. 2-6. Geological features in Leonardo's *Virgin of the Rocks*.
From Pizzorusso, "Leonardo's Geology."

intersecting the two layers with a sheet of hard rock, or "sill." As it cooled, the diabase contracted and produced vertical cracks with sharp edges, known as columnar fractures or joints.

The layer of sandstone below the diabase shows the same rounded weathering pattern as the top layer, and the contact surfaces between the two types of rock are clearly indicated by horizontal lines. Leonardo also observed correctly that no vegetation is growing out of the diabase, which is far too hard and too resistant to erosion to provide support for plants.

In the foreground of the grotto, the sandstone is not heavily weathered and has thus retained its original structure. The typical horizontal layers, or "beds," have been rendered with the utmost accuracy. In the distant background, finally, we can perceive the rocky pinnacles that appear in so many of Leonardo's drawings and paintings. They are remnants of processes of erosion in the distant past that have stripped away softer layers of rock, exposing the harder, more weather-resistant rock underneath.

Leonardo produced two versions of the *Virgin of the Rocks*. The dates and circumstances of the second version, which now hangs in the National Gallery in London, are not quite clear, and art historians believe that Leonardo may have let a fellow painter, Ambrogio de Predis, execute large parts of it.[35] This seems to be confirmed by Ann Pizzorusso's comparison of the geological details in both paintings. She finds that, even though the differences between the rock formations in the two versions may not be immediately obvious to the layperson, close evaluation shows that the rendering of geological details in the London version is clearly inferior to that in the Louvre.

"An observer with some knowledge of geology," Pizzorusso writes, "would find that the rock formations represented in the National Gallery work do not correspond to nature." With a detailed geological analysis, she demonstrates that the rocks in the London version are "synthetic, stilted characterizations," and she concludes: "It seems unlikely that the same person could have portrayed geological formations so accurately in the Louvre work and so incongruously in the National Gallery painting."[36] A similar conclusion was reached by botanist William Emboden on the basis of a comparison of botanical details in the two versions of the *Virgin of the Rocks* (see p. 105).[37]

The Fossil Enigma

In his extensive studies of sedimentation, Leonardo did not fail to notice that the sedimentary rocks of the Apennine mountains in northern Italy contained numerous fossils, and he correctly recognized them as the traces of organisms that had lived in the distant past:

> Between the various layers of the rocks are still to be found the tracks of the worms which crawled upon them when they were not yet dry, [and] all the marine clays still contain shells, and the shell is petrified together with the clay.[38]

The existence of fossils had been known since antiquity. Several ancient and medieval authors had commented on them without being aware of their true nature and age. Leonardo was familiar with these classical and medieval texts,[39] and in addition he had plenty of opportunities to observe well-preserved specimens of fossilized plants and animals. In fact, he tells the story that large numbers of fossil shells in perfect condition were brought to him in Milan:

> In the mountains of Parma and Piacenza multitudes of shells and corals filled with worm-holes may be seen still adhering to the rocks. When I was making the great horse at Milan a large sack of those, which had been found in those parts, was brought to my workshop by some peasants, and among them were many that had been conserved in their original condition.[40]

Marine fossils represented an enigma that natural philosophers had debated intensely since antiquity. If fossil shells were indeed remnants of marine organisms, how did they end up in sedimentary strata that now lie in the high mountains, several thousand feet above the current sea level? Leonardo was keenly aware of the problem:

> How will you explain the infinite number of kinds of leaves frozen into the high rocks of these mountains, among them seaweed that one finds mingled with shells and sand? And you will also see all kinds of petrified things, such as sea crabs, broken into fragments, scattered and interspersed with those shells.[41]

To solve the enigma, Leonardo studied a wide variety of body fossils (remains of organisms) and trace fossils (tracks of worms and snails) with the utmost care. On the basis of his detailed and sophisticated observations, he then presented his conclusions in a systematic exposition that covers several pages in the Codex Leicester and, with a series of lucid and original arguments, refutes the theories that were current in the Renaissance. Leonardo's exposition is an example of brilliant scientific reasoning, resulting in the first correct explanation of the nature of fossils. This understanding would not appear again in science until the end of the eighteenth century.

During the Middle Ages and the Renaissance, the most common belief about the origin of fossil shells in the mountains was that they had been carried there from the sea on the high waters and strong currents of Noah's flood, the Great Deluge of the Bible. A second, more esoteric theory was that these fossils were not remains of marine organisms but were special kinds of minerals that had been created right there in the mountains by divine intervention, or had grown in the rocks as a result of "celestial influences." Leonardo looked on both of these theories with disdain. He wrote of "the stupidity and simple-mindedness of those who imagine that these creatures were carried to such places distant from the sea by the Deluge," and "how another sect of ignoramuses maintain that nature or the heavens have created them in these places through celestial influences."[42]

In the Codex Leicester, Leonardo presents five powerful arguments against the diluvial theory. He notes that fossils appear in several layers of sedimentary rocks that were deposited at different sequential times, which is inconsistent with the biblical story of a single flood. "If you should wish to say that there were several deluges to produce these layers and the shells within them," he argues, "you would also have to affirm that such a deluge took place every year."[43] Moreover, he reasons, according to biblical tradition, the Deluge rose to "seven cubits above the highest mountain," and hence the shells would have been deposited at the very top of these mountains. But instead they are found predominantly near the mountain bases, "and always at the same level, layer upon layer."[44]

In addition to these two basic points, Leonardo uses sophisticated observation as the basis for three further arguments. Fossil strata deposited by strong currents, he explains, would not have preserved the tracks and trails of marine organisms that he found in sedimentary rocks. To him, these tracks are important evidence for the fact that the marine creatures were alive in the environment in which their fossilization occurred, rather than being carried there as dead animals by the Deluge.

Moreover, "if the Deluge had to carry shells for distances of 300 and 400 miles from the sea, it would have carried them mixed with various other species, all heaped together." But instead, "we see at such distances oysters all together, and the shellfish, and cuttlefish, and all the other shells that congregate together."[45]

Leonardo then notes that he has seen some fossils of bivalves (oysters, clams, and the like) with their two shells still joined. Since these shells are not cemented together in life but linked by an elastic ligament that quickly decays after death, any extensive transport in a flood would have separated the two valves. Hence, these animals must have been buried where they lived:

> On the edges of the rocks, we find a few with paired shells, like those that were left behind by the sea, buried alive in the mud, which later dried up and, in time, became petrified.[46]

Having established that the marine fossils could not have been transported to the mountains as dead animals by the currents of the Deluge, Leonardo then counters the argument that they might have migrated there through the high waters. He explains in great detail that the cockle, a bivalve mollusk often seen in fossils, could not have moved so far from the sea in the forty-day time period traditionally associated with the biblical Deluge:

> The cockle is an animal of no more rapid movement than the snail, when out of water, and is even somewhat slower because it does not swim but makes a furrow in the sand; and, using this furrow to support itself, it will travel between 3 and 4 *braccia** a day. Therefore this creature with such motion could not have traveled from the Adriatic sea as far as Monferrato in Lombardy, which is 250 miles away, in 40 days.[47]

In the same pages of the Codex Leicester, Leonardo also dismisses the belief that fossils are special minerals that have grown in the mountains under the influence of celestial forces. He argues that such minerals would grow in all strata of rocks, not only in those showing clear evidence of marine origins. "Where the valleys have not been covered by the salt waters of the sea," he points out, "there the shells are never found."[48]

In addition, Leonardo argues that the fossil shells must have been parts of living organisms in the past because they display features that are present in other life forms. In particular, he observes that the age of a

---

* One *braccio* equals approximately 2 feet.

fossil shell can often be determined from growth rings that record cycles of months and years:

> In these places . . . we [can] count in the crusts of cockles and snails the years and months of their life, as we do in the horns of oxen and sheep, and in the branches of trees that have never been cut in any part.[49]

This statement is a lucid example of the systemic thinking that is pervasive in Leonardo's science. Understanding a phenomenon, for him, always meant connecting it with other phenomena through a similarity of patterns. That he was able to associate the annual rings in the branches of trees with the growth rings in the horns of sheep is remarkable enough. To use the same analysis to infer the lifespan of a fossilized shell is extraordinary. As biologist Stephen Jay Gould has pointed out, this analysis of periodicities in growth became a rigorous and important subject in paleobiology only in the twentieth century.[50]

Implicit in all of Leonardo's arguments is his attempt to explain the origin of fossils in terms of natural processes that can be observed in the present, rather than by some miraculous or catastrophic events in the past. This idea was formulated explicitly as the so-called uniformitarian principle by the Scottish geologist James Hutton in the eighteenth century and was firmly established in the nineteenth century by Charles Lyell. The twin ideas that past changes in the Earth's surface can be understood in terms of forces and processes still operating today, and that these processes took place over extremely long periods of time, have since become cornerstones of modern geology. Both ideas were first set forth by Leonardo da Vinci and were integral to his science of the living Earth.

Leonardo's explanations of the nature of fossils are accompanied by precise descriptions of the specific sites in which he observed these deposits. According to historian of geology François Ellenberger, "we are justified, without being anachronistic, to talk about Paleoecology. . . . A modern geologist can only express surprise when he reads these texts in which there is nothing either erroneous or of no utility."[51]

In addition to recording the fossils' depositional environments, as they are called by geologists today, Leonardo describes the process of fossilization itself in great detail. "As befitted a sculptor who was urgently concerned with bronze casting," writes Martin Kemp, "Leonardo provided wonderfully vivid reconstructions of the casting of shells, exoskeletons and whole organisms within their moulds of enveloping mud."[52]

"When nature is on the point of creating stones," Leonardo explains, "it produces a kind of sticky paste which, as it dries, congeals into a solid mass together with anything it has enclosed in itself. However, it does not change it into stone but preserves it within itself in the form in which it has found it."[53] In Manuscript F, he devotes three pages to detailed descriptions of this process for the fossils of mollusks ("animals that have their bones on the outside"), fish ("animals that have their bones within the skin"), and for the impressions of leaves.[54] The description of the fossilization of fish, in particular, is a dazzling display of Leonardo's vivid and accurate imagination:

> As the rivers abated over time, these creatures, embedded and shut in with the mud, with their flesh and organs worn away and only the bones remaining, but having lost their natural arrangement, fell to the bottom of the mold formed by their impression. And as the mud rose above the level of the river, it dried up and formed first a sticky paste, and then changed into stone, completely sealing up what it contained, filling every crevice.
>
> And having found the hollow animal's imprint, it penetrated gradually through the tiny fissures in the earth from where the air escaped sideways. It could not escape upward as a result of the sediment which had fallen into the cavity, nor downward because the sediment that had already fallen had blocked up the porosity. There remained only the side openings, from which the air, condensed and under pressure from the action of the descending sediment, escaped with the same slowness as the sediment settled there; and in drying, this paste became stone, devoid of graininess, and it preserved the shapes of the creatures that had left their imprint, and enclosed their bones within it.[55]

The Formation of Mountains

With his meticulous observations and compelling reasoning, Leonardo demonstrated that the marine fossils found in mountain rocks had been formed in the fluvial and oceanic environments where these creatures lived in the distant past. His brilliant arguments invalidated the theories that were current in his time, but they did not solve the fossil enigma, as he well recognized. He still had to show how those layers of marine sediments ended up in the high mountains. In other words, he needed a theory of how mountains were formed during the extremely long periods of geological time. Leonardo did not hesitate to take on this formidable challenge.

The origin of mountains and, more generally, the formation of the Earth were topics that had been discussed by philosophers since antiquity.[56] Leonardo was familiar with many of the classical texts, including Ptolemy's celebrated *Cosmographia* and Pliny's encyclopedic *Natural History*, which contained summaries of the writings of almost five hundred Greek and Roman authors.[57]

In Greek philosophy, there were two schools of thought regarding the formation of mountains. Eratosthenes, a contemporary of Archimedes in the third century B.C., believed that the level of the Mediterranean Sea had been much higher in the past, when both the Strait of Gibraltar and the Bosphorus Strait had been closed, and that at some time the "Pillars of Hercules" (Strait of Gibraltar) had burst open, thus lowering the sea level and making mountains appear. The proofs of the former sea level, according to Eratosthenes, were found in the occurrence of fossils in those mountains. The other school of thought was promoted by the geographer Strabo in the first century B.C. According to him, mountains originated in earthquakes, volcanic eruptions, and other natural catastrophes.

Neither of these two theories was entirely satisfactory, and their relative merits were debated for centuries. The inhabitants of the Mediterranean countries were quite familiar with volcanoes, and volcanic eruptions were featured in many legends, such as that of Atlantis. But volcanic eruptions could not explain the deposits of shells and other fossils, as Strabo himself acknowledged. Eratosthenes, on the other hand, did not explain what kind of catastrophe had opened the Pillars of Hercules.

As far as the more general question of the formation of the Earth was concerned, there were also several opinions. Aristotle taught that the Earth was eternal, and that it continually repaired the degradations caused by erosion with the creation of new mountains. The Stoics, on the other hand, expounded the view that the visible signs of decay were proofs that the Earth would eventually perish. However, they believed that it would then recreate itself, and that in this way successive worlds would arise again and again.

Both of these were purely philosophical views, expounded without any attempts to describe corresponding geological processes. An interesting and much more detailed account was given by the Roman poet and philosopher Lucretius in the first century B.C. He maintained that the Earth was formed out of a chaotic mass of atoms of all kinds. As matter condensed, Lucretius explained, the ground began to sink. However, in some places this process was obstructed by the accumulation of rocks, and thus mountains were formed in those areas.

In the Middle Ages, the knowledge of Greek and Roman antiquity had been largely forgotten in Western Europe, but it was assimilated by Islamic scholars who translated the classic texts into Arabic and added their own commentaries and innovations.[58] The Arab natural philosophers revived the classical debate between the schools of Eratosthenes and Strabo, which they renamed as the theories of "neptunism" and "plutonism."

According to neptunism, named after the Roman god of the sea, land is continually worn down by erosion and the eroded earth is carried into the sea, where hills and mountains gradually build up from the deposits. This causes the water level to rise, while the height of the mountains on land slowly decreases. At a certain moment, the sea overflows its basin, invades the land, and exposes mountains that were formerly under water. In this way, plains change into oceans, and oceans into plains and mountains. These changes, the neptunists believed, occur periodically in vast cycles of 36,000 years.

To our modern minds, this theory sounds strange, as it is clearly inconsistent with the basic laws of gravity and the flow of water. Yet the neptunist belief that all mountains were formed by layers of deposits at the bottom of the oceans was favored by naturalists until the nineteenth century.

Opposed to neptunism was the theory of plutonism, named after the Greek god of the underworld, which postulated vertical forces in the Earth's interior that were capable of lifting or lowering land. Proponents of plutonism in the tenth and eleventh centuries included the great physician Avicenna (Ibn Sina).[59] Two centuries later, the scholastic philosopher Albertus Magnus, who was the teacher of Thomas Aquinas, proposed a similar theory, according to which mountains were uplifted by vapors released from the interior of the Earth.

In spite of the reputations of these scholars, however, the theory of plutonism (or "vulcanism," as it also came to be called) was not generally accepted in subsequent centuries, because it required the assumption that there was fire at the center of the Earth. Today we know that the core of the planet does consist of hot molten rock, known as magma.* But during the Middle Ages and the Renaissance the idea of a central fire in the Earth was inconceivable because it contradicted the accepted Aristotelian cosmology.

According to Aristotle, the four elements naturally arranged themselves in concentric spheres with earth at the center, surrounded successively by the spheres of water, air, and fire (or light).[60] He explained that the elements were constantly disturbed and pushed into neighboring spheres, but would then naturally return to their proper places. Hence, the medieval philosophers could accept the occasional existence of fire within the Earth, as evidenced by volcanoes, but could not conceive of fire occupying a permanent place at the Earth's center. Aristotle's authority was so powerful that the idea of subterranean fire, or heat, was seriously considered only in the eighteenth century when Hutton used it to explain the igneous origin of granite.

Christian theologians during the Middle Ages had no unanimity about cosmology. In the Old Testament, the Earth is described as a flat disk floating on the ocean, with the firmament held up by pillars above it. However, the modern notion that everybody in medieval times believed that the Earth was flat is no more than a popular cliché. In actual fact, most scholars, from antiquity to the Renaissance, knew that the Earth is spherical.[61]

A far greater obstacle to the proper understanding of geological processes was the biblical teaching that the world was created a mere 6,000–8,000 years ago. Today we know that even the much longer period of 36,000 years proposed by Arab scholars in their theory of neptunism is far too short to account for significant geological changes.†

The first to expand the notion of geological cycles was the French scholastic philosopher Jean Buridan in the fourteenth century. According to Buridan, the Earth consists of two hemispheres, one entirely aqueous and the other terrestrial, which periodically interpenetrate each other, so that oceans turn into land and vice versa in enormously long cycles. Leonardo

---

* Strictly speaking, the Earth's core consists of a small solid inner core and a larger molten outer core.

† The ancients believed that 36,000 years was the period of precession of the Earth's axis, which they had observed from the movement of the positions of the equinoxes along the ecliptic relative to the fixed stars. The actual number is approximately 25,800 years. Periods of geological time, by contrast, are measured in millions of years.

learned about Buridan's theory through the writings of Albert of Saxony, a disciple of Buridan, and he used the notion of the two hemispheres as an important element of his own theory of the formation of mountains.[62] But whereas Buridan and Albert believed that the Earth's asymmetry was created and regulated by God, Leonardo attempted to explain the complex dynamics of the Earth's shifting balance and the ensuing uplift of mountains entirely in terms of natural causes.

Leonardo's Tectonic Theory

Leonardo was well acquainted with the principal texts on the formation of the Earth by Islamic and Christian medieval scholars, and he used some of their key ideas to formulate his own theory. It is an elaborate blend of Aristotelian and medieval ideas combined with his own observations, and, like all of his geology, it includes some ideas that sound quite modern. In view of the fact that a proper understanding of the uplift of mountains in terms of plate tectonics was achieved only in the twentieth century, Leonardo's tectonic theory is indeed exceptional.

Leonardo begins his arguments by noting that, if the world were composed entirely of water, it would form a perfect sphere, like a drop of dew: "The surface of the sphere of water does not move from its perimeter around the center of the world which it surrounds at an equal distance."[63] Then he introduces Buridan's idea of the two interpenetrating bodies, one aqueous and the other terrestrial. However, in contrast to the scholastic philosopher, Leonardo reasons like a modern scientist, first citing an experiment and then using empirical evidence to construct a simple geometric model of the Earth:

> A drop of dew, perfectly round, affords us an opportunity to consider . . . how the watery sphere contains within itself the body of earth without the destruction of the sphericity of its surface. If you take a cube of lead the size of a grain of millet, and by means of a very fine thread attached to it you submerge it in this drop, you will see that the drop will not lose any of its original roundness, although it has been increased by an amount equal to the size of the cube which has been enclosed within it.[64]

Instead of a small cube, Leonardo then reasons, we could also imagine a terrestrial pyramid, or tetrahedron, immersed within the sphere of

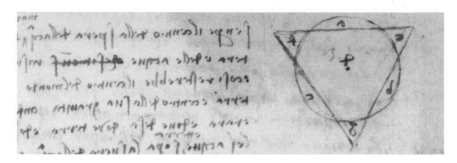

FIG. 2-7. Geometric model of the Earth.
Codex Leicester, folio 35v (detail).

water, with its corners protruding into the air. This is his model of the
Earth, presented in a simple sketch in the Codex Leicester (fig. 2-7), a mass
of land partially surrounded by water.

What is remarkable here is not only Leonardo's ingenious use of a
simple theoretical model—a technique that would become an integral
part of the scientific method in subsequent centuries—but also his utterly dynamic image of the Earth as a mass of land floating in water. The
conception is not unlike that in modern plate tectonics, except that in the
modern theory the land masses are conceived as drifting on a plastic layer
of partially molten rocks (see p. 93).

Having illustrated the dynamic relationship between land and water
with his geometric model, Leonardo then proceeds to discuss the actual
shape of the Earth.[65] He explains that the Earth has a geometric center,
which he calls the "center of the world." If the mass of land were homogeneous and equally distributed, it would form a perfect sphere around
that geometric center, and so would the sphere of water surrounding the
element earth.

In reality, however, the mass of land is far from homogeneous. The
interior of the world, in Leonardo's view, is a complex conglomeration of
solid earth and rocks, water running through various conduits, and large
caverns hollowed out by erosion. Because of this unequal distribution of
land and water, one hemisphere of the Earth will always be heavier than
the other, and hence the Earth's "center of gravity" will not coincide with
the "center of the world."* Dividing the sphere horizontally with the

---

* In modern science, the term "center of mass" is used instead of "center of gravity," which was the
term used by Leonardo and which is still used colloquially today.

heavier hemisphere at the bottom, Leonardo pictures the center of gravity as lying below the geometric center, within the heavier hemisphere.

However, the relative positions of the two centers of the Earth are not static. As in the model of the terrestrial pyramid within the sphere of water, the land masses of the Earth can slide deeper into the sphere of water or emerge farther from it. To explain the forces causing such movement, Leonardo introduces an Aristotelian argument. As a living organism, he reasons, the Earth will naturally move toward a state of balance, and hence will strive to bring its center of gravity closer to the geometric center.

As Gould has pointed out, this movement may be likened to that of a seesaw.[66] Just as two people of unequal weight can balance a seesaw if the heavier person moves inward and the lighter person moves outward, so the solid masses of the heavier hemisphere will gradually sink toward the center of the world, while the rocks of the lighter hemisphere will rise out of the water. And this is how, according to Leonardo, layers of sedimentary rock emerge from the sea to form mountains:

> And so the lightened side of the Earth is continually raised, and the antipodes draw nearer to the center of the Earth, and the ancient beds of the sea become chains of mountains.[67]

In Leonardo's view, water has continually shaped the forms of the Earth, and thus erosion is the major cause for the unequal distribution of mass in the two hemispheres. He saw the uprising of mountains and their erosion as different stages of the same cycle of transformation that took place over enormous periods of geological time. In the Codex Atlanticus, he noted how the uplift of the mountains is always followed by their erosion:

> I maintain that . . . the mountains, the bones of the Earth, with their wide bases penetrated the air and rose up into it, covered over and clad with much high-lying soil. Subsequently the frequent rains and the swelling of the rivers by repeated washing stripped bare part of the lofty summits of these mountains, so that the rock finds itself exposed to the air and the earth has departed from these places.[68]

And in Manuscript F, Leonardo explains how the erosion of mountains, in turn, causes them to rise up farther:

> The Earth is always growing lighter in some part, and the part that becomes lighter pushes upwards, and submerges as much of the opposite part as is necessary for it to join its . . . center of gravity to the center of the world.[69]

Today, the enormous cycle of erosion, sedimentation, uplift, and renewed erosion is known to geologists as the rock cycle.* The idea was central to Leonardo's conception of the Earth as dynamic and continually changing, and it is one of several reasons why his geology sounds so modern.

In addition to the gradual erosion of the mountains and their concurrent gradual uprising, Leonardo's tectonic theory also includes catastrophic events that take place in the interior of the Earth. Since antiquity, it was believed that there were numerous water veins inside the Earth (see p. 26), and Leonardo imagined that this water, over time, had carved out huge subterranean caves. As their erosion continued, the caves would eventually become unstable. Parts of the Earth's crust would collapse into them, and the movement of these enormous masses of rock toward the center of the world would lighten the upper hemisphere and thus contribute to the uplift of its mountains. In the Codex Leicester, Leonardo provides a clear description of this imaginary tectonic process:

> The very large space of the Earth that was filled with water, that is the immense cavern, must have had a considerable portion of its vault fall toward the center of the world, finding itself detached by the subterranean water courses that continually wear away the place through which they pass. . . . Now this great mass could fall . . . and it made the Earth lighter at the point where it broke off, and that part of the Earth immediately moved away from the center of the world and rose to the height where one sees the layered rocks, produced by the orderly action of the waters, at the summits of the high mountains.[70]

---

* The actual rock cycle is more complex, involving several additional processes that result in cyclical transformations of igneous, sedimentary, and metamorphic rocks into one another (see Wicander and Monroe, *Essentials of Geology*, p. 10).

*Il grand'amore nasce dalla gran
cognitione della cosa che si ama.* (*Trattato della pittura*, chapter 80)

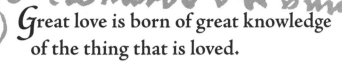

Great love is born of great knowledge
of the thing that is loved.

The theme of catastrophic collapse in subterranean caves was taken up again in subsequent centuries and was discussed under the name of "catastrophism" until the emergence of plate tectonics in the twentieth century.[71] The modern theory of plate tectonics provides a unifying conceptual framework for understanding the composition, structure, and internal processes of the Earth.[72] Its key idea is that the lithosphere, the hard outer layer of the Earth, is composed of giant plates that move over a plastic layer of partially molten rocks, known to geologists as the asthenosphere. At their boundaries, these plates may move apart, slide sideways past each other, sink beneath one another, or collide head on. These extremely slow interactions of massive plates generate the forces that create, respectively, mid-oceanic ridges, earthquakes, volcanic island arcs, and mountain ranges. The force that drives the movements of the plates has to do with circular currents of hot magma in the Earth's mantle underneath the lithosphere. The details of these thermodynamic processes are still not fully understood.

In Leonardo's tectonic theory, the large masses of land float in water rather than in hot magma (the element of fire). Nevertheless, the general features of his explanation of the origin of mountains have much in common with the modern theory. In both cases, the rising up of mountains is a consequence of catastrophic events that involve enormous masses of rock shifting inside the Earth. And even Leonardo's idea of land masses slowly rising out of the water when they become lighter, in order to keep land and water in balance, has an equivalent in modern geology. According to the so-called isostatic theory, parts of the upper crust that float on the layer of molten rocks will rise up vertically when they become lighter, just as they do in Leonardo's theory.[73] Modern geologists have confirmed, for example, that northern Europe is still rising because of such isostatic adjustment, triggered by the melting of the ice cap that covered the region a million years ago. Once again, it is astonishing to realize the prescient nature of Leonardo's scientific conceptions, especially when they concern processes of enormous scale that take place over immensely long periods of time.

The Distant Geological Past

During the years 1502–4, Leonardo traveled widely in Tuscany and Romagna in connection with various projects of military and hydraulic engineering.[74] While he drew beautiful and ingenious maps of the regions he visited, he also undertook extensive studies of their geological formations, observing and recording them with remarkable accuracy. "Everyone, and especially the field geologist," notes Ellenberger, "will appreciate the inestimable value and be amazed at the freshness and perspicacity of these observations and their interpretation."[75]

Many of these observations took place in the Arno basin, both upstream and downstream from Florence. In one entry in the Codex Leicester, for example, Leonardo noted:

> Near Monte Lupo, [the Arno] left gravel deposits, and these are still to be seen welded together, forming of various kinds of stones from different localities and of varying color and hardness one solidified mass. And a little further on, where the river turns toward Castel Fiorentino, the hardening of the sand has formed tufa stone. Below this, it deposited the mud in which the shells lived; and this has risen in layers acccordingly, as the floods of the turbid Arno were poured into this sea . . . as is shown in the cutting of the Colle Gonzoli, laid open by the Arno, which wears away its base.[76]

As he had depicted the differences between igneous rock and sandstone in the *Virgin of the Rocks* twenty years earlier, Leonardo now clearly perceived the differences between mountains formed from layers of hard rock (known to modern geologists as turbidite) and layers of clay, full of fossils, at the foot of these mountains. A sketch in the Codex Leicester (fig. 2-8) illustrates the difference between these two types of sediment, which date from different geological periods: deposits of clay at the foot of a mountain of horizontally layered and much older turbidite.

Leonardo used his observations in the Arno valley to speculate about the basin's geological history. He had known since his childhood that downstream of Florence, near Vinci, the plain of the Arno is intersected by a transverse chain of hills, Monte Albano and Gonfolina, and he speculated that in the distant past, the basin of Florence and that of Arezzo, further upstream, were the sites of two giant lakes at different levels:

> In the great valley of the Arno above Gonfolina, a rock was united since antiquity with Monte Albano in the form of a very high bank. This kept the river dammed up in such a way that, before it could empty itself into the sea ... it formed two large lakes, the first of which is where we now see the city of Florence flourish together with Prato and Pistoia. ... In the upper part of the Val d'Arno, as far as Arezzo, a second lake was formed, and this emptied its waters into the aforementioned lake.[77]

In subsequent years, the image of interconnected mountain lakes flowing into one another became an important visual metaphor in Leonardo's paintings, illustrating the continual flows and transformations that have shaped the surface of the Earth over immense periods of geological time. These lakes are clearly visible in the backgrounds of his mature masterpieces, the *Mona Lisa* and the *Madonna and Child with Saint Anne* (see plates 7 and 11).

FIG. 2-8. Sketch of clay deposits at the foot of a mountain of horizontally layered sandstone. Codex Leicester, folio 36r (detail).

In both paintings, Leonardo portrays his vision of the distant geological past—the forms of the Earth in ceaseless movement and transformation. As the mountains rose up from the primeval ocean, pockets of inland seas and lakes were created that would eventually find their way back to the ocean as their waters cut narrow gorges into mountains and hills. Gradually, these waters would carve out valleys and deposit in them masses of gravel and sand that eventually would become fertile soil. In other words, the mythical rock formations in these two master paintings represent Leonardo's meditations on the birth of the living Earth.

3

The Growth of Plants

Leonardo's science is a science of living forms that are continually shaped by underlying processes, whether he studied the rocks and sediments of the Earth, shaped by water, or the organic forms of plants, animals, and the human body, shaped by their metabolism. Invariably, he would begin with the outward appearance of these living forms and then proceed to investigate their intrinsic nature. Thus, at the core of his botanical studies, we find the two grand themes that appear again and again in other branches of his science—nature's organic forms and the patterns of metabolism and growth underlying them.

Leonardo's outstanding work in botany, as well as his original contributions to landscape and garden design, are discussed in great detail in the magnificent volume *Leonardo da Vinci on Plants and Gardens*, by botanist William Emboden.[1] This chapter is greatly indebted to Emboden's analysis.[2]

Unlike most of his other scientific studies, Leonardo's work in botany began relatively late in his life (see Chronology, p. 326). During the earlier years, his drawings of plants and trees were made mainly as studies for paintings. Notes on plants and landscape, often dealing with colors and light in addition to botanical accuracy, appear in his manuscripts most frequently after 1500, when he was forty-eight years old. The skill in his botanical drawings reached its culmination around 1508–10, and it was only after 1510, when Leonardo was in his sixties, that his botanical texts turned into purely scientific inquiries unrelated to paintings.

Plants were frequently used by Renaissance painters to decorate the geometrical and abstract spaces that were typical of the paintings of the time, especially in the Florentine school. These plants were usually arranged in formal decorative motifs. Some were rendered accurately, while others were purely imaginary. Botticelli's celebrated *Primavera*, for example, pictures a complex allegory in a garden setting with a glorious abundance of flowers. According to Emboden, "thirty of the forty plant species are identifiable and some are very well figured. Others are imaginary and seem included for mere decorative value."[3]

Beyond decoration, many plants in Renaissance art served another important purpose, especially in religious paintings. They were often associated with religious stories well known to the public, and thus had the "iconographic" function of conveying meaning through symbolic imagery. Leonardo exploited these additional layers of meaning in many of his paintings while depicting the plants that embodied the appropriate symbols with high botanical accuracy and masterful renderings of light and shade.

In addition, Leonardo was careful to represent the plants within their proper habitats and with seasonal accuracy. He did this to such an extent that a modern botanist today can recognize, for example, the setting of the *Virgin of the Rocks* as a Tuscan scene between March and April.[4] All these characteristics are what make the plants in Leonardo's masterpieces so unique. In the words of Emboden:

> Free from formal definitions that characterize Mantegna, Ghirlandaio, Perugino, and Botticelli in their use of plant motifs, Leonardo introduced an enormous vitality into his plant figurations. Beyond the iconographic use of plants, [he] considered the ecological context, seasonal correctness, and careful botanical depiction .... Leonardo's forceful environments seem charged and infused with an organic unity.[5]

Spiral Movement

When Leonardo investigated "all the forms of nature" in various branches of his science, he always looked for the processes and patterns of organization they had in common. One particular pattern that fascinated him throughout his life was that of spiral movement (*moto elico*). As mentioned earlier, Leonardo saw the spiral form as an archetypal code for the ever-changing and yet stable nature of living forms (see p. 62). He observed and drew it repeatedly in swirling vortices of water and air, in growth

FIG. 3-1. Andrea del Verrocchio and Leonardo da Vinci,
"Study for the *Giostra*," 1475. Uffizi Gallery, Florence,
Gabinetto Disegni e Stampe.

patterns of plants and animals, in curling locks, and in human movements
and gestures.

In his botanical drawings, Leonardo sometimes pictured stylized spiraling foliage to express his strong sense of the dynamic nature of organic
forms. He adopted this habit very early on; in fact, it is evident in one of
the very first botanical drawings from his hand that has come down to us.

In the mid-1470s, when Leonardo had just been certified as a master painter but was still working in Verrocchio's *bottega* (workshop), his
teacher collaborated with him on the preliminary design for a triangular tournament standard made of cloth, showing a winged Cupid and
a recumbent nymph (fig. 3-1). The standard had been commissioned by
the Medici family for a lavish pageant preceding a jousting tournament
(*giostra*) in 1475, and the drawing, now in the Uffizi Gallery in Florence,
is known as "Study for the *Giostra*." Its origin and authorship were long
controversial, but scholars now agree that Verrocchio made a preliminary
sketch in black chalk, barely visible in the original, and then let Leonardo
fix and elaborate it in pen and ink.[6] Being well aware of his pupil's special
talent for rendering natural forms, Verrocchio also left the drawing of the

landscape entirely to Leonardo, who added the plants and the rocky ledge on which the nymph reclines.

The long-stemmed plants from which Cupid emerges are rendered so accurately that they could be identified by botanists as a species of tall grass known as broomcorn millet (*Panicum miliaceum*). However, Leonardo chose to give the lower leaves a swirling, spiraling movement that is not characteristic of that plant. This spiral form of the leaves was an original creation of the young Leonardo that had no precedent in Renaissance art.[7] It seems that from the beginning of his career, Leonardo must have viewed patterns of plant growth as manifestations of a more general pattern embodied in various forms of organic life.

Ten years later, the stylized spiraling foliage appeared again in Leonardo's first finished masterpiece, the *Virgin of the Rocks* (plate 8), which contains an entire ecosystem of exquisitely rendered plants. In the lower left corner, Leonardo painted a tall species of iris known as yellow flag iris (*Iris pseudoacorus*). He accurately pictured its characteristic flowers and flat, sword-like leaves. But whereas in nature its leaves emerge from the ground in a fan-like arrangement in a single plane, Leonardo introduced a spiral movement into the foliage that is very similar to that of the broomcorn millet he drew in his youth. In this masterpiece, now in the Louvre, the spiral form of the plant's lower leaves is much more pronounced, conveying not only a strong sense of growth and vitality but also an impression of compelling elegance.

Leonardo's drawings of spiraling foliage reached a climax around 1506–8 in his studies for *Leda and the Swan*, in which the artist's central theme was the mystery of life's inherent procreative power (see p. 320). A preliminary study, now in the Rotterdam Collection, shows us a sensuous female nude kneeling in a moist swamp and turning toward the swan at her side with a gesture of great tenderness (fig. 3-2). The erotically charged composition is heightened by the phallic reed mace (*Typha latifolia*; also called bulrush or cattail) silhouetted against the sky and by the swirling grasses at her feet. The spiral movement of Leda's body is repeated in the swan and in the spiraling foliage surrounding them—all symbolizing the abundance of life's generative forces.*

---

* A second study for the kneeling Leda, now in the Devonshire Collection at Chatsworth, is similarly spiraled.

FIG. 3-2. Study for *Leda and the Swan*, c. 1505–10.
Museum Boijmans Van Beuningen, Rotterdam.

Leonardo made several studies of individual wetland plants in preparation for *Leda and the Swan*, including his celebrated drawing of the Star of
Bethlehem (*Ornithogalum umbellatum*, plate 6), the species botanists have
identified as the grasses at the feet of the kneeling Leda in the Rotterdam
study. The foliage in this highly stylized drawing forms the most exaggerated spirals. Indeed, the whole composition is strikingly reminiscent of
a water vortex, another of Leonardo's archetypal living forms. During
the same period, Leonardo also produced a study for the head of Leda

(fig. 3-3) in which the same swirling movement appears in her hair—one more manifestation of the spiral as symbol of nature's fecundity and pro-creative power.

"Many flowers portrayed from nature"

Plants and trees play important roles in nearly all of Leonardo's paint-ings. They have symbolic meanings and convey metaphoric messages while displaying the artist's profound knowledge of botanical forms and their underlying processes. Flowers seem to have been Leonardo's first subject when he showed great talent for drawing as a young boy in his native Vinci. In his Notebooks many years later, he listed "many flowers portrayed from nature" among the works he had produced in his early youth.[8]

Leonardo's sophisticated botanical and ecological understanding is fully displayed in his early masterpiece, the *Virgin of the Rocks* (plate 8). As mentioned earlier, the painting has been called "a geological tour de force" because of the artist's astonishingly accurate representation of com-plex geological formations (see p. 77). It might be called a botanical tour de force with equal justification. The luxuriant plants filling the rocky grotto are not scattered about the painting in a decorative pattern but are shown to grow only in places where weathered sandstone has decomposed suffi-ciently to allow their roots to take hold (see p. 78). Only species appropri-ate to the moist environment of the grotto are portrayed, each in a specific habitat and a seasonally accurate stage of development.[9]

Within those botanical and ecological constraints, Leonardo selected specific plants that evoked for his contemporaries multiple layers of subtle symbolic meanings associated with the religious themes of his composi-tion.[10] Behind the Virgin's left shoulder is a graceful columbine (*Aquile-gia vulgaris*). Its Latin name is derived from *aquila* (eagle), as the flowers were thought to resemble an eagle's talon. In antiquity, the plant was also known as "lion's herb" and its common name, columbine, alludes to the resemblance of the flower to a cluster of doves. To the Renaissance mind, these associations were rich in religious symbolism. The eagle and the lion were the symbols of the evangelists John and Mark, the dove personified the Holy Spirit, and the columbine's tripartite leaves were a perfect sym-bol of the Trinity.

Just above the Virgin's left hand one can barely see a cluster of tiny whirls, formed by the leaves of a plant known as Our Lady's bedstraw (*Galium verum*). According to legend, Joseph used the dried straw of this

FIG. 3-3. Study for the head of Leda (detail), c. 1505–10.
Windsor Collection, *Figure Studies, Profiles, and Caricatures*, RL 12516 (detail).

plant in the manger to make a bed for Mary, and its white blossoms turned to radiant gold when Jesus was born.

The rosette of leaves above the knee of the Christ Child has been identified by Emboden as belonging to the primrose *Primula vulgaris*, which was considered an emblem of virtue because of its pure white flowers. Emboden points out that the purity of Christ was usually represented by a white rose, but that Leonardo chose the white primrose instead because a rose would have been inappropriate for the given setting and season.

Several plants in the painting allude to various stages in the Passion of Christ. The palm leaves above the infant Saint John, identified with the genus *Raphis*, were an ancient symbol of immortality and evidently are meant here to herald Christ's entry into Jerusalem, just as Saint John would herald Christ as the Messiah. The three clusters of leaves behind the infant Saint John could belong to several plant species. However, in view of the implied time of year, they have been identified by Emboden as representing the anemone known, because of its trifoliate leaves, as herb trinity (*Anemone hepatica*). A related small cluster of anemones (*Anemone hortensis*) can be seen under the seated Christ Child. The anemone represented the blood drops of Christ and was said to have blossomed under the cross on Calvary when the blood fell from Christ's wounds. Finally, the Resurrection of Christ is symbolized by the leaves of bear's breech (*Acanthus mollis*) between the right knee and left heel of Saint John. As Emboden explains, it was an Italian tradition to plant bear's breech over graves, where it came to symbolize the Resurrection because it dies back to the ground in autumn and reemerges rapidly with a wealth of green foliage in spring.

The elegant iris in the lower left corner of the painting, with its striking spiraling leaves, has already been mentioned (see p. 100). Emboden points out that this is not the species *Iris florentina*, which Leonardo often portrayed in his drawings, but instead the ecologically appropriate wetland species *Iris pseudoacorus*.[11]

Many more plant species are portrayed in the *Virgin of the Rocks*, all chosen for particular symbolic virtues. They include St. John's Wort (*Hypericum perforatum*), the plant consecrated to Saint John and believed to have protective powers; a cyclamen (*Cyclamen purpurascens*), which symbolizes love and devotion because of its heart-shaped leaves; several species of ferns, believed to be benevolent repositories of souls; and branches of oak (*Quercus robur*), which embodies a host of iconographic virtues.

As mentioned earlier, there are two versions of the *Virgin of the Rocks*, one now in the Louvre and the other, painted several years later, in the National Gallery in London. It is widely believed that Leonardo let his fellow painter Ambrogio de Predis execute large portions of the London version. We have seen that this seems to be confirmed by a comparison of the geological details in both paintings (see p. 80).

Emboden came to a similar conclusion after comparing details of the plants in the two paintings. He points out that there are fewer plant species in the London version and shows that many of them are rendered inaccurately and without the sophistication displayed in the Louvre version. This leads Emboden to the conclusion that in the London version, "most certainly the plant life is not from the hand of Leonardo. . . . It is impossible to believe that the same painter who, in the Paris version of the same painting, took such great care to render plants with seasonal and ecological accuracy, not to mention iconography, produced the precarious landscape with simplistic conventions of botanical presentation."[12]

Botany for Painters

During his earlier years, Leonardo drew individual plants mainly as studies for paintings. Later on, he also jotted down in his Notebooks instructions for painters on how to render the effects of light and shade and the diversity of colors he observed in nature. These notes on how to paint plants and trees became increasingly frequent after 1500. They are collected in Jean Paul Richter's classic selection of Leonardo's writings in a section titled "Botany for Painters."[13] The excerpts collected in this section, filling almost forty pages, contain detailed descriptions of subtle color variations and of the effects of light and shade on various parts of trees and plants. In the Codex Arundel, for example, Leonardo noted:

> The trees in a landscape are of various kinds of green, inasmuch as some verge toward blackness, as firs, pines, cypresses, laurels, box and the like. Some tend toward yellow, such as walnuts and pears, vines and verdure. Some are both yellowish and dark, as chestnuts and common oak.[14]

And in Manuscript G:

> Young plants have more transparent leaves and a more lustrous bark than old ones; and particularly the walnut is lighter in May than in September.[15]

Leonardo was indeed a master in rendering the appearance of trees under various light conditions. In the words of Emboden, Leonardo's trees manifest "an omnipresent mystical quality imparted by the juxtaposition of light and shadow."[16] A sheet in the Windsor Collection (fig. 3-4) contains a particularly elegant example of a single tree drawn in red chalk, in which these subtle optical effects are superbly displayed. Leonardo's accompanying note reads:

> That part of a tree which stands out against shadow is all of one tone, and where the trees or branches are thickest, there it is darker because light has less of an impression there. But where the branches are against other branches, there the luminous parts show themselves brighter, and the leaves shine as the sun illuminates them.[17]

FIG. 3-4. Study of a tree, c. 1508. Windsor Collection, *Landscapes, Plants, and Water Studies,* folio 8v (detail).

Only a very few of Leonardo's early studies of individual plants have come down to us. One of the finest, and perhaps the most famous, is his Madonna lily (*Lilium candidum*) from the early 1470s (fig. 3-5). This is the same species as the lily held by the angel in Leonardo's Uffizi *Annunciation,* but the two differ in the arrangement of their flowers, buds, and leaves. The study is an impressive testimony to Leonardo's unparalleled mastery of botanical drawing even in his early twenties. The renderings of the lily's six stamens, its six-part envelope of flower petals, and the arrangement of leaves on its stem are completely accurate. "The Madonna lily," writes Emboden, "is a masterpiece of botanical imagery in every detail."[18]

Leonardo must have produced many more studies of flowers and plants than those we know today in order to be able to paint the *Virgin of the Rocks,* the *Leda,* and the complex interlace of luxuriant foliage that covered the vault and ceiling of the Sala delle Asse in the Sforza Castle at

Milan.[19] Indeed, art historian Jane Roberts estimates that several hundred studies of plants and flowers from his hand must have been lost.[20]

The high point of Leonardo's plant studies was reached in the drawings he produced around 1508–10. As Emboden explains, the unique quality of these works is that they depart strongly from the artist's earlier studies for paintings and take on the characteristics of independent scientific illustrations.[21] For example, a drawing of an anemone (*Anemone nemorosa*) and a marsh marigold (*Caltha palustris*) in the Windsor Collection (fig. 3-6) represents a fine comparative botanical study in which the flowers of the two species are similar but the forms of their leaves are diverse. On the subsequent folio (fig. 3-7), a rush (*Scirpus lacustris*) is contrasted with a sedge (*Cyperus monti*). Both are aquatic plants and are somewhat similar in appearance but belong to different families.

FIG. 3-5. Madonna lily, c. 1472–75. Windsor Collection, *Landscapes, Plants, and Water Studies*, folio 2r.

In the accompanying text, Leonardo notes the differences between the two species, pointing out in particular the angularity of the sedge's stem.

The transition of Leonardo's botanical drawings from studies for paintings to scientific illustrations was accompanied by a series of texts that represent his first purely scientific inquiries into the nature of botanical forms and processes. To appreciate the significance of this evolution in Leonardo's thought we first need to have some idea of the history of botany since antiquity, which formed the intellectual context within which he operated.

Botany from Antiquity to the Renaissance

Throughout antiquity and in the centuries that followed, the study of the living world was known as natural history and those who pursued it were known as naturalists.[22] The ideas of the ancients about plants and animals were represented in great detail in the encyclopedic works of four

FIG. 3-6. Marsh marigold (left) and anemone (right), c. 1506–8.
Windsor Collection, *Landscapes, Plants, and Water Studies*, folio 23.

masters—Aristotle, Theophrastus, Pliny the Elder, and Dioscorides—all
of which were available to the Italian humanist scholars in printed Greek
and Latin editions.

Aristotle was the classical author most widely available to Renaissance
scholars. His numerous works included several treatises on animals, including the *Historia animalium* (History of Animals). While Aristotle's
observations of plants were less accurate than his observations of animals,
his disciple and successor Theophrastus was a keen botanical observer.
His treatise *De historia plantarum* (Of the History of Plants) was a pioneering work that made him famous as the "father of botany." However,
while Theophrastus was a master of botanical categories, his botany remained purely descriptive. He never inquired into any root causes, and his
discussions of environmental influences were scarce and faulty. "He was a
great figure in his time," comments Emboden, "but we must not compare
him to any Renaissance botanist, [let alone] to Leonardo da Vinci."[23]

In the first century A.D., the Roman naturalist Pliny the Elder wrote
a monumental encyclopedia titled *Historia naturalis* (Natural History),
comprising thirty-seven books, which became the favorite scientific encyclopedia in the Middle Ages and the Renaissance. In this massive

153

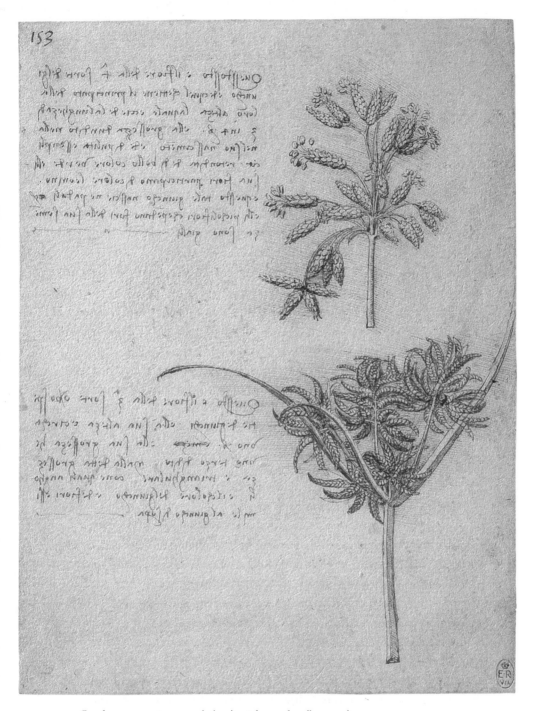

FIG. 3-7. Study contrasting a rush (top) with a sedge (bottom), c. 1510. Windsor Collection, *Landscapes, Plants, and Water Studies*, folio 24.

compendium, Pliny mentions more than one thousand plants, a number not to be equaled in any book until the Renaissance. However, according to Emboden, "there is no evidence of understanding or inquiry" in any of these numerous entries.[24]

In the subsequent centuries, botany was often considered a subdiscipline of medicine, since plants were mainly studied for their use in the healing arts. For centuries, the authoritative text in this field was the *De materia medica* (Regarding Medical Materials) by the Greek physician Dioscorides, who was a contemporary of Pliny. It contained references to six hundred plant species, arranged in three categories: aromatic, alimentary, and medicinal. The work was soon translated into Arabic and Latin, and some editions were lavishly illustrated. An exquisite example is the edition known as the Juliana Codex, named after the daughter of a Roman emperor, to whom it was given as a gift. An almost perfect facsimile in the National Library in Vienna remains one of the most beautifully illustrated manuscripts in history.

The *Materia medica* remained the sole authority for physicians until the Renaissance. No drug was considered legitimate that was not found in it. This doctrinaire use greatly impeded original botanical thought and established botany as a discipline almost exclusively in the service of medicine. Until the sixteenth century, plants would not be investigated as entities in themselves, but merely as accessories to healing and the medical arts. The only other writings on plants discussed their culinary uses or their roles as decorative elements in gardens.

One of the few independent botanical scholars of the Middle Ages was the Abbess Hildegard of Bingen, an extraordinary twelfth-century mystic and polymath who wrote theological treatises as well as botanical and medical texts, composed liturgical songs and visionary poems, and created superb illuminated manuscripts. Two of her books deal with plants in the region of Bingen. They contain original descriptions of seventy plant species, which are identified by their German vernacular names and discussed in relation to medicine.[25]

Another important medieval botanist was the German scholastic philosopher Albertus Magnus (Albert, Graf von Bollstädt) who lived about one hundred years after Hildegard. Albert collected his botanical observations in a volume titled *De vegetabilibus* (Of Plants),* which contains sev-

---

* *Vegetabilia* (later *plantae*) was the Latin term for "plants" in the Middle Ages.

eral original insights and came to be considered the most important work on botany since Theophrastus. "As a recorder of nature, a morphologist, and the originator of the proper mode of thought in science," writes Emboden, "Albert is unequalled until we encounter Leonardo."[26]

The fifteenth century was the age of the Renaissance herbals—botanical books containing descriptions and illustrations of herbs and plants and their medical properties. With the newly invented printing press, numerous copies of standard texts could be produced, and the use of woodcuts and copper plates made it possible for the first time to reproduce illustrations with complete accuracy.[27] Soon large numbers of herbals, patterned after the *Materia medica*, appeared from presses all over Europe and became extremely popular. Most fifteenth-century herbals went into multiple editions, often under several titles. A single work might be known under many names, which has caused considerable confusion among historians of botany and medicine.[28]

The scholarship involved in the production of most of these herbals was quite dismal.[29] Their main purpose was to show local examples of medical drugs referred to in the classical texts, and misidentifications were very common. The compilers were not concerned about the fact that species of plants found, say, in the Mediterranean were not those of northern Europe. If a plant could not be identified because it did not grow in their region, they would often have a woodcut made of some plant that looked remotely similar and ascribe to it the medicinal properties mentioned in the manuscript from which their book was derived. The vast majority of herbals in the early Renaissance were endlessly repetitive, drawing from one another and from antiquity, but not from nature.

Leonardo the Botanist

At the beginning of the sixteenth century, when Leonardo began his advanced botanical studies, botany was still in a purely descriptive phase and was considered merely an accessory to the healing arts. Even at the great universities of Pisa and Padua, whose professors included some of the leading botanists of the time, no genuine science of botany was taught in which plants were studied for their own sake.

As in so many other fields, Leonardo took his scientific work in botany far beyond that of his contemporaries. Like his fellow humanists, he was very familiar with the texts of the classical naturalists, but he refused to repeat their teachings uncritically.[30] Indeed, he despised the established scholars who merely quoted the classics in Latin and Greek. "They strut about puffed up and pompous," he wrote scornfully, "decked out and adorned not with their own labors but with those of others."[31] Leonardo always studied the classical texts carefully and then tested them by subjecting them to rigorous comparisons with his own direct observations of nature.

In contrast to his contemporaries, Leonardo not only depicted plants accurately but also sought to understand the forces and processes underlying their forms. In these studies, often based on observations that were astonishing for their time, he pioneered the emergence of botany as a genuine science. Emboden concludes his extensive analysis of Leonardo's corpus of botanical ideas with the following assessment:

> Collectively, the astute observations by this great "disciple of nature" argue compellingly for his position as one of the greatest thinkers in botany, and a man in advance of his time. Botany as a descriptive science was taken by Leonardo into realms of thought characteristic of the late 17th century and even into the sphere of some 20th-century concepts.[32]

Leonardo's botanical notes are scattered throughout the Codices, and in addition there is a major section on botany in Part Six of the *Trattato della pittura* (Treatise on Painting), the famous anthology compiled after Leonardo's death by his disciple Francesco Melzi.[33] As Emboden and other historians have noted, less than half of the material in the *Trattato* can be found among Leonardo's remaining manuscripts, indicating that substantial portions of his writings on botany have been lost. In fact, Leonardo scholar Carlo Pedretti has concluded from his thorough analysis of the *Trattato*'s chronology that Melzi must have copied its botanical sections from an entire lost manuscript on botany written by Leonardo.[34]

Emboden also points out that the presentation and botanical notation on the sheet depicting a rush and a sedge (see fig. 3-7) suggest a leaf from a treatise on plants, and Pedretti has suggested that Leonardo may have referred to such a treatise on another sheet of the Windsor Collection where he mentions a planned "discourse on herbs."[35] The format of such a manuscript may well have been that of the classical handbooks, but its contents would have gone far beyond those of a traditional herbal. "It would appear," writes Emboden, "that Leonardo had every intention of writing, or actually executed, a treatise that would explain every aspect of plant growth known to him."[36] Such a book would have been far ahead of its time. The first studies of plants for their own qualities were not published until several centuries later.

At the core of Leonardo's botanical theory we find, as mentioned earlier, the two grand themes that also appear in the other branches of his science—nature's organic forms and patterns, and the processes of metabolism and growth underlying them. In subsequent centuries, the investigations of these two themes gave rise to two major branches of modern botany: plant morphology and plant physiology. The term "morphology" was coined in the eighteenth century by the German poet and scientist Johann Wolfgang von Goethe, and its subject, the study of biological form, became the primary concern for biologists in the late eighteenth and early nineteenth centuries.[37] The development of plant physiology was triggered by the great advances in chemistry in the eighteenth century. A century later, the perfection of the microscope gave rise to a new branch of botany, plant anatomy, dedicated to the study of the structures and parts of plants, including features invisible to the naked eye. Plant anatomy subsequently expanded into all fields of biology, including molecular biology and genetics.

The development of these branches of botany during the past three centuries reflects a tension that has been present in Western science and philosophy from their beginnings.[38] It is the tension between the study of matter (or substance, structure, quantity) and the study of form (or pattern, order, quality). The study of matter was championed by Democritus, Galileo, Descartes, and Newton; the study of form by Pythagoras, Aristotle, Kant, and Goethe. Leonardo clearly followed the tradition of Pythagoras and Aristotle in developing his science of living forms, their patterns of organization, and their processes of growth and transformation.

In his morphological studies, as we shall see in more detail, Leonardo observed and recorded various growth and ramification patterns of flowers and trees. In particular, he noted different arrangements of branches and leaves around the stem—a field of study known in modern botany as phyllotaxis. In his plant physiology he was especially interested in the nourishment of plants by sunlight and water, and the transport of the "vital sap" (*umore*, or "humor"; sugars and hormones, in modern language) through a plant's tissues. He correctly distinguished between the two types of vascular tissues known today as phloem and xylem, and he made astute observations about the movement of sap when a tree is injured. Leonardo was also the first to recognize that the age of a tree corresponds to the number of rings in the cross section of its trunk, and that the width of the rings correlates with the wetness or dryness of those years.

Not all of Leonardo's botanical observations were original, but he always articulated them much better than his contemporaries. Indeed, the botanical sections in the *Tratatto della pittura* amount to genuine studies in theoretical botany. In the words of William Emboden,

> The text was transformed . . . into a true volume of scientific inquiry which is not paralleled in any similar treatises of that time. Whatever borrowings from the works of others appear, and although there are errors in interpretation in several instances, the work remains an extraordinary document in its experiential verification of phenomena.[39]

Branching Patterns

Most of Leonardo's notes in Part Six of the *Tratatto della pittura* are instructions for painters on how to render plants, trees, and landscapes under varying atmospheric conditions. A good third of this section, however, deals with his morphological studies. In particular, he describes and illustrates various patterns of phyllotaxis that are characteristic of plants and trees.

Leonardo correctly identified the three basic types of ramifications: alternate (branches switching from side to side), opposite (two branches growing in opposite directions from the same node), and spiraled (successive branches rotating through equal angles around the stem). In a sketch in the *Trattato* (fig. 3-8), he illustrated these three types with the branching patterns of an elm (*olmo*), an elder (*sambuco*), and a walnut (*noce*), respectively.

In view of Leonardo's lifelong fascination with the spiral as an archetypal pattern of life (see p. 22), it is not surprising that he paid special attention to the branching patterns known today as "spiraling phyllotaxis." He identified several different types of these spiraling arrangements of leaves on the stem, noting that, in each case, an exact number of rotations around the stem is completed after a certain number of branchings.

For example, he pointed out that "nature has arranged the leaves of the last branches of many plants in such a way that the sixth leaf is always above the first, and so it follows successively if the rule is not impeded."[40]

While studying these branching patterns, Leonardo observed different patterns of flowering. "Some of the flowers that grow on the branches of shrubs bloom first at the very top of these branches," he noted in the Trattato, "and oth-

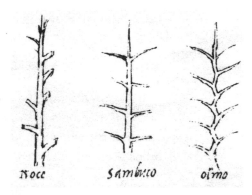

noce   sambuto   olmo

FIG. 3-8. Basic ramification types, illustrated with an elm (right), an elder (center), and a walnut (left). Trattato, chapter 890.

ers open the first flower at the very lowest part of the stem."[41] While being direct and simple, this observation defines a basic principle* that is still used in botany today to establish taxonomic categories.[42]

Having identified the basic types of branching patterns, Leonardo proceeded to study the processes underlying their formation. To begin with, he correctly observed that branches and fruits always sprout from the lateral buds located just above the attachment points of leaves. "When the shoot and the fruit of the following year spring from the bud," he noted in Manuscript G, "the eye lies above and in close contact with the insertion of the leaf."[43] Further on in the same Notebook, he added a lovely metaphor about the ways in which he saw the leaf nourishing and protecting the bud in its axil (the angle formed by the stalk of the leaf and the main stem):

> Every shoot and every fruit originates above the insertion of its leaf, which serves as its mother, giving it water from the rains and moisture from the dew that falls at night from above, and often it shields them from excessive heat of the rays of the sun.[44]

Leonardo's inquisitive mind was not content with the description of the morphology of branching in terms of axils and lateral buds. He wanted to know what causes these buds to grow in specific places, generating specific

---

* The principle is known to botanists as determinate versus indeterminate inflorescence.

sequences of branching. He answered this question with a remarkable hypothesis. He suggested that branching patterns have to do with the "humor," or "vital sap," that nourishes the plant's tissues:

> Between one ramification and the other, if there are no other particular branches, the tree will be of uniform thickness. And this takes place because the whole sum of the sap that feeds the beginning of this branch continues to feed it until it produces the next branch. And this nourishment, or equal cause, produces equal effect.[45]

This assertion, interlinking the morphology of branching patterns and the physiology of nutrient flow, is indeed quite extraordinary. "It is no small suggestion," comments botanist Emboden, "that there is a 'humor' establishing a cause and effect relationship between the sequencing of branching. What we now know to be hormonally conditioned activity is related to the inactivation of new branches in an area that . . . supports an existing branch of a high level of activity, and it is here suggested by Leonardo, albeit obliquely."[46] With this suggestion, Leonardo was far ahead of his time. "The distancing between branches . . . went unexplained until the 20th century," writes Emboden, "when the centers of inactivation generated by hormonal activity became known to botanists."[47]

Leonardo's prescient intuition of the causal link between the flow of sap and the patterns of phyllotaxis led him to another highly original observation concerning successive levels of branchings in a tree. At each level, he asserted, the total cross-sectional area of the branches must remain constant. In Manuscript M, he illustrated this rule clearly with two simple sketches (fig. 3-9), and he expressed it succinctly in a passage in Manuscript I. "All the branches of trees," he notes, "at every stage of their height, when put together, are equal to the thickness of their trunk." And then he adds: "All the ramifications of the waters, at every stage of their length, being of equal movement, are equal to the size of their parent stream."[48]

What makes Leonardo's assertion so remarkable is not so much its intuitive plausibility but the reasoning on which it is based. When a branch grows, Leonardo argues, its thickness will depend on the amount of sap it receives from the area below the branching point. In the tree as a whole, there is a constant flow of sap, which rises up through the trunk and then divides between the branches as it flows upward and outward through successive ramifications. Since the total quantity of sap carried by the tree

FIG. 3-9. Ramifications of a tree in which the total cross-sectional area of the branches remains constant at each level. Ms. M, folio 78v.

is constant, the quantity carried by each branch will be proportional to its cross-section, and hence the total cross-section at each level will be equal to that of the trunk.

Leonardo's argument is typical of the kind of systemic thinking that we find again and again in his science. Having established the conceptual link between the morphology of successive ramifications and the physiology of flowing sap, he then compares this flow of sap to the flow of water through the tributary branches of a river. In his extensive studies of flowing water, he had already discovered and clearly articulated the principle of continuity (see p. 45). Now it was quite natural for him to apply it to the flow of sap in a tree and to deduce corresponding rules of proportion. Moreover, Leonardo applied the same reasoning in his anatomical studies to the flow of blood through branching arteries and veins and to the flow of air through the ramifications of the trachea, comparing both to the branching patterns of rivers and trees.[49]

As far as the ramifications of trees are concerned, modern botany has shown that Leonardo's rule is not completely accurate, because the flow of sugars and hormones is not the only factor determining the thickness of the branches.[50] Nevertheless, Leonardo's intuitive understanding of the causal link between phyllotaxis and the flow of sap, long before the development of biochemistry, is truly impressive.

Plant Growth

For Leonardo, describing "all the forms of nature" with great accuracy and depicting them in magnificent drawings and paintings was not enough. He had to go deeper and understand the nature and causal roots of the processes that underlie living forms and continually shape them. Indeed, the exploration of these causal relationships is one of the main characteristics that distinguishes Leonardo's research from that of other Renaissance scholars and makes it look so modern to us. In his botany, this meant, as mentioned earlier, that he interlinked the disciplines now known as plant morphology and plant physiology.

In his studies of plant growth, Leonardo explored fundamental questions about many basic processes that are studied by plant physiologists today: how do plants acquire the energy and nutrients necessary for their growth? how do they grow in response to environmental stimulation? what are the pathways of nutrient flow through the plant tissues? how do plants regulate their growth? what are the stages of germination from seed to seedling? In modern botany, these questions are answered in the language of biochemistry and of cellular and molecular biology, involving concepts like photosynthesis, tropism, metabolic pathways, and plant hormones. Leonardo, of course, did not have access to these levels of scientific explanation. But his meticulous observations and great intuition for the nature of organic forms led him to many insights that are remarkably close to modern botanical knowledge.

The ancients believed that plants grew by literally ingesting earth to nourish themselves and increase their mass. Leonardo examined the traditional teachings critically, and to do so he tested them with a simple experiment. He unearthed the roots of a small squash plant and brought it to maturity by supplying it only with water. "I made the experiment . . . of leaving only one small root on a gourd," he recorded in Manuscript G, "and this I kept nourished with water, and this gourd brought to perfection all the fruits it could produce, which were about 60 of those long gourds."[51] From this experiment, Leonardo drew the remarkable conclusion that "the sun gives spirit and life to the plants, and the earth nourishes them with moisture."[52]

To appreciate the originality of this statement and the way Leonardo arrived at it, we must remember that botanical experiments were unheard of in the early sixteenth century. As Emboden has noted, it was not until the mid-seventeenth century that an experiment similar to Leonardo's was carried out. In the 1640s, the Belgian physician Jan Baptista van Helmont planted a small willow tree in an earthenware pot to which he added only water. After five years, Helmont recorded that the weight of the tree had increased dramatically but that the earth had lost only a few grams. He concluded from this that all of the additional plant body had been produced from the water alone.[53]

Today we know that Helmont's conclusion was incorrect, since most of the mass produced in plant growth comes from the air. The roots take in water and mineral salts from the earth, and the resulting sap rises up to the leaves, where it combines with carbon dioxide (CO_2) from the air to form sugars and other organic compounds. In this marvelous process, known as photosynthesis, solar energy is converted into chemical energy and bound in the organic substances, while oxygen is released into the air. The bulk of the plant body—including the cellulose and other compounds produced through photosynthesis—consists of heavy carbon and oxygen atoms, which plants take directly from the air in the form of CO_2. Thus, while many people today still tend to believe that plants grow out of the soil, in actual fact most of the plant mass comes from the air.

Both Leonardo and Helmont lived long before the advent of chemistry and hence were unable to recognize the complex processes involved in photosynthesis. However, as Emboden points out, Leonardo came closer to our modern understanding "in suggesting that the sun as well as the moisture from the earth were responsible for the mass of the plant body."[54] The critical role of sunlight in photosynthesis was discovered by the Dutch plant physiologist Jan Ingenhouz toward the end of the eighteenth century, and a full understanding of its complex biochemistry was not reached until the twentieth century.

Manuscript G contains another remarkably prescient passage in which Leonardo seems to intuit the role of the atmosphere in the process of photosynthesis. A few pages after the description of his botanical experiment, he notes:

> The lower branches, after they have formed the angle of their separation from the parent stem, always bend downward so as not to crowd against the other branches which follow above them on the same stem and to be better able to take in the air which nourishes them.[55]

This passage is noteworthy not only because of the brilliant (and correct) suggestion that plants receive nourishment from the air, but also because it is an example of Leonardo's observation of tropism, the tendency of plants to orient themselves in response to environmental stimuli. In addition to noting the bending of branches in response to gravity, known to botanists today as geotropism, Leonardo observed the phenomenon of phototropism, that is, the orientation of plants toward light. "The extremities of the branches of plants," he noted in the *Trattato della pittura*, "unless they are overcome by the weight of fruit, turn toward the sky as much as possible."[56] Both phototropism and geotropism were rediscovered and studied in detail by Charles Darwin at the end of the nineteenth century.

To understand how plants orient themselves and grow in certain ways, Leonardo turned to the flow of sap through the plant tissues, as he had done for the explanation of branching patterns. He used the term "vital sap" for the essential life fluid of plants, and he believed that it nourishes the plant tissues and also regulates their growth. Today we know that the sap contains sugars and hormones and that the latter indeed affect various aspects of plant growth. As Emboden points out, these effects of hormonal activity on the growth of plants were not understood until the twentieth century.[57] That Leonardo described several of them qualitatively in the early sixteenth century is truly exceptional.

In his studies of trees, Leonardo correctly distinguished between the dead outer layer of the tree's bark, also known as cork, and the living inner bark, known to botanists as phloem, which he called very aptly "the shirt that lies between the bark and the wood."[58] He recognized that the function of this vascular tissue is to transport sap throughout the plant and that it is therefore of critical importance for keeping the plant alive. "In the bark and shirt is the life of the plant," he noted in the *Trattato*.[59] However, Leonardo did not recognize that the transport of water and minerals takes place through the xylem (or wood) inside the phloem, although he identified the inner bark and the wood (the phloem and the xylem) as two distinct tissues. The transport system of the xylem was not known until the late seventeenth century.[60]

A fine example of Leonardo's precise botanical observations is that of the so-called secondary growth (the increase of a tree's diameter), in which new cells are created in the phloem and some of them differentiate into cork, which becomes part of the bark while the bark's outermost layers split apart to accommodate the expansion. Leonardo's description of this rather complex process is completely accurate:

> The growth in thickness of trees is brought about by the sap which, in the month of April, is produced between the inner bark and the wood of the tree. At that time, the inner bark is converted into outer bark, and the outer bark acquires new cracks in the depths of the existing cracks.[61]

In his observations of secondary growth, Leonardo also noticed that some of the newly produced cells differentiate into wood, first turning into soft sapwood and eventually into the heartwood that provides the strength of the trunk. He discovered not only that this process generates the annual growth rings in the cross sections of a tree's branches and trunks, and that the approximate age of a cut tree can be determined by counting those rings, but also—remarkably—that the width of a growth ring is an indication of the climate during the corresponding year. "The rings on the cut branches of trees show the number of their years," Leonardo recorded in the *Trattato della pittura*, "and the greater or smaller width of these rings show which years were wetter and which drier." Then he added, almost as an afterthought: "Although this is of no importance in painting I want nevertheless to describe it in order to leave out as little as possible of what I know about trees."[62]

Dove si grida, non è vera scientia. *(Trattato della pittura*, chapter 33)

Where there is shouting, there is no true science.

Leonardo was keenly interested in how the flow of sap through a tree's inner bark affects its growth in various ways. As Emboden has pointed out, these studies are fascinating because several of Leonardo's observations refer implicitly to hormonal activity, centuries before the discovery of hormones.[63] To observe the flow of sap in the phloem, Leonardo paid special attention to injuries sustained by trees. He noted that in such cases, sap rapidly flows to the defense of the injured part, where it stimulates vigorous growth. The grafting of branches offered him ample opportunities to observe this effect. "If a branch of a tree is cut off and there be grafted or inserted one of its own twigs," he noted, "in time this twig will grow much larger than the branch which nourishes it, because the nourishment or vital saps rush to the defense of the injured place."[64] This phenomenon, well known to botanists, is associated today with the collective activity of several types of hormones.[65]

In a passage in the Codex Atlanticus, Leonardo describes the same effect in the case of a tree that has lost part of its bark:

> When a tree has had part of its bark stripped off, nature . . . diverts to the stripped portion a greater quantity of nutritive moisture than to any other part; so that . . . the bark there grows much more thickly than in any other place. And this moisture has such power of movement that, having reached the spot where its help is needed, it makes various buddings and sproutings, not unlike water when it boils.[66]

As before, Leonardo's observation of the powerful movement of "nutritive moisture" to the injured area refers implicitly to the migration of sugars and hormones, which produce more tissue in the process of regeneration than in the rest of the bark. The rapidity of the flow of sap serves to seal the area before bacteria and fungi can cause it to rot. As Emboden explains, Leonardo's "buddings and sproutings after the manner of water when it boils" have been confirmed by modern botanists and are understood today as resulting from high hydrostatic pressures in the special elongated cells of the phloem known as sieve tubes.[67]

In view of his accurate location of the flow of sap in the phloem, or "inner bark," it is not surprising that Leonardo was well aware of the lethal effect of girdling, in which a tree dies when an entire ring of bark is removed. "If you take away a ring of bark from the tree," he recorded in Manuscript B, "it will wither from the ring upward, and all below will remain alive."[68] As Emboden points out, this statement shows that Leonardo understood that sap is stored in the roots and lower portions of a tree, and that girdling, by preventing any stored sap from flowing upward through the phloem, cuts off the tree's vital nourishment.[69] Indeed, the storage of sap in the roots is mentioned explicitly in a note in Manuscript G: "The trunks of the trees have a bulging surface that is caused by their roots, which carry nourishment to the tree."[70]

In the view of Emboden, Leonardo's most remarkable observation regarding the effects of the flow of sap on plant growth is expressed in the following note from the *Trattato della pittura*:

> The sap of the branch, when it does not absorb the heat of the sun, falls to the lower part of its branch; and the sap nourishes more where it is in greater abundance.[71]

According to Emboden, a modern botanist would interpret this passage as describing the migration of growth hormones, known as auxins, along a branch in such a way that they accumulate on the lower side of the branch owing to the effect of gravity.

Emboden points out that Leonardo's statement actually includes two distinct observations, both of which are "most extraordinary in [their] correctness."[72] One is that some part of the sap can be inactivated by sunlight, which may indeed occur with auxins. The other observation is that the sap, when it is not struck by the sun, "falls to the lower part of the branch." This migration of auxins from the light to the dark side of a stem has been recognized in modern botany as the fundamental reason that plants bend toward the sun.

We can only marvel at the fact that, long before the discovery of hormones and the advent of biochemistry, Leonardo was able to use his tremendous powers of observation and his great intuition to arrive at a correct qualitative understanding of branching patterns, secondary growth, annual growth rings, phototropism, and the responses of trees to injuries. Like modern plant physiologists, he explained these phenomena in terms of specific peculiarities in the flow of the life fluid of plants through their vascular tissues.

During the years 1508–12, while Leonardo was engaged in his most intensive studies in theoretical botany, he also began to organize his Notebooks, mapped out several comprehensive treatises, and undertook advanced anatomical studies.[73] It is therefore not surprising that he often established conceptual links between his botanical observations and his investigations of the "qualities of forms" in other areas. He compared the branching patterns of trees to those of rivers and of blood vessels. He drew spiraling foliage reminiscent of spiraling water vortices. He likened the flow of sap through a plant's vascular tissues to the flow of blood through human arteries and veins.

One of his most sophisticated comparisons of organic forms in different living systems is to be found in his studies of plant seeds. A sheet in the Windsor Collection contains the following note:

> All seeds have an umbilical cord, which is broken when the seed is ripe. Likewise, they have a matrix and secundina, as herbs and all seeds that are produced in pods demonstrate. But those which are produced in nutshells, like hazelnuts, pistachios and the like, have a long umbilical cord, which shows itself in their infancy.[74]

As Emboden explains, Leonardo observed correctly that in flowering plants the seed develops from a structure within the flower's ovary, known today as the ovule, which remains attached to the ovary wall by a stalk, known to botanists as the funiculus, until it develops into a seed after fertilization.[75] The "matrix" and "secundina" in Leonardo's note refer to the outer layers of the ovule, known today as integuments.

What is most remarkable in Leonardo's observation is his identification of the funiculus (the stalk that attaches the seed to the ovary wall) with the umbilical cord, which attaches the mammalian embryo to the placenta. Having studied the development of the human fetus in great detail and pictured it in superb drawings (see pp. 312ff.), he could not help but be impressed by the delicate structural similarity in the developments of plant seeds and mammalian embryos. It was for him a compelling testimony to the unity of life at all scales of nature. Modern botanists completely agree. They call the tissue within the ovary to which the ovule is attached the placenta and, like Leonardo, they view the funiculus as the plant equivalent of the umbilical cord.

Leonardo's highly sophisticated observations of intricate botanical forms and his ability to understand them in terms of the underlying processes of metabolism and development puts him far above the natural philosophers of his time. In recognition of this fact, physiologist and Leonardo scholar Filippo Bottazzi concluded his classic essay "Leonardo as Physiologist" with the following homage:

In art he was supreme among the great; in the mechanical sciences, he was the first and foremost restorer. But the story of modern biology begins with him.[76]

PART II

Form and Transformation
in the Human Body

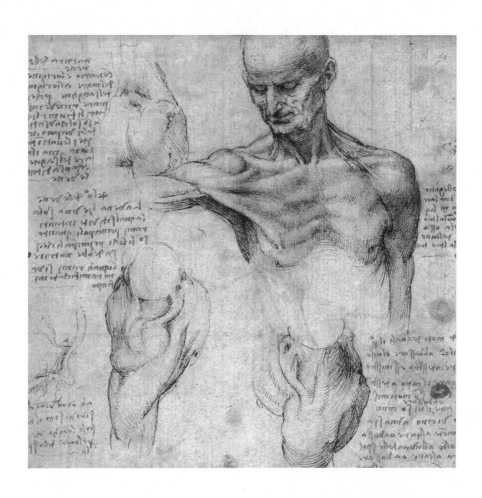

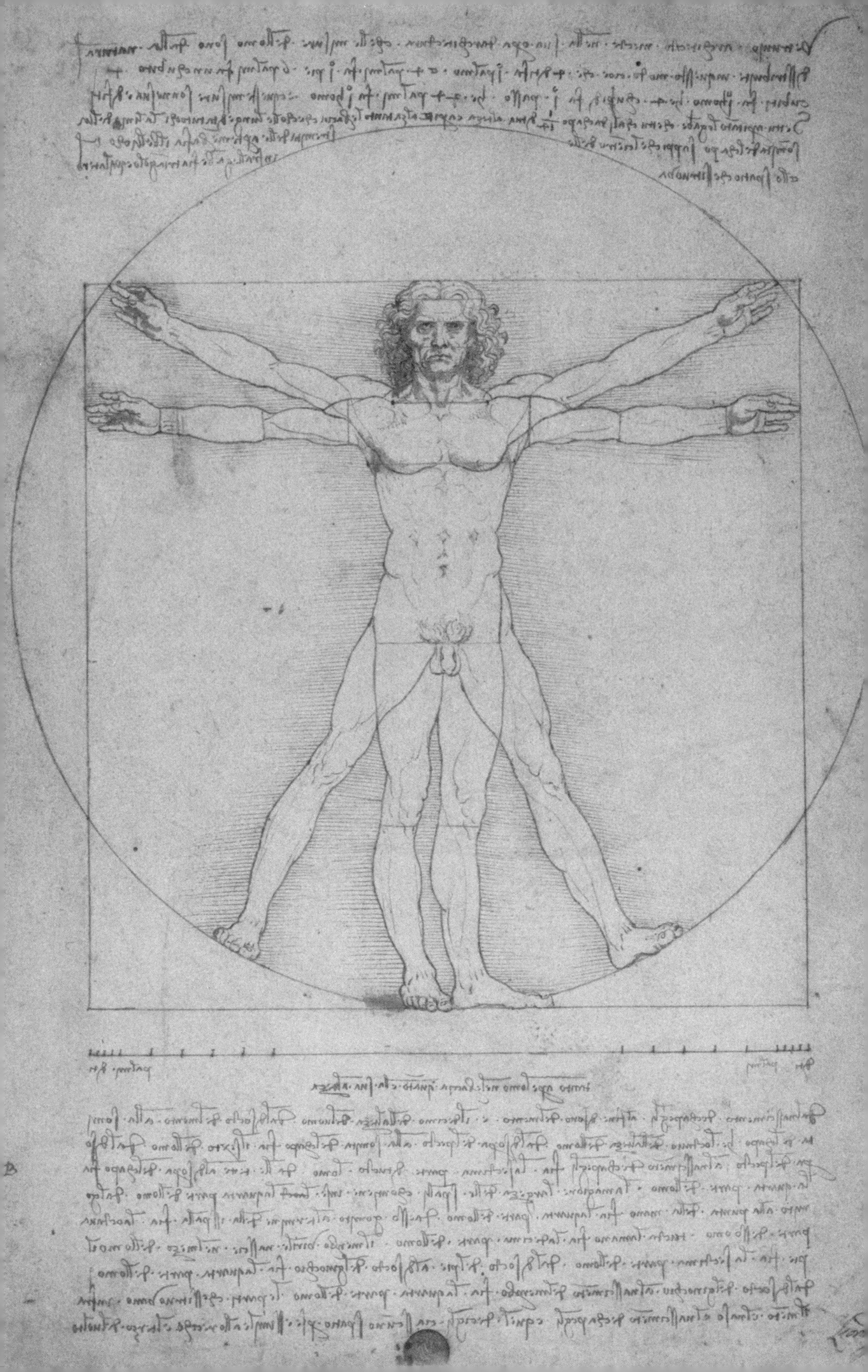

4

The Human Figure

In the preceding three chapters, I have analyzed Leonardo's science of life in the macrocosm—his studies of the flow of water, life's medium and vital fluid; his explorations of the Earth's living body and its transformations over enormous periods of time, in which rocks are gradually worn down, turning into gravel and eventually into the fertile soil that provides life's sustenance; and his observations of the growth patterns of plants and their interactions with the air, water, and soil of the living Earth.

For Leonardo, these manifestations of life, which are analyzed today in the separate sciences of fluid dynamics, geology, botany, and ecology, were all threads in one seamless fabric. And whenever he explored the forms of nature in the macrocosm of the living Earth, he looked for similar qualities and patterns in the microcosm of the human body (see p. 25). Indeed, from his first extended anatomical studies in Milan in his thirties to his most sophisticated research in cardiology and embryology in his old age, his work in anatomy always went hand in hand with explorations of related phenomena now studied in physics, geology, and botany.[1]

Chronology of Anatomical Research

From the days of his apprenticeship in Florence, Leonardo had been familiar with the dissection of muscles, which was practiced by many Renaissance painters. However, Leonardo's first extended anatomical studies and systematic dissections were not concerned with muscles but with the pathways of the sensory nerves, in particular the optic nerve, in the human skull. In the late 1480s, he recorded his discoveries in several stunning

PRECEDING Studies of the muscles of the neck and shoulder, c. 1509–10 (detail, see fig. 6-4).
FACING *The Vitruvian Man*, c. 1490 (detail, see plate 5).

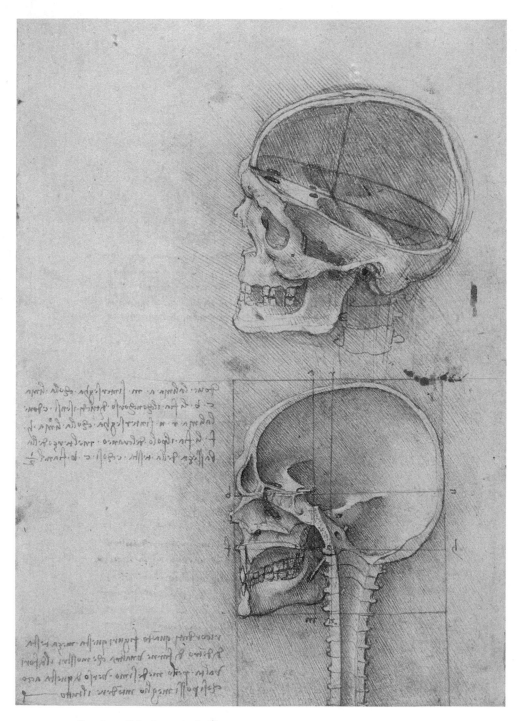

FIG. 4-1. Section of the human skull, c. 1489.
Windsor Collection, *Anatomical Studies*, folio 43r.

pictures of the skull (for example, fig. 4-1), which are famous for their delicate renderings of light and shade and their masterful application of visual perspective.[2] The trained eye of the physician sees in these drawings amazingly accurate depictions of the skull's cavities and nerve endings—the eye socket, its neighboring sinuses, the tear ducts, and the foramina (openings) for the optic and auditory nerves.[3]

This phase of Leonardo's anatomical work was part of an elaborate research program dedicated to the entire process of perception and knowledge. It began with the study of perspective, the geometry of light rays (known today as geometrical optics), and the interplay of light and shadow. From there, Leonardo proceeded to explore the very nature of light, the anatomy of the eye and physiology of vision, and the pathways of sensory impressions along the cranial nerves to the center of the brain, where he located the "seat of the soul." I have discussed Leonardo's wide-ranging examinations of the process of perception at length in my previous book.[4] They not only resulted in many remarkable anatomical discoveries but led him to formulate highly original ideas about the relationship between physical reality and cognitive processes, which have re-emerged only recently in the new interdisciplinary field of cognitive science.

The next phase of Leonardo's anatomical research was concerned with the human body in motion. However, detailed anatomical investigations of this grand theme had to wait for over fifteen years. The main reason for this long gap was that in the early 1490s, Leonardo became fascinated with mathematics, especially geometry, and realized that a thorough analysis of how nerves, muscles, tendons, and bones work together to move the body requires a basic understanding of the principles of mechanics. "Nature cannot give movement to animals without mechanical instruments," he observed,[5] and so he immersed himself in prolonged studies of how to apply geometry to elementary problems involving weights, force, and movement—the branches of mechanics now known as statics, dynamics, and kinematics. From the beginning, this research was concerned both with the workings of machines and with the understanding of the human body and its movements. Indeed, Leonardo began with a series of experiments to determine how much force various parts of the body could generate in various positions, and he meticulously recorded his results (see p. 213).

Around 1506, at the beginning of his second period in Milan, Leonardo must have felt that he had acquired sufficient knowledge and skills in mechanics to take up his anatomical studies once again. Two years later, he

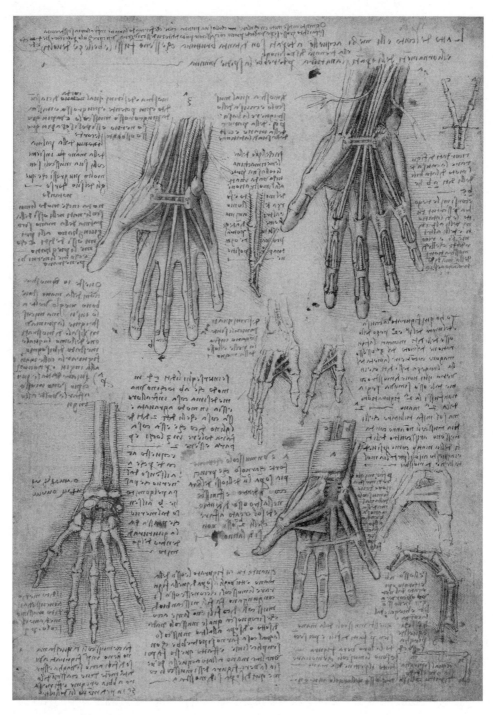

FIG. 4-2. The mechanisms of the hand, c. 1510.
Windsor Collection, *Anatomical Studies*, folio 143r.

outlined his plans for a vast treatise on anatomy titled *De figura umana* (On the Human Figure). It was to contain 120 "books" (chapters) with demonstrations of every conceivable aspect of the combined actions of nerves, muscles, tendons, and bones.

"My configuration of the human body will be demonstrated to you just as if you had the natural man before you," Leonardo announced at the outset of his detailed outline, and then explained why this would require numerous dissections.[6] We do not know how many chapters of his ambitious treatise on the human figure Leonardo composed, but the large number of superb drawings of the body's nerves, blood vessels, muscles, and bones he produced during the years 1506–10, which are now in the Windsor Collection, make it evident that his promises were not exaggerated.

On one of these folios, showing the anatomy and actions of the muscles and tendons of the hand (fig. 4-2), Leonardo wrote the following reminder to himself, prominently placed at the center of the page:

> Arrange it so that the book on the elements of mechanics with its practice shall precede the demonstration of the movement and force of man and other animals, and by means of these you will be able to prove all your propositions.[7]

The first decade of the sixteenth century was the period of Leonardo's most intensive anatomical research. It resulted in the most prodigious output of his scientific drawings, covering almost all aspects of the human anatomy. His skills in anatomical dissections were now far superior to those in the first period and his observations were often stunningly accurate. At the same time, this was the period of his most sophisticated research in geology, recorded in the Codex Leicester and in Manuscript F (see chapter 2). It seems that investigating the bones, tissues, and blood of the human body and the rocks, soil, and water of the living Earth were scientific activities for Leonardo that reinforced one another and confirmed his belief in the fundamental unity of life at all scales of nature.

Leonardo always considered the human body an animal body, as we do in biology today, and he frequently compared various parts of the human anatomy to the anatomy of animals. This was also natural for him because his access to human cadavers was limited, and he often had to rely on dissections of animals.* One of the main themes of Leonardo's comparative

---

* It is noteworthy that his great compassion for animals prevented Leonardo from practicing vivisection.

anatomy was the juxtaposition of the mechanisms of movement in humans and animals, in particular the flight of birds, which fascinated him throughout his life (see pp. 251ff.).

The third phase of Leonardo's anatomical research was triggered by his encounter with the young anatomist Marcantonio della Torre in 1510.[8] His discussions with this brilliant medical scholar inspired Leonardo to expand his anatomical work far beyond the areas involved in physical movement. During the subsequent years, he delved deep into the body to study the functions of the internal organs, respiration, and the flow of blood.

The central theme of this third and last phase was Leonardo's persistent quest to understand the nature of life. This is evident in all his anatomies of internal organs, and especially in his investigations of the heart—the bodily organ that has served as the foremost symbol of human existence and emotional life throughout the ages. Leonardo's careful and patient studies of the movements of the heart and the flow of blood, undertaken in Milan and Rome when he was over sixty, are the culmination of his anatomical work. He not only understood and pictured the heart in ways no one had before him but also observed subtleties in its actions and in the flow of blood that would elude medical researchers for centuries.

During the last decade of his life, while he was engaged in his most advanced studies of the human heart, Leonardo became intensely interested in another aspect of the mystery of life—its origin in the processes of reproduction and embryonic development. That he had always considered embryology an integral part of his studies of the human body is evident from an early outline of his planned treatise on the human figure, written in 1489 during the first phase of his anatomical studies. This outline begins with the following sweeping declaration:

> This work should begin with the conception of man, and should describe the nature of the womb, and how the child lives in it, and to what stage it resides in it, and in what way it acquires life and food; its growth, and what interval there is between one degree of growth and another, and what it is that pushes it out of the body of the mother.[9]

Leonardo's embryological studies, based largely on dissections of cows and sheep, included most of the topics he had listed and led him to many remarkable observations and conclusions. He described the life processes of the fetus in the womb, including its nourishment through the umbilical cord, in astonishing detail, and he also made a series of measurements on animal fetuses to determine their rates of growth. Leonardo's embryological drawings are delicate revelations of the mysteries surrounding the earliest stages of human life.

Beauty and Proportion

During the first phase of his anatomical research, in which he investigated the pathways of the cranial nerves, Leonardo also developed a keen interest in the proportions of the human body, and he attempted to associate idealized anatomical forms with the mathematical perfection of geometric figures. This clearly can be seen in his drawings of the human skull. In the two drawings shown in figure 4-1, the skull is inscribed in a circle with intersecting lines suggesting a coordinate system (upper drawing), and in a rectangle on which various marks indicate the skull's proportions (lower drawing). Leonardo's studies of the ideal proportions of the human skull during this period are reflected in the regular proportions of heads in several of his portraits, while deviations from these ideal proportions are clearly visible in his famous drawings of "grotesques."[10]

Leonardo's associations of human anatomy with the perfection of mathematical proportions and geometric forms arose from his intense fascination with mathematics during this period. The circle and the square, in particular, were the forms of perfect symmetry in the Platonic tradition, together with the five Platonic solids (the tetrahedron, cube, octahedron, dodecahedron, and icosahedron), which Leonardo explored in a series of drawings during his studies of geometry and proportion with the mathematician Luca Pacioli.[11]

With his interest in proportion, Leonardo followed a long tradition that originated in classical antiquity. Artists, including the Greek sculptor Polyclitus and the Roman architect Vitruvius, had made detailed studies of ideal human proportions, and throughout the Middle Ages and the Renaissance there had been many attempts to establish a canon of proportions for the human figure. The most sophisticated of these canons was the one proposed shortly before Leonardo's birth by the great architect and humanist Leon Battista Alberti in his treatise *De statua* (On Sculpture).[12]

In the Renaissance, artists and philosophers alike were fascinated by the study of proportions. It not only helped them to correctly represent the human figure in paintings and sculptures but also was the very foundation of Renaissance aesthetics. Proportion in painting, sculpture, and architecture was seen as the essence of harmony and beauty, in the microcosm of the human body as well as in the macrocosm at large. Accordingly, architects endeavored to design the dimensions of their buildings in such a way that they reflected the proportions of the ideal human body, expressing them, whenever possible, in terms of the classical symbols of perfection—the square, the circle, and the golden section.*

During the years 1489–90, Leonardo threw himself into a systematic study of human proportions with his usual vigor and attention to detail. He took a wealth of measurements to establish a comprehensive system of correspondences between all parts of the body and recorded them in a series of detailed drawings (for example, fig. 4-3).[13] Leonardo's meticulous studies of the proper relationships among various parts of the body, and of the relationships of the parts to the body as a whole, were one of the foundations of the amazing accuracy of his anatomical drawings. "Applying the science of proportion to his anatomical research," writes architectural historian and Leonardo scholar James Ackerman, "Leonardo was able to produce vivid images of the body in which each part could be represented in its proper place, and of the proper size in relation to other parts, and he and his contemporaries developed schemata of anatomical illustration which required only minor changes over the centuries."[14]

In addition to human proportions, Leonardo undertook detailed studies of the ideal proportions of the horse (fig. 4-4). Throughout his life he had a special fondness for horses and, like many Renaissance artists, he regarded the proportions and movements of horses as most "noble," second only to those of humans.†

In contrast to other Renaissance artists, Leonardo's approach to the study of proportion was organic and physical.[15] Unlike Albrecht Dürer a few years later, Leonardo never adopted a fixed regular grid into which the

---

* The golden section (also known as "golden ratio" or "divine proportion") was defined by Euclid as the division of a line segment into two parts in such a way that the ratio of the whole segment to the larger part is equal to the ratio of the larger part to the smaller. In the Renaissance, the golden section was considered the proportion most pleasing to the eye, and many artists and architects incorporated approximate golden ratios into their works (see Livio, *The Golden Ratio*).

† Leonardo's extensive studies of horses are assembled in a special volume of the Royal Collection at Windsor Castle; see p. 355 below.

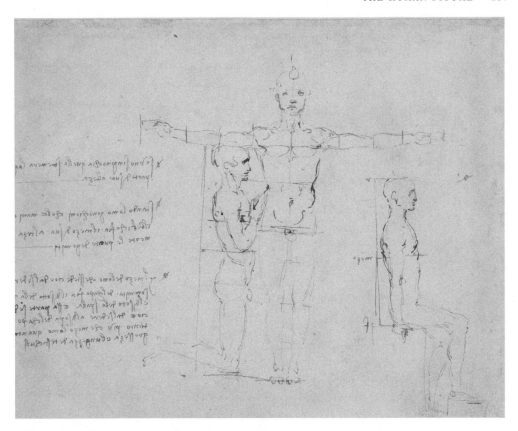

FIG. 4-3. Studies of the proportions of the body
when standing, kneeling, and sitting, c. 1490.
Windsor Collection, *Anatomical Studies*, folio 27r.

body or its parts would be inscribed, but rather preferred to first draw the
human body as he saw it. Then he would take his measurements and enter
them on the sketch he had drawn from nature. As art historians Frank
Zöllner and Johannes Nathan point out, even in such an elaborate study
as that of a horse's leg shown in figure 4-4, the proportions of the drawing
do not always agree with the measurements he recorded for the horse.[16]
It seems that for Leonardo, these sketches and the numbers recorded on
them were always works in progress, tentative approximations of an ideal
canon of beauty and proportion.

Another reason Leonardo could never confine the body to a rigid
geometric grid was his utterly dynamic view of the human figure. The
representation of the human form in terms of its movements and development—the second grand theme of his anatomical research—is already

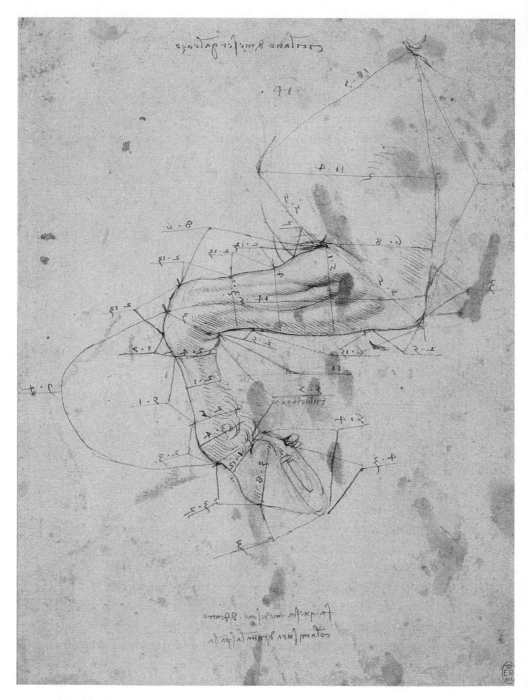

FIG. 4-4. Studies of the proportions of a horse's leg, c. 1485–90. Windsor Collection, *Horses and Other Animals*, folio 94.

foreshadowed in his studies of proportion. Human proportions are never static for Leonardo; they change when the body moves and adjusts its position (as shown in fig. 4-3), and they also change over its lifetime.

According to science historian Domenico Laurenza, this attempt to combine the search for mathematical perfection with careful attention to change and transformation is a highly original aspect of Leonardo's studies of proportion and, indeed, is characteristic of his science as a whole:

> Variability and constancy of form, the physical world and geometrical perfection, the qualitative dimension and the quantitative dimension: Leonardo's reflection is torn between these opposite poles, trying to reach a synthesis. . . . In his investigation of forms, both natural and geometric, Leonardo tries to achieve a synthesis between these two aspects—the existence of laws of order and geometric constancy in the midst of continuing variability and transformation of physical forms.[17]

In the classical tradition, the contrasting ideas of qualitative, ever-changing forms and of eternal mathematical perfection were expounded in the philosophies of Aristotle and Plato, respectively. The Italian humanists eagerly studied both schools. Florence under the Medici was the center of Platonism, while Milan, under the influence of the universities of Padua and Bologna, was predominantly Aristotelian. Leonardo, who spent many years in both cities, was well aware of these philosophical debates. His science, with its emphasis on qualities and continual change and transformation, was much more Aristotelian than Platonist. But he was also fascinated with mathematical precision, and the tension between the two schools of thought is often evident in his writings.

The Vitruvian Man

Leonardo's most famous and most complete drawing of human proportions is his iconic image of a man's body in two different, superimposed positions—first standing inscribed in a square, and then with arms and legs outstretched inscribed in a circle (plate 5). The drawing, on a single sheet now housed in the Accademia Gallery in Venice, is known as the Vitruvian man because it is a faithful illustration of the classical canon of perfect human proportions established by the Roman architect Vitruvius. In the accompanying text, Leonardo summarizes the relevant passages from the third volume of Vitruvius' treatise *De architectura* (On Architecture), in which the precise proportional relationships between various parts of the ideal human body are defined.

"Vitruvius, the architect," Leonardo begins his summary, "says in his work on architecture that the measurements of the human body are distributed by nature as follows: 4 fingers make 1 palm, and 4 palms make 1 foot; 6 palms make a cubit; 4 cubits make a man's height." This establishes a system of units of length expressed as fractions of the height of a man: 1 cubit = ¼ of the height, 1 foot = ¹/₆, 1 palm = ¹/₂₄, and 1 finger = ¹/₉₆. These units are marked off on a calibrated scale beneath Leonardo's drawing.* Leonardo then proceeds to state Vitruvius' prescription of how to inscribe the ideal human body in a circle:

> If you open your legs so much as to decrease your height by 1/14, and spread and raise your arms so that your middle fingers touch the line of the crown of the head, you should know that the navel will be the center of the extremities of the outspread limbs; and the space between the legs will form an equilateral triangle.

The inscription of the figure into a square follows from the rule that "the span of a man's outstretched arms is equal to his height." And finally, Leonardo lists a long series of further proportions established by Vitruvius, for example:

> From the roots of the hair to the bottom of the chin is the tenth part of a man's height; from the bottom of the chin to the crown of the head is the eighth of the man's height; from the top of the breast to the crown of the head is the sixth of the man . . . ; the maximum width of the shoulders contains in itself the fourth part of the height . . . ; the complete hand is the tenth part of the man.[18]

The famous Venetian folio is unique among Leonardo's studies of proportion. Unlike his other drawings, it is not a working record of his own observations and measurements but a definitive visual representation of a classical canon for the ideal human figure. Its careful layout and precise execution suggest that the folio may have been intended for reproduction

---

* The four units of length defined by Vitruvius and expressed in terms of fractions with even denominators corresponded to the duodecimal system that was in use in Europe until the introduction of the meter in the nineteenth century (see Zöllner and Nathan, *Leonardo da Vinci*, p. 348). The cubit was an antique unit of length equal to the length of the forearm and hand.

by an engraver, and it has also been suggested that Leonardo may have planned to use the drawing as the starting point for a treatise on human proportions. If so, he would have followed a practice he adopted in many other areas of his scientific studies. He usually started from commonly accepted concepts or explanations, and often summarized what he had gathered from the classical texts before proceeding to verify it with his own observations.[19] Indeed, Leonardo did not hesitate to correct the Vitruvian canon. On the Venetian folio, he referred to the foot as "the seventh part of the man" instead of one-sixth, as Vitruvius had it, and in later anatomical drawings he recorded the face as one-ninth of the figure's height instead of Vitruvius' one-tenth.

Even though Leonardo's famous drawing is a strict illustration of the rules of proportion set out by Vitruvius, it is nevertheless a highly original contribution. Whereas Vitruvius described first the configuration of the body in a circle and then in a square, Leonardo visually unified the two descriptions. As Laurenza explains, "Only an image could demonstrate the two possibilities in a simultaneous—and hence harmonious—manner. Leonardo, in so doing, presents a faithful reading of Vitruvius, but seems to understand that concept with greater depth than Vitruvius himself."[20]

Before Leonardo, other Renaissance artists tried to superimpose the two Vitruvian positions but failed to do so. Most of them ended up representing the two positions in separate drawings. Others tried to combine the two images but could not achieve the seemingly natural harmony of the Venetian folio. In their drawings, not all the body parts would touch the circle and square, as prescribed by Vitruvius, or the human figure would be elongated in an unnatural way to achieve the correct result.[21]

Leonardo realized that, if the circle and square are concentric, the Vitruvian prescriptions cannot be represented correctly without distorting the body unnaturally. Hence, he chose two different centers for his two figures. The navel is the center of the figure in the circle, as described by Vitruvius, while the pubic bone is the center of the figure in the square. This adjustment is Leonardo's principal innovation, which clearly distinguishes his drawing from earlier illustrations.*

* In Leonardo's drawing, the relationship between circle and square is determined by the position of the navel, which divides the height of the figure approximately in the proportion of the golden section. It is unlikely, however, that Leonardo actually constructed the circle from the square in this way. In 1490, his mastery of Euclidean geometry would not have been sophisticated enough to do so. Besides, Leonardo's starting point would have been the body, not a geometrical construction (see Laurenza, "L'uomo geometrico").

With this choice, Leonardo achieved a perfect interdependence of the two figures. Only the head is exclusively related to the square, as specified by Vitruvius. The fingertips of the outstretched arms touch both circle and square, while the feet do so when the legs are closed. The entire image is one of a single body with two positions coexisting harmoniously. It is Leonardo's unique synthesis of the Vitruvian canon of proportions, unequalled by any artist before him or since.

Leonardo's Vitruvian man may also be seen as a unique integration of the Aristotelian emphasis on change and transformation with the Platonic ideal of eternal geometric perfection. As the two positions of the human figure coexist and interact harmoniously, so do the two geometric figures. When the body's position changes, circle and square are "transformed" into each other.[22] The Vitruvian man is an early example of Leonardo's lifelong fascination with the transformation of geometric figures. Fifteen years after he drew the Venetian folio, this interest would lead him to develop his special "geometry done with motion," which we recognize today as a distant forerunner of the branch of modern mathematics known as topology.[23]

Proportion and Harmony

The ultimate goal of Leonardo's extensive studies of human proportions was not to establish a definitive geometric scheme, but rather to enable him to represent an individual human figure in such a way that each part was integrated into a harmonious whole. "The beautiful proportions of an angelic face in painting," he wrote in the *Trattato*, "produce a harmonious concord, which reaches the eye simultaneously, just as [a chord in] music affects the ear."[24] This search for harmony or beauty is apparent not only in his paintings but also in his anatomical drawings. As Laurenza explains,

> Leonardo's anatomy, too, is subjected to a search for harmonious configurations. . . . He attempts systematically to realize representations that reconstitute the extreme variety and variability of the individual anatomical parts (displayed in the course of the dissections) into a harmonious composition. . . . Around 1506–1508, he creates anatomical representations which not only show the organs as an ensemble, in their reciprocal relationships, but also emphasize the symmetry of some of them and the topological centrality of others.[25]

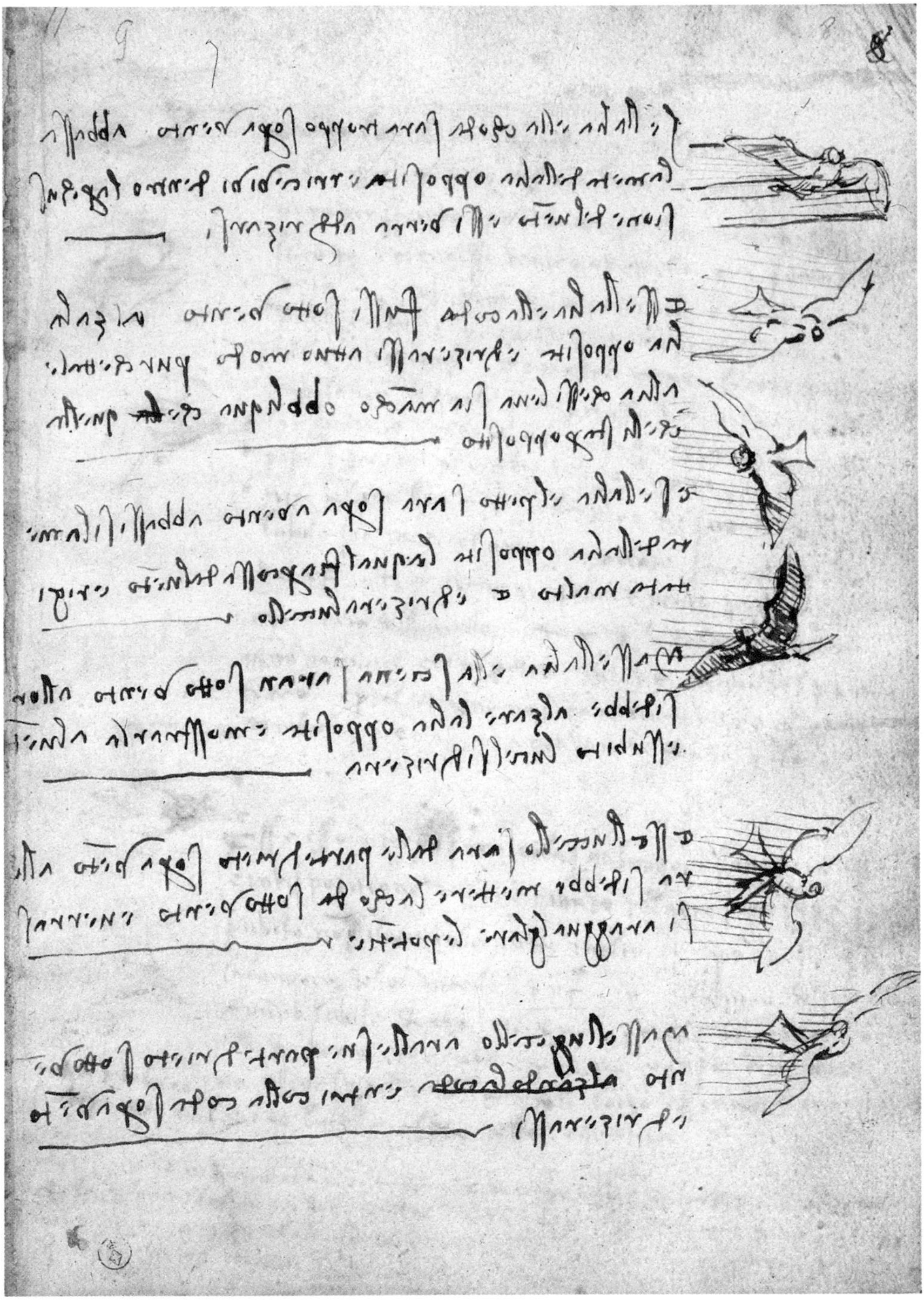

PLATE 1. Birds in flight, 1505.
Codex *Sul Volo*, folio 8r.

PLATE 2. Andrea del Verrocchio and Leonardo da Vinci,
The Baptism of Christ, c. 1470–75.
Uffizi Gallery, Florence.

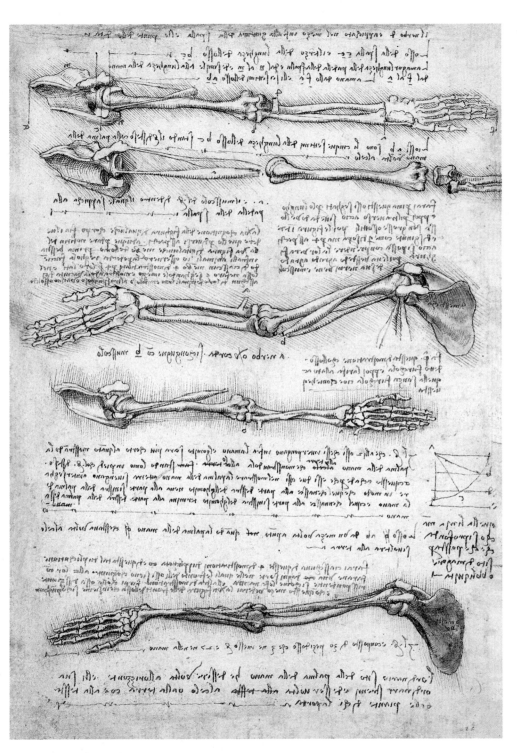

PLATE 3. Study of the rotation of the arm, c. 1509–10.
Windsor Collection, *Anatomical Studies*, folio 135v.

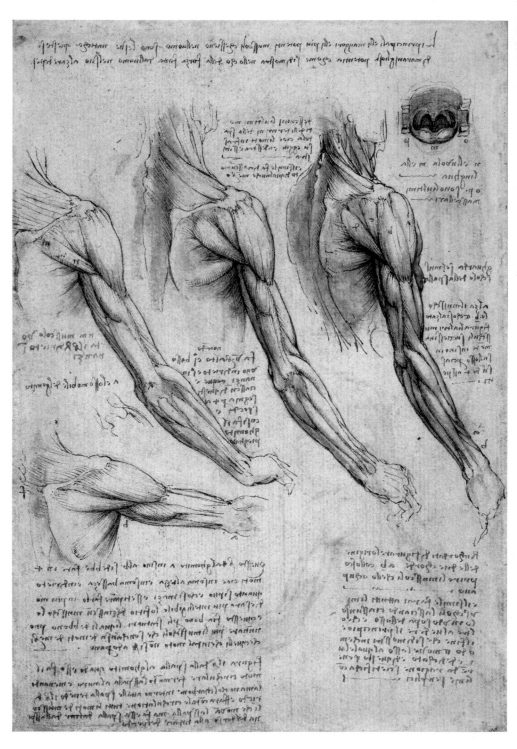

PLATE 4. Rotated views of the muscles of the shoulder and arm, c. 1509–10.
Windsor Collection, *Anatomical Studies*, folio 141v.

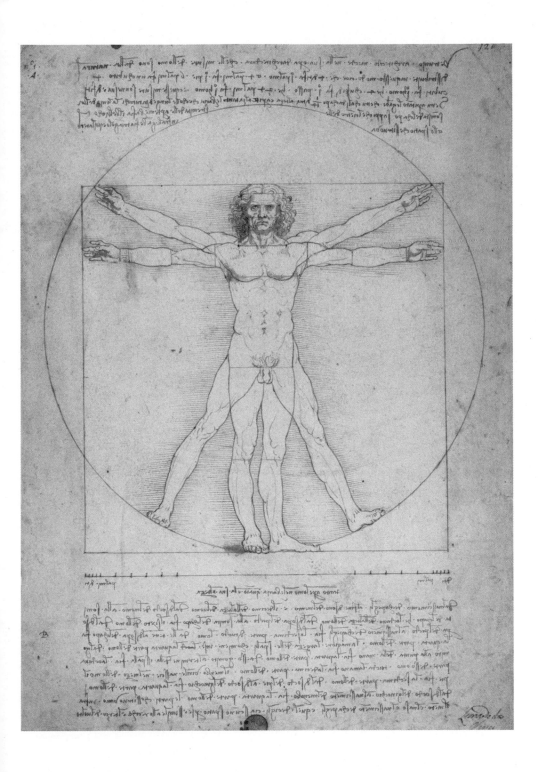

PLATE 5. *The Vitruvian Man*, c. 1490.
Galleria dell'Accademia, Venice.

PLATE 6. Star of Bethlehem, c. 1508.
Windsor Collection, *Landscapes, Plants,
and Water Studies*, folio 16r.

PLATE 7. *Madonna and Child with Saint Anne*, c. 1506–15.
Musée du Louvre, Paris.

PLATE 8. *Virgin of the Rocks*, c. 1483–86.
Musée du Louvre, Paris.

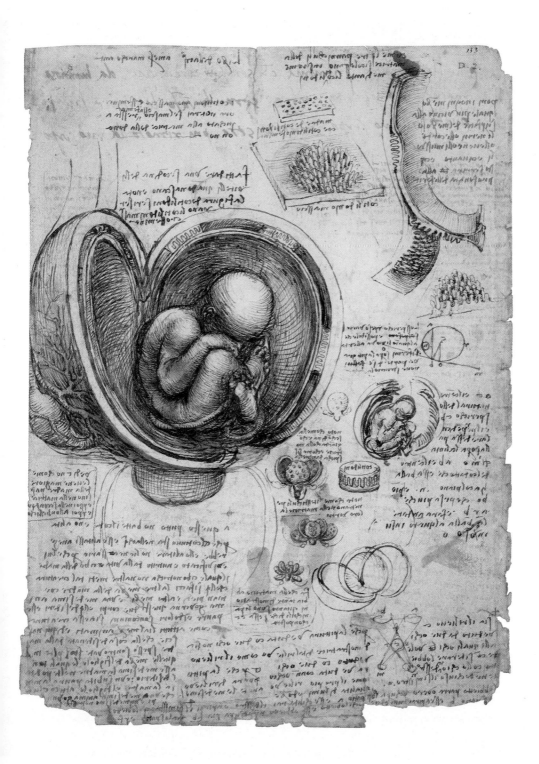

PLATE 9. The fetus in the womb, c. 1510–12.
Windsor Collection, *Anatomical Studies*, folio 198r.

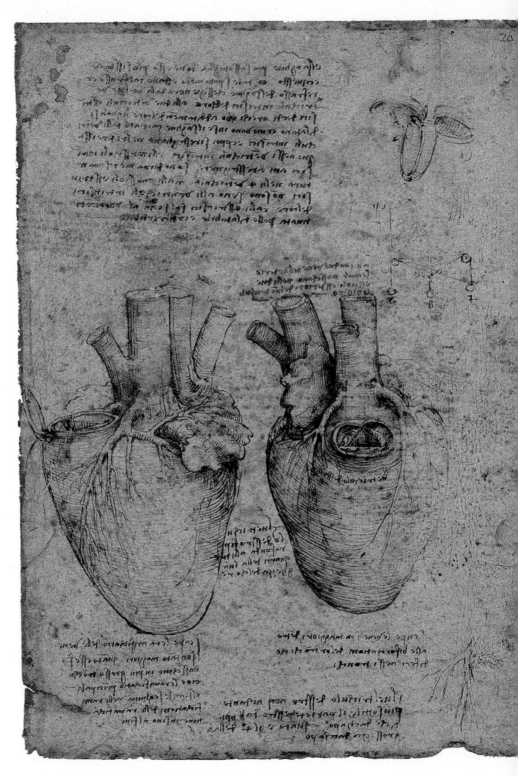

PLATE 10. The heart and its blood vessels, 1513.
Windsor Collection, *Anatomical Studies*, folio 166v.

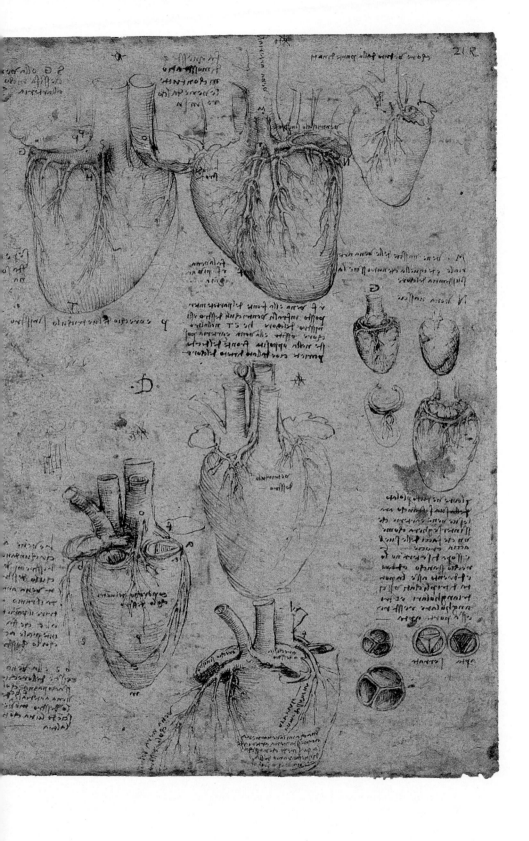

PLATE II. *Mona Lisa*, also known as *La Gioconda*, c. 1503–15.
Musée du Louvre, Paris.

The culmination of these integrated anatomical representations is Leonardo's famous drawing of the anatomy of a woman's body on a double sheet in the Windsor Collection (fig. 4-5). It is a superb composite image of many anatomical systems—the heart and blood vessels, the trachea and bronchi; the liver, spleen, and kidneys; and the urinary and reproductive systems—all shown in relation to one another within the integrated whole of the female body. The harmonious order is emphasized further by the strong bilateral symmetry among the organs. The

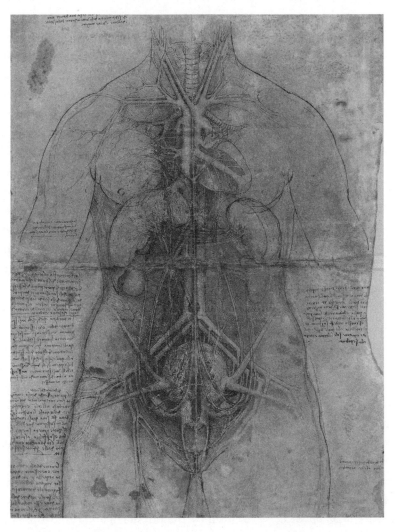

FIG. 4-5. Composite view of the internal organs of a woman's body, c. 1508. Windsor Collection, *Anatomical Studies*, folio 122r.

drawing contains several anatomical errors, in part because it is largely based on dissections of animals.[26] However, as Nathan points out, "this fact pales in comparison with [Leonardo's] ability to convey the complexity of the human body with such clarity, even while providing an enormous wealth of information."[27]

Leonardo's search for anatomical harmony and synthesis is complicated by the fact that the science of anatomy, by definition, is based on the analytical separation of the body's parts. Indeed, throughout his anatomical drawings there is a tension between composite representations, in which several parts of the body are portrayed in a single image within an integrated whole, and dissected representations that depict isolated parts.

This tension is explored in a fascinating monograph by Domenico Laurenza titled *La ricerca dell'armonia: rappresentazioni anatomiche del Rinascimento* (*The Search for Harmony: Anatomical Representations in the Renaissance*).[28] Laurenza shows that the contrast between composite and dissected representations is characteristic of Renaissance anatomy as a whole and, in fact, was already apparent in antiquity. Aristotle, the classical author most widely available to Renaissance scholars, wrote several biological treatises, including *Historia animalium* (History of Animals) and *De partibus animalium* (On the Parts of Animals). In the latter work he discussed the internal organs of animals, but always within a strong composite view. Aristotle also introduced the distinction between simple parts of the body (bones, tendons, etc.) and composite parts (head, arms, legs, etc.), which was widely adopted in subsequent centuries.

The culmination of anatomical knowledge in antiquity was reached in the second century A.D. with Galen, who wrote more than one hundred treatises, summarizing the medical knowledge of his time in accordance with his own theories, which were partly based on dissections of animals.[29] The distinction between composite and dissected representations was apparent in two of Galen's works. *De usu partium* (On the Usefulness of the Parts) reflected the synthetic approach, while his *De anatomicis administrationibus* (On Anatomical Procedures) represented the analytical approach. However, as Laurenza points out, the latter work was not available to the Italian humanists.

The medical bible throughout the Middle Ages and the Renaissance was the *Canon of Medicine*, written by the physician and philosopher Avicenna (Ibn Sina) in the eleventh century, a vast encyclopedia that codified the complete Greek and Arabic medical knowledge.[30] As expected from one of the foremost medieval interpreters of Aristotle, Avicenna's approach to anatomy was also largely synthetic and composite.

A third influential text on anatomy in the Renaissance was the *Anatomia corporis humanis* by Mondino de Luzzi, a professor at Bologna in the early fourteenth century who was one of the few medieval teachers who actually performed anatomical dissections himself. Although Mondino's *Anatomia* was a practical guide to dissections and thus a different kind of book, he nevertheless emphasized that dissection had to begin with the knowledge of the body as a whole.

Leonardo studied the works of Galen, Avicenna, and Mondino, the three principal medical authorities of his time.[31] Since he always tried to understand the organic forms of nature in terms of their relationships, patterns, and contexts, it is not surprising that he was drawn to the composite approach to anatomy emphasized by these authorities. As Laurenza demonstrates with a wealth of examples, the tension between analysis (dissection) and synthesis (recomposition) appears in the entire corpus of Leonardo's anatomical work. Again and again, the detailed analytical representations of bones, muscles, and nerves are subsequently reconfigured in more complete synthetic images. On many occasions, verbal reminders referring to future reconfigurations are added in the margins of dissective drawings.

FIG. 4-6. Study of the blood supply to the upper leg, c. 1508. Windsor Collection, *Anatomical Studies*, folio 112r (detail).

Sometimes Leonardo produces images of a sequential composition, in which a limb is successively "dressed" with its anatomical elements, from the bones to the skin. Figure 4-6, a study of the blood supply to the upper leg, is a typical example. The femoral blood vessels and their branches are pictured in relationship to the femur, and the ensemble of the femur and its vessels is shown within the outlines of the thigh and in relationship to the complete lower leg. At other times, a sequence of bones, muscles, nerves, and blood vessels may be integrated into a single representation.

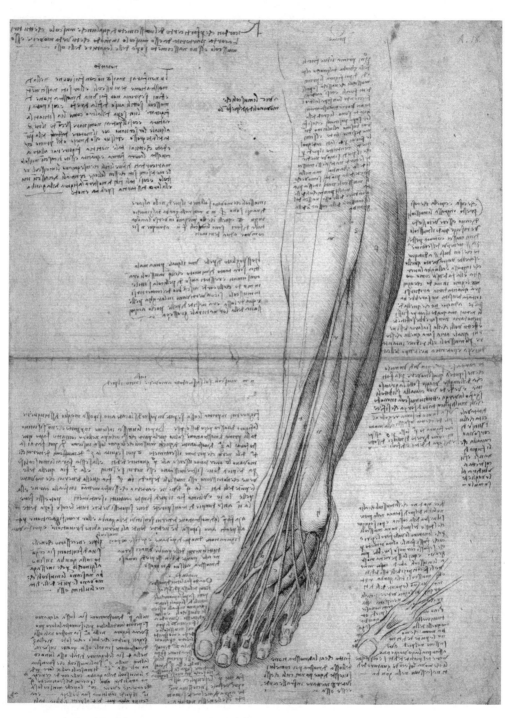

FIG. 4-7. Study of the anterior muscles of the leg, c. 1510.
Windsor Collection, *Anatomical Studies*, folio 151r.

Figure 4-7, for example, shows the superficial anatomy of the foot and lower leg in relationship to the complete calf (see pp. 239–40 for a more detailed discussion of this drawing).

Leonardo and Vesalius

During the first half of the sixteenthth century, the emphasis on composite anatomical representations, which had reached its peak in Leonardo's drawings, continued with several Italian anatomists,[32] but toward the mid-century the balance shifted from composite-synthetic to dissective-analytical representations. The culmination of the analytical tradition was reached in 1543, twenty-four years after Leonardo's death, with the publication of the famous work *De humani corporis fabrica* (On the Fabric of the Human Body) by the Flemish anatomist Andreas Vesalius.[33]

The *Fabrica*, as it came to be known, is a magnificent volume of more than two hundred anatomical illustrations based on dissections of human and animal bodies and organized into seven "books" (chapters) according to anatomical categories—bones, muscles, blood vessels, nerves, and so on. The illustrations are accompanied by extensive descriptions and an elaborate system of cross references between the anatomical features pictured and the printed text. The book is considered a landmark in the history of printing and typography. The skillful blending of images and text set a new standard for scientific illustration that would last for centuries.

The plates of the *Fabrica* are outstanding examples of sixteenth-century woodcut. Vesalius employed several artists to produce them, possibly from the studio of the famous Venetian painter Titian, which was not far from the University of Padua where Vesalius taught anatomy at the time.[34] In addition, some of the drawings were undoubtedly the work of the author himself. Upon its publication in Basel in 1543, the book rapidly gathered fame throughout Europe. While Leonardo's anatomical drawings lay hidden, the *Fabrica* was praised as the first major anatomical work based on actual dissections, and Vesalius became famous as the father of modern anatomy.

As Laurenza demonstrates in detail in his monograph, Leonardo and Vesalius personify extreme points in the tension between composite and dissective anatomical representations.[35] Whereas Leonardo shows us, say, a system of blood vessels in relationship to the adjacent bones and in proportion to the body as a whole, Vesalius concentrates on the analysis of

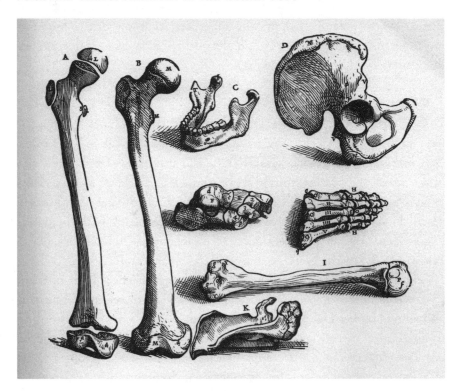

FIG. 4-8. Isolated representations of the human skeleton,
Andreas Vesalius, *De humani corporis fabrica*, book I.
From Saunders and O'Malley, *The Illustrations from the
Works of Andreas Vesalius.*

the anatomical parts per se, treating every anatomical system in isolation.
In the *Fabrica*, individual bones are completely isolated from the other
systems (fig. 4-8), as are the muscular system (fig. 4-9) and the blood vessels (fig. 4-10). The system of arteries is shown in isolation from the heart,
which is represented together with the lungs in another part of the work.
The lungs, in turn, are isolated from the trachea and bronchi. The brain
is shown in isolation from the nerves, which are discussed in a preceding
section.*

In all these illustrations, the external boundary of the body, which is

---

* To respond to the powerful Renaissance ideal of harmonious composite representations, Vesalius
decided to add a few composite, synthetic images in an appendix (*Epitome*) to the *Fabrica*. However,
as Laurenza points out, there too the order of the plates is dissective, from the superficial to the
deeper layers. The primary emphasis is always on analysis and dissection (see Laurenza, *La ricerca
dell'armonia*, pp. 91–92).

fundamental to the synthetic view, has been completely dissolved. "The result, in the end," Laurenza concludes, "is an analytical, piecemeal, fragmented vision. The visual synthesis is entirely virtual, left to the imagination and goodwill of the reader."[36]

As a consequence of this extreme analytical approach, the ideas of beauty, symmetry, and proportion are almost completely absent in the *Fabrica*. In order to attenuate the cold and macabre effect of his figures, Vesalius sometimes instructed the artist to present them in classical poses and to put them into pleasing landscapes (see fig. 4-10). Leonardo did not

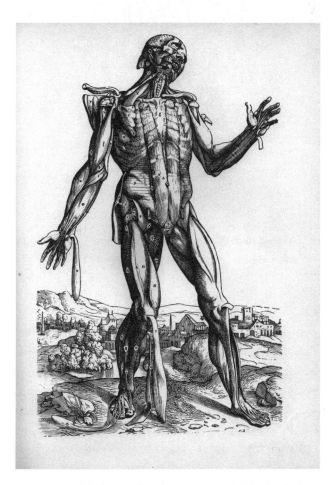

FIG. 4-9. The muscular system, presented in dissective order, Andreas Vesalius, *De humani corporis fabrica*, book 2. From Saunders and O'Malley, *The Illustrations from the Works of Andreas Vesalius*.

FIG. 4-10. Isolated representation of the portal system of veins,
Andreas Vesalius, *De humani corporis fabrica*, book 3.
From Saunders and O'Malley, *The Illustrations from the Works
of Andreas Vesalius.*

need such embellishments. Harmony, beauty, and a sense of "aliveness"
are intrinsic to all his anatomical figures.[37] In my opinion, this is a consequence of the fact that Leonardo's science as a whole is a science of living,
organic forms. The grand unifying theme that underlies his scientific work
in any field is his persistent quest for understanding the nature of life.[38]

Vesalius' *Fabrica* had a decisive influence on subsequent anatomical illustrations. In the years after its publication, numerous plagiarisms and
crude copies of the plates appeared in anatomy books all over Europe,
and henceforth the dominant style of anatomical representations would
be Vesalian: analytic rather than composite and synthetic. This trend has
continued to the present day. As Laurenza observes,

Composite anatomical consideration and representation correspond to today's topographical anatomy, a branch of anatomical
studies which, for medical students, is often merely the subject of a
secondary exam. The actual model for anatomical studies and representations is of the analytical type. [Accordingly], modern medicine and surgery understand illness as the disease of a particular
organ, rather than as an illness of the organism as a whole.[39]

Leonardo, amazingly, seems to have foreseen the dangers of such re-
ductionism. In his *Anatomical Studies*, he repeatedly emphasized the need
for "integral knowledge" (*cognizione integrale*) and denounced isolated
depictions of individual anatomical parts as a "monstrous thing" (*cosa
mostruosa*).[40] Especially revealing is his famous polemic against the "abbreviators," which was directed at all those who made summaries, or epito-
mes, of large and complex texts, and which he also addressed to certain
anatomists of his time. The polemic is found on a folio of anatomical notes
in the Windsor Collection,[41] and on an earlier folio Leonardo specifically
refers to "students who are obstructionists of anatomies and abbreviators
of them."[42] In other words, he criticized the way certain professors and
students conducted dissections, castigating them for the "abbreviations,"
or lack of integration, in their anatomical studies.

For the modern reader, the relevance of Leonardo's "discourse against
the abbreviators" goes far beyond the field of anatomy. "The abbreviators
of works," he declares, "do injury to knowledge and to love. . . . Of what
value is he who, in order to abbreviate the parts of those things of which
he professes to give complete knowledge, leaves out the greater part of the
things of which the whole is composed? . . . Oh human stupidity! . . . You
don't see that you are falling into the same error as one who strips a tree
of its adornment of branches full of leaves, intermingled with fragrant
flowers or fruit, in order to demonstrate that the tree is good for making
planks."[43]

This statement is not only revealing testimony of Leonardo's systemic
thinking and deep sense of harmony, but is also ominously prophetic. Reducing the beauty of life to mechanical parts and valuing trees only for
their lumber is an eerily accurate characterization of the mindset that still
dominates our world today. As I shall argue in the Coda, this makes Leonardo's legacy all the more relevant to our time.

5

The Elements of Mechanics

As Leonardo studied, drew, and painted "all the forms of nature," he investigated not only their external qualities and proportions but also the forces that had shaped and continued to transform them. He saw similar patterns in the macro- and microcosm, but his careful investigations of these patterns of organization made him realize that the forces underlying them were quite different.

In his extensive studies of flowing water, Leonardo recognized correctly that gravity and the fluid's internal friction, or viscosity, were the two principal forces operating in its movements (see p. 39). In his detailed observations of rock formations, he identified water as the chief agent in the formation of the Earth's surface (see p. 69). Moreover, he speculated about the nature of the tectonic forces that caused layers of sedimentary rock to emerge from the sea and to form mountains (see p. 89).

In his studies of plants and animals, Leonardo identified the soul as the vital force underlying their formation and growth. Following Aristotle, he conceived of the soul as being built up in successive levels, corresponding to levels of organic life. The first level is the "vegetative soul," which controls the organism's metabolic processes. The soul of plants is restricted to this metabolic level of a vital force. The next higher form is the "animal soul," characterized by autonomous motion in space and by feelings of pleasure and pain. The "human soul," finally, includes the vegetable and animal souls, but its main characteristic is reason.

The "Noble" Role of Mechanics

The autonomous, voluntary movements of the human body fascinated Leonardo and became a major theme in his anatomical work. From their

FACING Studies of power transmission, c. 1495.
Codex Madrid I, folio 123v (detail).

origin in the center of the brain (the "seat of the soul"), he traced the transmission of the forces underlying various bodily movements through the central and peripheral motor nerves to the muscles, tendons, and bones (see p. 211). Leonardo argued that these muscles, tendons, and bones were nature's "mechanical instruments," which were essential to bodily movements and were best analyzed in terms of the laws and principles of "mechanical science" (see p. 131).

From the early days of his apprenticeship as artist and engineer in Florence, Leonardo had been familiar with basic principles of mechanics applied to simple machines—levers, screws, wedges, pulleys, balances, and the like. Later on, he applied these principles in his inventions of more complex machines and mechanical devices. But it was in Milan, when he developed a keen interest in mathematics and made the transition from engineering to science, that he became motivated to study mechanics with more sustained effort and in much greater depth.[1]

At that time, Leonardo had begun to investigate the movements of the human body and had discovered a broader and more "noble" role for the science of mechanics. "The instrumental or mechanical science," he would write fifteen years later in his Codex on the Flight of Birds,* "is very noble and most useful above all others, because by means of it all animated bodies that have movement perform all their operations."[2] The contrast of this statement with the pronouncements of the architects of the mechanistic worldview in the subsequent two centuries is rather remarkable. For Descartes, Bacon, and Newton, the ultimate value of the science of mechanics was the human domination of nature. For Leonardo it was the understanding and imitation (in the case of the flight of birds) of the animal body in motion.

To understand in detail how nature's "mechanical instruments" work together to move the body, Leonardo immersed himself in studies of problems involving weights, forces, and movements. He showed how joints operate like hinges, tendons like cords, and bones like levers. While he studied the elementary principles of mechanics in relation to the movements of the human body, he also applied them to the design of numerous new machines, and as his fascination with the science of mechanics grew, he explored ever more abstract topics, often struggling with conceptual problems that would be fully understood only centuries after his death.

---

* *Codice sul volo degli uccelli*, or Codex *Sul Volo* for short.

The Body—a Machine?

In view of Leonardo's brilliant achievements in mechanical engineering and his extensive applications of principles of mechanics to the body's "mechanical instruments," it is tempting—but, in my view, erroneous—to believe that Leonardo saw the entire human body as a machine. Many Leonardo scholars have, implicitly or explicitly, taken this view. Kenneth Keele, for example, in his thorough analysis of Leonardo's entire corpus of anatomical studies, published in 1983, repeatedly referred to "the human machine," seemingly paraphrasing Leonardo.[3]

A decade later, Paolo Galluzzi echoed this perception in his superb volume on *Renaissance Engineers: From Brunelleschi to Leonardo da Vinci,* when he titled a section about Leonardo's anatomical studies "The Human Body as a Wonderful Machine." In the opening paragraphs of this section, Galluzzi quoted Leonardo's statement that "nature cannot give movement to animals without mechanical instruments" (see p. 131), and then added the following comments:

> Leonardo sought to demonstrate the close analogy between the machine and the body. He saw both as wonderful achievements of Nature, where iron laws govern not only mechanical instruments but also the motions of animals.[4]

I disagree with the assessment by Keele, Galluzzi, and other historians that Leonardo pursued a mechanistic approach in his anatomical studies and saw the human body as a machine. I believe that such a conclusion is based on an unwarranted Cartesian interpretation of Leonardo's writings. It amounts to projecting a reductionist, mechanistic model of the human body onto Leonardo's scientific views of the macro- and microcosm—views that were utterly organic and unmarred by the mind-body split introduced by Descartes more than one hundred years after Leonardo's death.

René Descartes based his view of nature on a fundamental division into two separate and independent realms: that of mind and that of matter.[5] The material universe was a machine and nothing but a machine. Nature worked strictly according to mechanical laws; everything in the material world could be explained in terms of the arrangement and movements of its parts. Descartes extended this mechanistic view of matter to living organisms. Plants and animals, for him, were simply machines.

Human beings were inhabited by a rational soul, which was of divine origin. But the human body was a mere automaton, indistinguishable from an animal-machine. The body was completely divorced from the mind, the only connection between the two being by the intervention of God.

The difference between Descartes's and Leonardo's views of living organisms is profound. For Leonardo, nature enables animals to move with the help of "mechanical instruments," but this does not mean that animals are machines. The crucial difference is that, in Leonardo's view, the soul, which is inherent in all living organisms, is the origin of bodily movement and is also the body's "composer" (*desso corpo compositore*), or formative force.[6] Leonardo's concept of the soul is quite different from that of Descartes. For him, the soul forms an indivisible whole with the body, controlling both its voluntary and involuntary movements. The range of the soul's activities is different in plants, animals, and humans; it increases in stages with increasing biological complexity (see p. 153). I have argued in my previous book that Leonardo's integrative concept of the soul—as the agent of perception, and as the vital force underlying the body's formation and movements—while strikingly different from Descartes's, comes very close to our modern concept of cognition.[7]

"Why nature cannot give movement to animals without mechanical instruments," Leonardo writes in his *Anatomical Studies*, "is demonstrated by me in this book on the active movements made by nature in animals."[8] It seems quite clear that he refers here to the voluntary movements of animals, which are achieved by means of mechanical instruments but originate in and are controlled by the animal's soul—a far cry from Descartes's animal-automata.

In many passages in Leonardo's manuscripts, he marvels at the beauty and grace that arise from subtle interactions between animals' bodies and souls. For example, he observes that the delicate cognitive processes (as we would say in modern scientific language) of a bird in flight will always be superior to those of a human pilot steering a flying machine:

> It could be said that such an instrument designed by man is lacking only the soul of the bird, which must be counterfeited with the soul of the man . . . [However], the soul of the bird will certainly respond better to the needs of its limbs than would the soul of the man, separated from them and especially from their almost imperceptible balancing movements.[9]

At times one can even discern a fine sense of humor in Leonardo's insistence that the movements of animals are not purely mechanical. "Nature does not go in for counterweights when she makes organs suitable for movement in the bodies of animals," he muses on another folio of his *Anatomical Studies*, "but she places inside the body the soul, the composer of this body."[10] More important, this passage is a clear statement on the integrative nature of the soul as both mover and composer of the body.

From the passages quoted above and from the general nature of his science as a science of organic forms, qualities, and transformations, it seems evident that Leonardo's approach to anatomy was not mechanistic, at least not in the Cartesian sense. He fully realized that that the anatomies of animals and humans involve mechanical functions, and this became the principal motivation for his extensive studies of "mechanical science." But these studies were always embedded in a broader organic conceptual framework.

Occasionally Leonardo referred to complex anatomical systems as "machines." On two folios of the *Anatomical Studies*—one showing the set of muscles controlling the complex movements of the head, and the other the superficial muscles of the thigh—he uses the term "this machine of ours" in sudden outbursts of awe and wonder about the body's complexity, interjected between detailed anatomical descriptions.[11] In both statements the term "machine" is applied to a complex system of mechanical functions, rather than to the body as a whole.

On other occasions, Leonardo uses the term "machine" to refer to phenomena in the macrocosm. He speaks of water as "the vital humor of the terrestrial machine,"[12] and of the ebb and flow of the tides as "the breathing of this machine of the Earth."[13] It is clear that here again "machine" should not be given a Cartesian meaning but should be understood to refer to a complex living system, nourished by water, with tides moving rhythmically like the breath and the flow of blood in the human body, and animated by a vital force of growth, or "vegetative soul" (see Leonardo's description of the living Earth, p. 67).

Leonardo da Vinci created a unique synthesis of art, science, and design.[14] He was a mechanical genius who invented countless machines and mechanical devices, and he maintained a lively interest in the theory of mechanics during most of his mature life. Yet his science as a whole was not mechanistic. He saw the world as an infinite variety of living forms continually shaped by underlying processes, and of patterns of organization recurring in the macro- and microcosm. He formulated mechanical models when he thought they would help him understand natural phenomena, but unlike scientists in subsequent centuries, he never considered the world as a whole, nor the human body, as nothing but a machine.

Leonardo's Machines

Throughout his adult life, and especially during his years at the Sforza court in Milan,[15] Leonardo was famous not only as an artist but also as a mechanical engineer. His duties as court painter and "ducal engineer" included, in addition to painting portraits and designing pageants and festivities, a variety of small engineering jobs that demanded ingenuity and skills in the handling of materials. Leonardo's many creative talents were perfectly suited for these tasks. He invented a large number of astonishing devices during this time, which firmly established his reputation as engineer-magician at court.

Among his inventions were doors that opened and closed automatically by means of counterweights; a table lamp with variable intensity; folding furniture; an octagonal mirror that generated an infinite number of multiple images; and an ingenious spit, on which "the roast will turn slow or fast, depending upon whether the fire is moderate or strong."[16]

Leonardo did not limit his engineering skills to these gadgets but invented numerous machines of a more industrial nature. These included a variety of textile machines for spinning, weaving, twisting hemp, trimming felt, and making needles, as well as machines for casting and hammering metal, shaping wood and stone, drawing strip and wire, coining and grinding—in short, machines for the basic industries of his time.

A special type of machine were the measuring instruments Leonardo invented and designed for his scientific experiments.[17] In particular, he made many attempts to improve clock mechanisms for time measurement, which was still in its infancy in his day. In the Codex Madrid I, Leonardo put forth a systematic exposition of the main components of a mechanical clock: the use of the spring as the driving force and of the fusee (a conical drum) to compensate for the lessening force of the spring; power transmissions through gear-trains; and various forms of regulation systems known as escapements. All these elements are discussed in detail and pictured in superb drawings covering several pages.[18]

In addition to mechanical engineering, Leonardo was also engaged extensively in civil and military engineering. He was known as one of Italy's leading hydraulic engineers and during his years at the Sforza court was probably in charge of all hydraulic works in Lombardy (see p. 32). He improved the existing systems of locks, invented special machines for digging canals, and skillfully inserted small dams into rivers to prevent damage to properties along their banks.

One of his most ambitious but unrealized hydraulic projects was a navigable waterway between Florence and Pisa. Leonardo imagined that this waterway would provide irrigation for parched land and could also serve as an "industrial" canal, providing energy for numerous mills that would produce silk and paper, drive potters' wheels, saw wood, and sharpen metal (see p. 23).

As a military engineer, Leonardo was frequently consulted about strategies of warfare, and he often responded with ingenious designs of new fortifications and grandiose plans to dam up or divert rivers to conquer enemy troops.[19] Most of his work for military rulers consisted in designing structures to defend and preserve towns and cities.[20] However, he also designed extravagant machines of destruction—bombards, explosive cannonballs, catapults, giant crossbows, and the like. At the same time, paradoxically, he was vehemently opposed to war, which he called a "most beastly madness" (*pazzia bestialissima*). Various explanations of this apparent contradiction can be put forward.[21] Leonardo was in constant need of a stable income that would allow him to pursue his scientific research, and he shrewdly relied on his great skills in mechanical engineering to secure financial independence by offering designs of impressive war machines. Moreover, he may have been aware that most of these fanciful designs would never be realized.

However, it is also clear from Leonardo's Notebooks that he was fascinated by the destructive engines of war, perhaps in the same way that natural cataclysms and disasters fascinated him. We may not be able to resolve the contradiction between his pacifist stance and his services as military engineer, but may have to accept it as one of many contradictions in the complex personality of a great genius.[22]

Leonardo's outstanding contributions to mechanical, civil, and military engineering are discussed extensively in several books, including the beautiful volume *Renaissance Engineers: From Brunelleschi to Leonardo da Vinci*, by science historian Paolo Galluzzi,[23] and the lavish catalogue of an exhibition at the Musée des Beaux-arts de Montréal, edited by Galluzzi, which covers both Leonardo's engineering and architecture in great detail.[24] His technical drawings are frequently exhibited around the world, often supplemented by wooden models that show in impressive detail how the machines work as he had intended.[25]

The combination of artist-engineer was not unusual in the Renaissance.[26] Leonardo's teacher Verrocchio, for example, was a renowned goldsmith, sculptor, and painter as well as a reputable engineer. The great Renaissance architect Filippo Brunelleschi first gained notice in Florence as a sculptor and later on, when he was famous as an architect, was also acclaimed for his inventive genius as an engineer. The young Leonardo admired him greatly and declared his indebtedness to the great architect by drawing several of Brunelleschi's renowned lifting devices and architectural plans.

What made Leonardo unique as an engineer, though, was that many of the novel designs he presented in his Notebooks involved technological advances that would not be realized until several centuries later. Even more important, he was the only one among the famous Renaissance engineers who made the transition from engineering to science. To know *how* something worked was not enough for him; he also needed to know *why*. Thus an inevitable process was set in motion that led him from technology and engineering to pure science. As art historian Kenneth Clark notes, we can see the process at work in Leonardo's manuscripts:

> First, there are questions about the construction of certain machines, then … questions about the first principles of dynamics; finally, questions which had never been asked before about winds, clouds, the age of the earth, generation, the human heart. Mere curiosity has become profound scientific research, independent of the technical interests which had preceded it.[27]

Leonardo's passage from the study of medieval empirical technology to theoretical mechanics began with the emergence of a strong interest in mathematics during his first period in Milan, when he was in his late thirties. An important event was a visit to the nearby city of Pavia in 1490. Leonardo went there together with the architect Francesco di Giorgio on behalf of the duke of Milan to inspect work on the city's cathedral.

In Pavia, Leonardo met the mathematician Fazio Cardano, a specialist in the "science of perspective," which in the Renaissance included geometry and geometrical optics. Leonardo's discussions with Cardano ignited a passion for mathematics that would remain with him until his old age.[28]

While his fellow architect and engineer Francesco returned to Milan as soon as their work was completed, Leonardo stayed in Pavia for another six months to consolidate his understanding of geometry with studies in Pavia's magnificent, world-famous library.[29]

Immediately after his return to Milan he began two new Notebooks, now known as Manuscripts A and C, in which he applied his new knowledge of geometry to a systematic study of perspective and optics as well as to elementary problems of mechanics. Leonardo's application of geometrical reasoning to the analysis of machines was highly original. Inspired, most likely, by his discussions of Euclid's celebrated *Elements of Geometry* with Cardano in Pavia, he began to separate individual mechanisms, or "elements," from the machines in which they were embedded. This conceptual separation did not arise again in engineering until the eighteenth century.[30]

In fact, Leonardo at that time planned (and may even have written) an entire treatise on *Elements of Machines*, in which he would use geometry to analyze basic mechanisms in terms of elementary principles of mechanics—the transmission of power and motion, measurement of forces, and so on. Such analysis was important to Leonardo not only for understanding and improving upon existing mechanisms but also for the "very noble" purpose of understanding the individual actions of muscles, tendons, and bones in generating bodily movements (see pp. 153–54).

Leonardo's treatise on *Elements of Machines*, if ever written, has been lost. But his Codex Madrid I contains extensive preparatory studies for such a treatise. In this manuscript, written in the late 1490s while he was also completing *The Last Supper*, Leonardo analyzed over twenty elementary mechanisms in countless variations—screws, levers, hinges, springs, couplings, gears, pulleys, and so on. As historian of technology Ladislao Reti has shown in his thorough and beautiful analysis of the Codex Madrid, Leonardo's elements of machines include all the mechanical devices described in the work of early nineteenth-century French scholars of the Ecole Polytechnique, which was traditionally considered to be the first systematic study of elementary mechanisms.[31]

The mechanisms described in the Codex Madrid were well known to Renaissance engineers, although Leonardo invented many new versions and combinations of them. However, none of his predecessors or contemporaries had analyzed in detail how they worked. For centuries, animals had been attached to carts or traction devices; men had turned cranks,

worked mechanical tools by hand, and operated treadmills. Simple machines were built and used according to tradition without asking how friction could be reduced or the transmission of muscle power improved.

Leonardo's approach was profoundly different. He never followed
traditional solutions without questioning them, but analyzed them according to mechanical rules and principles deduced from observation and
experiment. He paid special attention to the transmission of power and
motion from one plane to another, which was a major challenge of Renaissance engineering. An extremely complex example is his design of a
water-powered mill for simultaneously drawing and rolling cannon-barrel
segments. Leonardo's illustration of his design (fig. 5-1) shows a machine
assembly of fifteen links in which the initial power of the water turbine (at
bottom left of the assembly) is transmitted three times between vertical
and horizontal axes with the help of a combination of toothed wheels and
worm gears. Each time the power increases twelvefold while the turning
speed successively decreases until it reaches the sturdy solid wheel (at top
right of the assembly) that presses on the cannon segment beneath it.[32] The
transfer of power is clearly indicated by Leonardo in a small diagram below the main drawing in which the numerical power ratios are indicated
for each gear.

This drawing is a splendid illustration of Leonardo's extraordinary
capacity for coordinating complex mechanical functions, analyzing them
precisely, and presenting them visually with great clarity. From the time
when his interest in mathematics was kindled at the University of Pavia,
his work in mechanical engineering became inextricably linked to the
analysis of his machines in terms of geometry and principles of mechanics.

The Science of Weights

Leonardo began his theoretical studies of mechanics with the "science of
weights," known today as statics, which is concerned with the analysis of
loads and forces of physical systems in static equilibrium, such as balances
and levers. In the Renaissance as today, this knowledge was very important for architects and engineers, and the medieval science of weights com-
prised a large collection of works compiled in the late thirteenth and fourteenth centuries, which Leonardo studied extensively. Some of these were
treatises and fragments translated from Greek or Arabic, usually ascribed
to Euclid or Archimedes, while others were original writings of medieval
authors.[33]

The mathematical foundations of statics were established in antiquity by the great mathematician and scientist Archimedes in a treatise titled *On the Equilibrium of Planes*, which contains his exact determinations of centers of gravity and his proof of the general law of the lever. Archimedes's mathematical proofs were purely geometrical, which delighted Leonardo

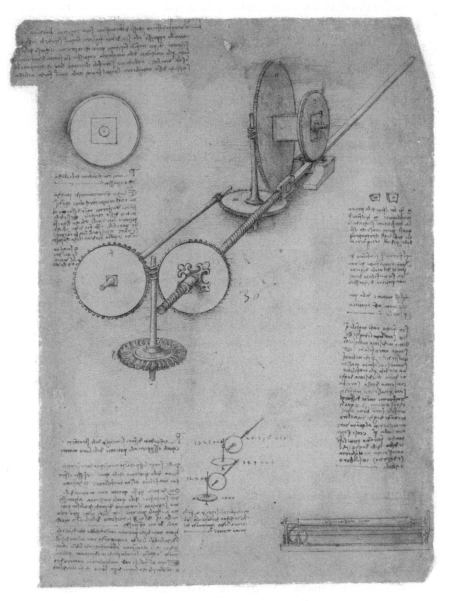

FIG. 5-1. Water-powered mill for rolling and drawing cannon-barrel segments. Codex Atlanticus, folio 10r.

and led him to declare enthusiastically that "mechanics is the paradise of the mathematical sciences."[34]

Of the medieval authors, Leonardo drew most heavily upon two works by Jordanus de Nemore, a thirteenth-century mathematician about whom almost nothing is known but who left several treatises on mathematics and mechanics that show considerable skill and originality. Leonardo read the two principal works by Jordanus on statics, *Elementa de ponderibus* (Elements of the Science of Weights) and *De ratione ponderis* (On the Nature of Weight), the latter being a greatly expanded and improved version of the *Elementa*.

In his usual fashion, Leonardo absorbed the key ideas from the best and most original texts in the corpus of the medieval science of weights, commented on some of their postulates in his Notebooks, verified them experimentally, and refuted some incorrect proofs. The Codex Atlanticus, in particular, contains several pages of his Italian translation of various postulates from Jordanus's *Elementa* and *De ratione ponderis*, probably from a single manuscript that contained both works.[35]

The centerpiece of Leonardo's mathematical treatment of statics is the classical Archimedean law of the lever. It states that a lever, or balance, will be in equilibrium when the ratio of the two weights (or forces) is the inverse of the ratio of their distances from the fulcrum. This law appears repeatedly in various forms in Leonardo's Notebooks. In the Codex Atlanticus, for example, he states:

> The ratio of the weights that hold the arms of the balance parallel to the horizon is the same as that of the arms, but is an inverse one.[36]

In the Codex Arundel, Leonardo expresses the law in terms of a formula that in modern algebraic notation would be written as $w_2 = (w_1 d_1)/d_2$:

> Multiply the longer arm of the balance by the weight it supports and divide the product by the shorter arm, and the result will be the weight which, when placed on the shorter arm, resists the descent of the longer arm, the arms of the balance being in equilibrium at the outset.[37]

Leonardo used the law of the lever to calculate the forces and weights necessary to establish equilibria in numerous simple and compound systems involving balances, levers, pulleys, and beams hanging from cords. In addition, he carefully analyzed the tensions in various segments of the cords, probably for the purpose of estimating similar tensions in the muscles and tendons of human limbs.

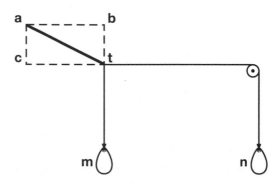

FIG. 5-2. Pivoting plank in equilibrium with two forces acting at different angles. Ms. E, folio 65r (as reconstructed by Clagett, "Leonardo da Vinci: Mechanics").

Science historian Marshall Clagett has discussed Leonardo's diagrams and analyses of the principles of statics in great detail.[38] Clagett emphasizes that Leonardo applied the law of the lever not only to situations where the forces act in a direction perpendicular to the lever arms, but also to forces acting at various angles. The Codex Arundel and Manuscript E contain numerous diagrams of varying complexities with weights exerting forces at different angles via cords and pulleys. Leonardo recognized that in such cases the relevant length in the law of the lever is not the actual length of the lever arm but the perpendicular distance from the line of the force to the axis of rotation. He called that distance the "potential lever arm" (*braccio potenziale*) and marked it clearly in many diagrams.

In a diagram in Manuscript E (fig. 5-2), for example, Leonardo shows a bar that is pivoted at one end at point *a* with a weight *m* suspended from its other end at point *t*. A second weight *n* exerts a horizontal pull via a cord running over a pulley. The problem is to determine the weights *m* and *n* necessary to keep the bar in equilibrium. In his solution, Leonardo identifies the two potential lever arms as *ab* and *ac*, and he states correctly that, at equilibrium, the weights *m* and *n* will be inversely proportional to the distances *ab* and *ac*.

In modern statics, the potential lever arm is known as the "moment arm" and the product of moment arm and force is called the "moment of force," or "torque." Leonardo clearly recognized the principle that the sum of the moments about any point must be zero for a system to be in static equilibrium. According to Clagett, this discovery was his most original contribution to statics, going well beyond the medieval science of weights of his time.

Fluids in Equilibrium

While Leonardo experimented with balances, levers, and pulleys to explore the laws governing mechanical systems in static equilibrium, he also studied the equilibrium of fluids, known today as hydrostatics. In the Codices Madrid, which contain most of his early investigations of the "science of weights," we also find comments on water pressure and references to the principle of Archimedes, as well as drawings of scales measuring the buoyancy of weights submerged in water.

Since antiquity, hydrostatics had been an independent discipline, unrelated to the study of the flow of water (now known as hydrodynamics). Its general principles had been enunciated clearly by Archimedes in his classical text *On Floating Bodies*. This treatise contains, in particular, the famous principle that now bears Archimedes' name. It states that the buoyant force on a submerged object is equal to the weight of the fluid displaced by the object.

Most of the theoretical work of Archimedes was so advanced that it was poorly understood by his contemporaries and succeeding generations.[39] Translations of various fragments on hydrostatics were reproduced in several medieval texts, generally without a clear understanding of the underlying principles. These texts could have given Leonardo only a very sketchy knowledge of Archimedean hydrostatics.

Leonardo certainly had a general knowledge of buoyancy. He knew that objects weigh less in water than in air and he even tried to determine the difference experimentally. Two similar drawings in the Codices Madrid show weights hanging from a scale, one weight in the air and the other submerged in water contained in a vessel.[40] In the Codex Madrid I (fig. 5-3), the text accompanying the illustration mentions several experiments of that kind and lists

FIG. 5-3. Experiment for determining the force of buoyancy. Codex Madrid I, folio 181r (detail).

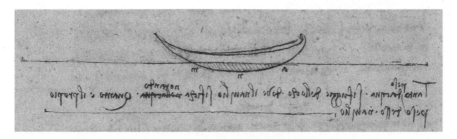

FIG. 5-4. Illustration of the amount of water displaced
by a floating boat. Codex Madrid I, folio 123v (detail).

quantitative results. Leonardo also knew the basic principle of floating
bodies. An earlier folio in the same Codex Madrid I shows an elegant little
sketch of a floating boat (fig. 5-4) together with the following comment:
"As much weight of the water leaves the place where a boat floats, as the
weight of that boat itself."[41] However, as far as we can tell from his existing manuscripts, Leonardo never fully stated the principle of Archimedes.
He knew the formula for floating objects but seems to have been unaware
that a similar formula holds for submerged objects. Leonardo's numerous
notes on hydrostatics make it evident that he reached only a partial understanding of the Archimedean laws of buoyancy.

It is interesting to examine in detail what prevented Leonardo from
fully understanding the principle of Archimedes. To understand the origin of the force of buoyancy, one needs to know that water pressure increases with depth, and that the increased pressure is exerted in all directions. As a consequence, there is a net upward force on the bottom of a
submerged object, as illustrated in figure 5-5.

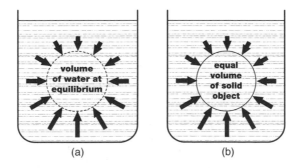

FIG. 5-5. Buoyant forces on a "water ball" in equilibrium
(a) and on a solid spherical object of equal volume (b).
From HyperPhysics.com, Georgia State University.

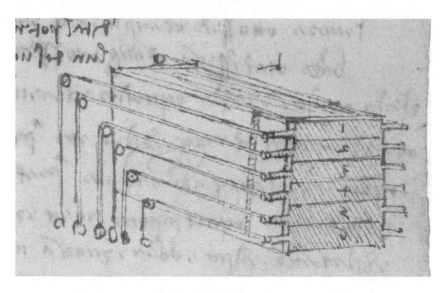

FIG. 5-6. Measurement of the variation of water pressure with depth by means of a series of moveable plates, sustained by counteracting forces that are generated by weights and transmitted to the plates via strings and pulleys. Codex Leicester, folio 6r (detail).

A hypothetical ball of water anywhere in the vessel will be in equilibrium because its weight is supported exactly by the force of buoyancy, or net pressure (see fig. 5-5a). If the water ball is replaced by a solid spherical object of equal volume, the distribution of pressure on the object will be the same (see fig. 5-5b). Hence, the buoyant force on the solid object is equal to the weight of the water displaced, as stated by Archimedes.

Leonardo knew, at least in his later years, that water pressure in a reservoir acts on vertical walls and increases from the surface to the bottom. In the Codex Leicester, which he wrote when he was in his late fifties, he depicts a very ingenious experiment designed to measure the increase of pressure with depth (fig. 5-6). As described in the accompanying text, the experiment involves a water tank "in which one of the walls is a loose parchment, sustained by a row of plates, as the drawing shows, at which plates you put so much opposite weight as to sustain precisely those plates in contact with the front of the aforementioned water tank."[42] If Leonardo had actually carried out this experiment, it would have shown him the correct linear increase of pressure with depth.

In spite of his knowledge of the variation of water pressure, it would have been difficult for Leonardo to reason that the weight of the water ball in figure 5-5a is balanced by the net upward pressure. Archimedes's sophisticated understanding of the distribution of water pressure had been lost over the centuries and would not be rediscovered by Blaise Pascal until 150 years after Leonardo. In addition, the concept of the weight of a portion of water, immersed in the surrounding water filling the vessel, was foreign to Leonardo. During most of his life, he accepted the Aristotelian view that "all the elements are without weight in their own sphere but possess weight outside their sphere."[43] This Aristotelian theory of gravity, which was commonly held in the Middle Ages and the Renaissance, made it difficult, if not impossible, to understand the hydrostatic equilibrium of a portion of water as the balance between its weight and the net upward pressure of the water surrounding it.*

The Aristotelian view of a natural weightless state of the elements also prevented Leonardo from developing a sophisticated conception of water pressure, which is essential for a full understanding of the principle of Archimedes. He was aware of the pressure generated by a piston, but he had great difficulties conceptualizing a pressure distribution of water in its "natural" state.

Pressure in fluids is a concept of modern hydrostatics where it is defined as force (or weight) per unit of area. As physicist and Leonardo scholar Enzo Macagno pointed out in his detailed studies of Leonardo's writings on hydrostatics,[44] this definition was hard to accept in the Middle Ages and the Renaissance because it requires the division of two quantities of different dimensions—a force by an area. Such divisions, or multiplications, were fully accepted only in the seventeenth century when geometry was replaced by algebra—a much more powerful mathematical language, capable of expressing relationships between physical quantities in terms of abstract equations.

Because of these mathematical limitations, Leonardo and his contemporaries never reached more than a qualitative understanding of fluid pressure. The first to give a full account of the pressure distribution in a fluid in hydrostatic equilibrium was the mathematician and philosopher Blaise Pascal in the seventeenth century. His formulation is now known as Pascal's principle. It states that pressure applied to a confined fluid at any point is distributed undiminished through the fluid in all directions and acts upon every part of the vessel's confining surface at right angles.

* See p. 40 above for a more detailed discussion of Leonardo's view of gravity.

Leonardo struggled with the concept of pressure for many years without ever fully understanding it. As historian of science Constantino Fasso has documented, some pages of the Notebooks show anticipations of essential insights that would be formulated clearly in subsequent centuries, while others contain confused and contradictory statements.[45] For example, the Codex Madrid I, in which Leonardo recorded his early investigations of hydrostatics, already contains an evident, though imprecise, anticipation of Pascal's principle. Next to a sketch of a weight placed on a wine skin, there is a marginal note: "Any part of the skin feels equally the pressure of the weight."[46] But elsewhere in the same Notebook, Leonardo gives a wrong description of a similar situation. He depicts a vessel filled with water on which pressure is exerted by a weight through an air cushion, and he comments that the increase in pressure "pushes . . . in all parts of that vessel in the way and proportion exerted before by the water alone."[47] In other words, he assumes that the increase in pressure is not constant throughout the vessel (as stated by Pascal's principle), but varies with depth in proportion to the hydrostatic pressure that existed before the application of the weight.

Even Leonardo's mature writings on hydrostatics in the Codex Leicester are not free from such contradictions. The description of his brilliant experiment to measure the variation of water pressure with depth (see fig. 5-6) is a good example. After stating correctly that the pressure on the vertical walls increases from the surface to the bottom of the vessel, and describing how to measure the increase, Leonardo adds a few lines in which he confuses the issue again. "The same rule," he asserts, "may be used at the bottom to see in which part of the bottom of the vessel water presses more on that bottom."[48] Clearly, at least at the time of writing these lines, Leonardo was not aware that the water pressure is distributed equally across horizontal planes.

On a folio in the same Codex Leicester, a few pages after the contradictory statements mentioned above, we find Leonardo's most mature discussion of hydrostatic pressure. Once again, he considers a vessel filled with water to which pressure is applied through an opening at the top:

Water, pressed through the mouth of the vessel," he writes, "acquires in its contact with that vessel a uniform pressure [*potentia*]. I intend that the water, because it is pressed, acquires that uniform pressure in addition to the unequal pressure that existed in this water before, since it is evident that water by itself exerts more weight on an orifice at the bottom of the vessel than on the surface, and for each degree of depth it acquires degrees of weight.[49]

Comparing this statement to the one in the Codex Madrid mentioned above, which describes the same experiment and was written about ten years earlier, makes it evident that Leonardo's thoughts on hydrostatic pressure had matured significantly during the intervening years. In the Codex Leicester, he states unequivocally and correctly that the pressure exerted on the water by a weight is distributed uniformly throughout the vessel, and that the resulting total water pressure is composed of two parts: the original hydrostatic pressure that is unequal, increasing from the surface to the bottom, and the constant pressure that is added through the application of the weight. Moreover, he correctly states that the original hydrostatic pressure increases linearly with depth.

It seems that the only flaw in Leonardo's analysis is the lack of an explicit definition of pressure as force (or weight) divided by area, and hence he shows a slight confusion between pressure and weight. Otherwise, this statement in the Codex Leicester is a clear anticipation of Pascal's principle. Since the passage is, as far as we know, chronologically Leonardo's latest discussion of hydrostatic pressure, we may take it as his definitive pronouncement on the subject.

Another hydrostatic phenomenon that puzzled Leonardo a great deal was the equilibrium of liquids in communicating vessels, which was well known during his lifetime. "The surfaces of all liquids at rest that are joined together below are always of equal height," he noted correctly in the Codex Atlanticus.[50] In the medieval "science of weights," questions of statics had always been treated by applying the laws of the equilibrium of the balance, and so it was natural for Leonardo and other Renaissance engineers to use the same approach in trying to explain the law of communicating vessels. The analogy of a balance in equilibrium with equal arms, loaded by the weights of the water in the two vessels, is correct only when the two vessels are equal. When one vessel is larger, the hypothetical balance can be in equilibrium only if the water surface is higher in the smaller vessel. This is contrary to the experimental evidence, as Leonardo did not fail to notice.

The resolution of this paradox—how a small amount of water on one side can balance a large amount on the other side—requires the full understanding of hydrostatic pressure reached by Pascal in the seventeenth century. According to Pascal's principle, the hydrostatic pressure must be the same at all horizontal planes for the water to be in equilibrium. This means that the height under the surfaces must be the same, but their areas are irrelevant because pressure equals weight per unit area.

Lacking this sophisticated conception of pressure, Leonardo never succeeded in completely solving the paradox of communicating vessels. However, he found an ingenious explanation during his first reflections on hydrostatics. In the Codex Madrid I, there is a sketch of communicating vessels of unequal size with water levels at equal height indicated correctly (fig. 5-7). In this drawing, Leonardo has divided the water in the larger vessel into several columns. In the accompanying text, he explains that not all the water in the larger vessel is active in counterbalancing the weight of the water in the smaller vessel, but only one column (a–n) with the same cross section as that of the smaller vessel (m–r). In view of Leonardo's very limited understanding of water pressure at the time, this reasoning is remarkable. "Leonardo's approach to the problem of communicating

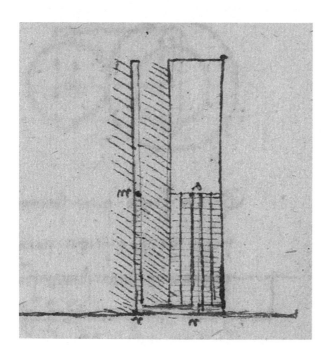

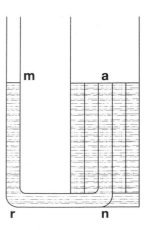

FIG. 5-7. Leonardo's pictorial explanation of the law of communicating vessels. Codex Madrid I, folio 150r (detail); reconstructed by Macagno, "Mechanics of Fluids in the Madrid Codices."

vessels," comments Fasso, "seems to me the most advanced that could be achieved at a time in which the concept of pressure had not yet been devised."[51]

Forces and Motion—A Conceptual Maze

During his studies of the "science of weights" and of hydrostatics, Leonardo became interested in the general relationships between forces and motion. In his attempts to outline what in subsequent centuries would be called a "science of motion," he encountered conceptual difficulties that were far greater than those in his works on statics. The geometrical reasoning he used for his analyses of machines was much harder to apply to the dynamic phenomena of bodies moving under the influence of forces and colliding with one another. Besides, the concepts required to describe these phenomena mathematically—concepts like energy, momentum, force, acceleration, and so forth—had not yet been fully developed. In fact, it would take another two centuries to clearly identify and define these basic concepts of mechanics. As science historian Domenico Bertoloni-Meli points out, even Galileo's early speculations on motion, a hundred years after Leonardo's, were entangled in "a conceptual and terminological maze at the intersection between Aristotelianism and a new science."[52]

According to the Aristotelian four-element theory, commonly held in the Middle Ages and the Renaissance, the movements of the elements arise from their natural tendencies to return, when disturbed, to their proper places within concentric spheres around the Earth.[53] Leonardo held on to this teleological explanation of forces and motion for most of his life, but on several occasions he questioned its basic premises, realizing that they were obstacles in his attempts to understand mechanical phenomena. For example, in a small notebook written during his early studies of mechanics and now known as Manuscript I, he listed a series of questions about motion that he intended to explore. Besides being a lively testimony to Leonardo's relentless curiosity, these questions clearly indicate his doubts about the Aristotelian scheme:

> What is the cause of motion? What is motion in itself? What is it that is most adapted for motion? What is impetus? What is the cause of impetus, and of the medium in which it is created? What is percussion? What is its cause? What is rebound? What is the curvature of straight motion and its cause?[54]

A few years later, he expressed similar doubts about the Aristotelian view of gravity in connection with the flow of water (see p. 40). In these struggles with the conceptual maze of classical and medieval mechanics, Leonardo sometimes showed an intuitive grasp of abstract concepts and relationships that was far ahead of his time, while at other times he could not free himself from the constraints of the traditional Aristotelian ideas.

In reviewing Leonardo's achievements in kinematics and dynamics in the following pages, I shall a few times compare them to those of Galileo, the other great scientist from Tuscany, born more than a century after Leonardo. Galileo published his early speculations on movement, *De motu antiquiora* (Older Works on Movement), in 1592, about a hundred years after Leonardo's early work on mechanics; and Galileo's mature work, the *Discorsi* (Discourses), was published in 1638, about 125 years after Leonardo's mature writings. The comparison between Leonardo's and Galileo's mechanics is fascinating because they performed similar experiments, struggled with similar conceptual problems, and used similar mathematical language, stating the regularities they discovered in terms of proportions and geometrical laws rather than algebraic equations. Galileo's mature work marks a kind of conceptual halfway point between his and Leonardo's early speculations and the publication of Isaac Newton's *Principia* (Principles) in 1687. Newton's grand opus, in which he used algebraic notation and his newly invented calculus, was the triumphant completion of the new "science of motion."

The Four "Powers" of Nature

In his attempts to bring some clarity to classical and medieval mechanics, Leonardo identified four basic variables—motion (or velocity), weight, force, and percussion (or "impact," as we would say today). He did not use the modern term "variable" but instead used the term "power" (*potentia*), which appears frequently, with a wide range of meanings, in his Notebooks. In his writings on hydrostatics, for example, *potentia* means "pressure"; in other texts it corresponds to our modern term "energy"; and Leonardo also describes the force exerted by living organisms as an "invisible power."

The fact that Leonardo's use of *potentia* is rather vague compared to modern scientific terminology is not surprising for the fifteenth century. Today we know that his four "powers" of nature all have different dimensions, and hence I believe that *potentia* in this context is best understood in the sense of our modern term "variable." This interpretation is reinforced by Leonardo's insistence that the outstanding common characteristic of the four "powers" is that they vary continuously (like the variables in our modern mathematical functions). Among Leonardo's early notes on mechanics in the Codex Madrid I is this emphatic statement:

> We will be telling the truth by affirming that it is possible to imagine all the powers capable of infinite augmentation or diminution. . . . They can grow from nothing to infinite greatness by equal degrees. And by the same degrees they decrease to infinity by diminution, ending in nothing.[55]

Having defined the basic variables of mechanics, Leonardo then tries to establish quantitative relationships between them. For most of his life, he believed that all such relationships could be represented as direct or inverse proportions—more generally, as what we now call linear functions. He called such proportional, or linear, algebraic relationships "pyramidal," using pyramids and isosceles triangles (i.e., triangles with two equal sides) to represent them geometrically.[56]

On a folio in the Codex Atlanticus, Leonardo illustrated the actions of the powers of nature with various examples, heading the page prominently with the words: "All natural powers . . . are to be called pyramidal inasmuch as they have degrees in continuous proportion toward their diminution as toward their increase."[57] In other words, all powers vary continuously and their variations can be expressed in terms of linear relationships.

Leonardo's belief that "pyramidal" (linear) relationships were universal in nature was derived from his familiarity with linear proportions in perspective. Like most mathematicians of his time, he frequently used geometrical figures to represent algebraic relationships, and he saw the pyramid, or isosceles triangle, as a powerful symbol of a conceptual link between optics and mechanics.

This belief in the universal nature of linear relationships prevented Leonardo at times from recognizing other types of algebraic relationships that are not so easily expressed in terms of geometrical figures. For example, he failed to recognize that the distances traversed by falling bodies increase with the squares of the times instead of linearly. Galileo made the same error in his early work, and for the same reasons. In a letter written in 1604, he stated as an "indubitable principle" that the speeds of falling bodies increase in proportion to the distances traversed, which is equivalent to saying that both speeds and distances increase linearly with time. According to science historian Bertoloni-Meli, "Galileo may have been led to (incorrectly) assuming the proportionality between speeds and distances by the intuitively privileged role . . . of direct proportionality over more complex relations."[58]

Galileo eventually corrected his error, and Leonardo, too, discovered more complex functional relationships between physical variables late in his life. But by then he was too busy with other projects to revise his statements on the universality of pyramidal relationships between the four powers of nature.[59]

Leonardo's writings on motion, weight, force, and percussion are widely dispersed in his Notebooks, from the beginning of his scientific notes in the Codex Trivulzianus and the early Codices Forster and Manuscript A to his notes in Manuscript E, compiled in old age, and throughout the large collections of notes in the Codices Arundel and Atlanticus.* In the following pages, I will be able to touch upon only a few highlights of Leonardo's observations, discoveries, and speculations in the field of mechanics.

The Nature of Motion

When Leonardo extended his studies of mechanics beyond statics, he realized that the nature of motion and its relationships to various forces would now be the central subject of his investigations. Thus, it is not surprising that he considered motion to be the most fundamental of his four powers of nature. "Speak first of motion," he wrote in a note to himself, "then of weight because it arises from motion; then of force, which arises from weight and motion; then of percussion, which arises from weight, motion, and often from force."[60]

The causal links between the four powers established in this statement seem somewhat arbitrary, and indeed Leonardo soon realized that motion, weight, force, and percussion were so tightly interconnected that it was impossible to single out any one of them as primary:

> Gravity, force, and percussion are of such a nature that each by itself alone can arise from each of the others . . . and all together and each by itself can create motion and arise from it.[61]

* For a complete list of the scholarly editions of Leonardo's Notebooks, see p. 355.

Nevertheless, it seemed to Leonardo that motion would be the best starting point to analyze the relationships between the four powers of nature. In the Codex Arundel, he reasserts that motion, like the other three powers, varies continuously, and in the same passage he correctly identifies the basic relationship between velocity, distance, and time:

> That motion is slower which covers less distance in the same time. And that motion is swifter which covers more distance in the same time. . . . It is in the power of motion to extend to infinite slowness and likewise to infinite velocity.[62]

Even more remarkable is the fact that Leonardo recognized the relativity of motion. "The motion of the air against a fixed thing is as great as the motion of the moving thing against the motionless air," he noted in the Codex Atlanticus. "And the same occurs in water, which in a similar circumstance has shown me the very same nature."[63] Indeed, on an earlier page in the same Codex, Leonardo accurately observed: "The action of a pole drawn through still water resembles that of running water against a stationary pole."[64]

Leonardo must have recognized the importance of this discovery, because he recorded it many times in various manuscripts. All these statements are clear and beautiful expressions of an important principle of modern mechanics. The relativity of motion was rediscovered and formulated mathematically in the late seventeenth century by the renowned physicist and mathematician Christiaan Huygens in connection with the laws of collision. It is the basis of the wind tunnel, the principal experimental tool of modern aerodynamics.

As we do in kinematics today, Leonardo distinguished between "straight" (linear) and curved motion, and he listed circular, spiral, and "irregular" movements as special types of curved motion.[65] In view of his great fascination with spirals (see p. 62), it is not surprising that he identified four distinct spiral movements corresponding to convex, plane, concave, and "columnar" (helical) spirals.

In his extensive work with machines, Leonardo had ample opportunity to study wheels rotating about various axes. He correctly distinguished between angular and linear velocity,* as we would say today. "That part

---

* The angular velocity of a rotating wheel is the rate of change of the angle through which the wheel turns; it is the same for each point on the wheel. The linear velocity of any point on the wheel is its tangential speed, which is proportional to its distance from the axis of rotation.

of a revolving wheel moves with less motion which is nearest to the center of that revolution," he noted in Manuscript E.[66] As mentioned before,
Leonardo contrasted the circular motion of wheels with the spiral motion
of water vortices in his pioneering studies of turbulent flows of water and
air (see p. 47). "The spiral or whirling motion of every liquid is so much
swifter as it is nearer to the center of its revolution," he observed with complete accuracy, "[whereas] the circular motion of a wheel is so much slower
as it is nearer to the center of the revolving object."[67]

Leonardo fully realized that with the distinction between angular
and linear velocity he had discovered a property that was characteristic
not only of rotating wheels but of all circular motion. We can find several
illustrations of this insight in his manuscripts, including the following,
involving hunting tools, in his early notes on mechanics in Manuscript I:

> Field-lances or hunting-whips have a greater movement than the
> arms . . . because, in moving the arm, the hand describes a much
> wider circle than the elbow; and in consequence, moving in the
> same time, the hand covers twice the pathway covered by the elbow;
> therefore, it may be said to be of a speed double that of the motion
> of the elbow.[68]

Leonardo did not fail to recognize the centrifugal force generated by
circular motion, and he also observed correctly that, when an object rotating on a string is released from its rotation, it will fly off tangentially:

> The weight that moves around the fixed point of a string where it is
> joined, pulls and stretches this string with great power, and if such a
> string is separated from its fixed point, the weight carries with it the
> said string along that line into which it was drawn at its separation
> from its fixed point.[69]

It took another 150 years for these characteristics of circular motion to be
rediscovered by Robert Hooke and Christiaan Huygens.

Force and Motion

In his analysis of the relationships between force and motion, Leonardo
stayed largely within the confines of the Aristotelian framework. He distinguished between "natural" motion—the spontaneous movement of an
element toward its natural state—and "violent" or "accidental" motion, in

which an element is displaced from its natural state by some force. "Gravity and levity are accidental powers," he explained in the Codex Atlanticus, "which are produced by one element being drawn through or driven into another. No element has gravity or levity within its own element."[70]

In this Aristotelian context "gravity" did not refer to a force, as it does in Newtonian physics, but rather to the "heaviness" that is created by displacing a solid object upward, away from the Earth and thus out of its natural place. Similarly, "levity" was thought to be created when air is displaced downward and submerged in water. In both cases, the displacements were "violent" motions for which forces were required, while the return of the element to its natural place was due to an inherent tendency rather than an external force.

Aristotle's distinction between natural and violent motion, and his assertion that these two types of motion were fundamentally different and could not be mixed, were accepted by natural philosophers throughout the Middle Ages and the Renaissance. In the early seventeenth century, the distinction between natural and violent motion was abandoned by Galileo but was still debated among his contemporaries.

The situation clarified gradually when the concept of inertia came into focus. Indeed, according to science historian Robert Lenoble, "Modern mechanics was born with the principle of inertia."[71] The final decisive step was made by Newton, who clearly recognized inertia as the tendency of a massive body to preserve its state of rest, or uniform straight motion, unless acted upon by a force. A consequence of this fundamental insight, now known as Newton's first law of motion, was that force henceforth was no longer associated with motion in general, but specifically with changes of a body's state of motion; in other words, with acceleration.

Like all medieval and Renaissance scholars, Leonardo accepted Aristotle's assertion that an object in violent motion would continue to move only as long as there was a force acting on it. "No inanimate thing can push or pull something without going along with it," he wrote in his early Manuscript A, "and what pushes it can only be force or weight."[72] An obvious problem with this Aristotelian position was the difficulty in explaining why a thrown stone, for example, continues to move after losing contact with the hand exerting a force on it, or an arrow after losing contact with the bow that propelled it. The medieval philosophers were well aware of this difficulty, and they found an ingenious solution. They postulated that the moving force impressed, or infused, an impetus into the moved object that kept it in motion until the impetus eventually dissipated, like heat in an iron after it is removed from the fire.

The medieval theory of impetus was formulated in its most elaborate form in the fourteenth century by the French scholastic philosopher Jean Buridan. Leonardo studied Buridan's theories, including his imaginative theory of tectonic movements in geological cycles, through the writings of Albert of Saxony (see p. 89). As he progressed with his investigations of mechanics, he used the concept of impetus with increasing frequency. In his very first Notebook, the Codex Trivulzianus, he described the phenomenon of impetus but did not use the term:

> Every moved or percussed body retains in itself for some time span the nature of that percussion or movement; and it will retain it so much more or less as the power and the force of that blow, or motion, is greater or smaller.[73]

Around the same time he carefully analyzed, again without using the term "impetus," how the disturbance caused by a stone thrown into a still pond is transported outward in circular ripples:

> The water, though remaining in its position, can easily take this tremor from neighboring parts and pass it on to other adjacent parts, always diminishing its power until the end.[74]

In subsequent years, definitions of "impetus" along the lines of Buridan appear frequently in Leonardo's Notebooks. In Manuscript G, for example, he notes:

> Impetus is the impression of motion transmitted by the motor to the moved object. Impetus is a power impressed by the motor in the moved object.[75]

Leonardo applied the concept of impetus to many mechanical phenomena that we now associate with inertia, such as the stability of a spinning top, various oscillating motions, and collisions.[76] However, while Buridan described impetus quantitatively as being proportional both to the quantity of matter and the velocity of the object, no such quantitative treatment is evident in any of Leonardo's extant Notebooks.

In his early work on mechanics, Leonardo maintained the medieval imagery of impetus as a power that was infused into a moving object and that subsequently dissipated by itself. But as he became increasingly interested in friction, he became keenly aware of the effects of viscosity in water and air (see p. 41) and realized that it was air resistance that gradually diminished the impetus of a projectile:

> The power of the motor . . . attaches itself to the moved body, and over time it is consumed in the penetration of the air, which is always compressed in front of the moving object.[77]

In addition, Leonardo recognized that in the motion of projectiles there was a continuous interplay between violent and natural motion. "The natural motion, conjoined with the motion of a motor, consumes the impetus of that motor," he observed late in his life in Manuscript E.[78] With this statement describing the interplay between the projectile's inertia and the force of gravity (as we would say today), Leonardo clearly transcended the Aristotelian framework in which natural and violent motions could never be mixed.

In the centuries after Leonardo, the medieval concept of impetus gradually evolved into the modern concept of momentum, defined as the product of an object's mass and its velocity and as a vector, that is, a quantity having both a magnitude and a direction. At the end of the sixteenth century, Galileo still used impetus in the sense of a self-dissipating entity.[79] Descartes abandoned this image by introducing the term "quantity of motion" to replace "impetus," but he saw it as being independent of direction. In the seventeenth century, finally, Huygens was the first to state the conservation of the quantity of motion and gave it the full meaning of our contemporary concept of momentum.

Conservation of Energy

Among the basic concepts of mechanics, energy was the one that took longest to be identified and precisely formulated. It is much more abstract than the concepts of mass, force, or momentum, and has become one of the most important concepts of modern physics. Its importance is due to the fact that the total energy in any physical process is always conserved. Energy can change into many different forms—gravitational, kinetic, heat, chemical, and so on—but the total amount of energy in a particular process, or set of processes, never changes. There is no known exception to the conservation of energy. It is one of the most fundamental and most far-reaching laws of physics.

The famous German philosopher and mathematician Gottfried Wilhelm Leibniz, a contemporary of Newton, is usually credited with being the first to recognize the conservation of energy. He called it a "living force" (*vis viva*) and defined it as the product of the mass of an object and its velocity squared.* Newton accepted the conservation of energy but is said to have disliked conservation principles and did not attach great significance to them. Interestingly, the discussion of conservation as a fundamental law of nature involved philosophical and theological concerns for both Newton and Leibniz.[80]

The term "energy" in its modern sense was first used in the early nineteenth century, and scientists and philosophers argued for many years about whether energy was some kind of substance or merely a physical quantity. It was only with the formulation of thermodynamics in the mid-nineteenth century that energy was defined as the capacity of doing work, and the conservation of the total amount of energy through multiple transformations in mechanical and thermodynamic processes was clearly formulated.

Considering the very gradual emergence of the concept of energy over more than three centuries, and the reluctance even by Newton to attach importance to energy conservation, it is truly remarkable that Leonardo da Vinci had an intuitive grasp of it as early as the late fifteenth century. In his science of living forms that undergo continual changes and transformations, Leonardo paid special attention to the conservation of certain quantities—mass and volume in particular—and developed his own original "geometry done with motion" to express these principles of mathematically.[81] He extended the concept of conservation even to the motion of solid objects in space. "Of everything that moves," he noted, "the space which it acquires is as great as that which it leaves."[82]

Leonardo saw the conservation of volume as a general principle governing all changes and transformations of natural forms, whether solid bodies moving in space or pliable bodies changing their shapes. He applied it to the flow of water and other liquids (see p. 42), as well as to various movements of the human body, especially the contraction of muscles.[83]

In view of Leonardo's special perspective on principles of conservation,

* This corresponds to the modern definition of kinetic energy, $E_{kin} = \frac{1}{2} mv2$.

it is perhaps not surprising that he intuitively recognized the conservation of energy in his studies of mechanics. A striking example of this intuition is presented in the Codex Madrid I, where he describes and sketches an experiment involving a container from which water is tapped at various heights (fig. 5-8).

Leonardo begins the discussion of this experiment by clearly stating the problem. "Here the question is asked," he writes, "which of these four waterfalls has more percussion and power in order to turn a wheel: fall *a* or *b*, *c* or *d*?" In the subsequent analysis, he notes that the initial spouting speed of the jets increases as the tapping level is lowered because of the increase in water pressure. However, he then makes the following conjecture.

> I have not yet experimented, but it seems to me that [the four jets] must have the same power . . . [for] where the force of percussion is lacking, it is compensated by the weight of the waterfall.[84]

In modern terminology, we would say that Leonardo reasons as follows. As the kinetic energy (or "force of percussion") generated by the free fall of the water particles diminishes with the decreasing height of the spouts, this decrease is compensated for by the increasing potential energy (or "weight of the waterfall"), which results in the increase of water pressure and initial spout velocity. As a result of this compensation, the total energy (or "power") of the jets at their impact on the ground remains constant regardless of the height of their spouts.

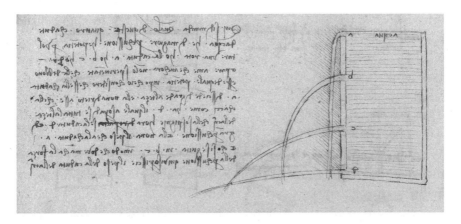

FIG. 5-8. Water jets falling from a container at four different heights. Codex Madrid I, folio 134v (detail).

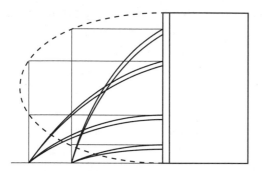

FIG. 5-9. Correct relationships between horizontal distances reached by jets and heights of spouts. Adapted from Fasso, "Birth of Hydraulics During the Renaissance Period."

It is worth noting that Leonardo's sketch of the four "waterfalls" is not accurate. Had he actually carried out the experiment, he would have noticed that the horizontal distances traveled by the jets are pictured incorrectly. Although the jets' spouting speed does increase with decreasing height of the taps, the time it takes for them to hit the ground becomes shorter, which affects the horizontal distances they can reach.

Today we can easily calculate the time it takes the freely falling water particles in each jet to reach the ground from the height of the spout, and the horizontal distance they can reach during that time with their initial spouting speed.[85] This calculation shows that the horizontal distance increases as the tapping level is lowered, reaches a maximum for the tap at half the height of the container, and then decreases again symmetrically for the taps in the container's lower half. The curve obtained by plotting the horizontal distances reached by the jets against the heights of their spouts is an ellipse (fig. 5-9).

Without our modern terminology and quantitative formulations, such a calculation was far beyond Leonardo's reach. These limitations make it all the more impressive that he intuited and correctly formulated the conservation of energy for flowing water. This important conservation law was rediscovered and precisely formulated only in the mid-eighteenth century by the mathematician Daniel Bernoulli, and is now known as Bernoulli's theorem.

Movements of Consumption

When physicists in the nineteenth century developed the science of thermodynamics they formulated two fundamental principles, known today as the first and second laws of thermodynamics. The first law is that of the conservation of energy. The second law states that, while the total energy involved in a process is always conserved, the amount of useful energy diminishes, dissipating into heat, friction, and so on. It is truly amazing

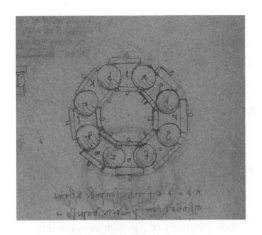

FIG. 5-10. Rotary ball bearing.
Codex Madrid I, folio 20v (detail);
model by Muséo Techni, Montreal, 1987.

that Leonardo da Vinci anticipated both of these fundamental laws of physics and that his thorough understanding of the dissipation of energy led him to deep insights about change, transformation, and the nature of time, foreshadowing similar insights in modern physics by more than three centuries.

The loss of machine power through friction was well known to Renaissance engineers. Their hoists, cranes, and other large machines were made of wood, and the friction between movable parts was a major problem. Leonardo invented numerous sophisticated devices for reducing friction and wear, including automatic lubrication systems, bearings made of semiprecious stones, and mobile rollers of various shapes—spheres, cylinders, truncated cones, etc.[86] Figure 5-10 shows an elegant example of a rotary bearing composed of eight concave-sided spindles rotating on their own axes, interspersed by balls that can rotate freely but are prevented from lateral movements by the spindles. When a platform is put on this ball bearing, friction is reduced to such an extent that the platform can be turned easily even when it is carrying a heavy load.

All the great Renaissance engineers were aware of the effects of friction, but Leonardo was the only one who undertook systematic empirical studies of its nature and properties. He investigated the frictional forces between various solids, as well as those involving water and air, and he designed experimental equipment for these studies that was far ahead of his time.[87]

Leonardo found by experiment that when an object slides against a surface, the amount of friction is determined by three factors: the roughness of the surfaces, the weight of the object, and the slope of an inclined plane:

> In order to know accurately the quantity of the weight required to move a hundred pounds over a sloping road, one must know the nature of the contact which this weight has with the surface on which it rubs in its movement, because different bodies have different frictions. . . .

> . . . Different slopes make different degrees of resistance at their contact; because, if the weight that must be moved is upon level ground and has to be dragged, it undoubtedly will be in the first strength of resistance, because everything rests on the earth and nothing on the cord that must move it. . . . But you know that, if one were to draw it straight up, slightly grazing and touching a perpendicular wall, the weight is almost entirely on the cord that draws it, and only very little rests upon the wall where it rubs.[88]

Leonardo's conclusions are fully borne out by modern mechanics. Today, the force of friction is defined as the product of the frictional coefficient (measuring the roughness of the surfaces) and the force perpendicular to the contact surface (which depends both on the object's weight and the slope of the surface). Leonardo not only analyzed the forces of friction correctly but also obtained reasonably accurate quantitative results two centuries before the modern study of friction began and three centuries before the subject was fully elaborated by the physicist Charles Coulomb.

Leonardo extended his keen interest in friction to his extensive studies of fluid flows. The Codex Madrid contains meticulous records of his investigations and analyses of the resistance of water and air to moving solid bodies, as well as the resistance of water and fire moving in air.[89] Well aware of the internal friction (viscosity) of fluids, he dedicated numerous pages in the Notebooks to recording its effects on fluid flow (see p. 41). "Water has always a cohesion in itself," he wrote in the Codex Leicester, "and this is the more potent as the water is more viscous."[90]

Air resistance was of special interest to Leonardo because it played an important role in one of his great passions—the flight of birds and the design of flying machines (see pp. 250ff.). "In order to give the true science of the movement of birds in the air," he declared, "it is necessary first to give the science of the winds."[91]

Careful reading of Leonardo's notes on mechanics makes it evident that he recognized that all the different kinds of friction he studied—the grinding of axles and pivots in machines, the friction between colliding objects, and the resistance encountered by bodies moving in water and air—had the same net effect. They all resulted in a "consumption of power," or dissipation of energy, as we would say today. The fundamental observation that in any physical process some energy is dissipated and cannot be recovered, which would become a cornerstone of the science of thermodynamics 350 years later, is illustrated repeatedly in Leonardo's manuscripts.

In all mechanical engineering, Leonardo points out, "you have to deduct as much . . . from the power of the instrument as that which is lost by the friction in its bearings";[92] and his experiments with rebounding objects led him to this conclusion: "I have learned from percussion that the falling movement exceeds the reflex movement."[93] A particularly elegant demonstration of energy dissipation is recorded in Manuscript A, which contains some of Leonardo's earliest investigations of mechanics. In a simple sketch (fig. 5-11), he compares the trajectory of a ball flying freely through the air with one where the ball bounces repeatedly, losing a certain amount of "power" in each bounce.

In his studies of mechanical engineering, Leonardo early on investigated the medieval belief that power could be harnessed through perpetual motion machines. At first he accepted this idea. He designed a host of complex mechanisms to keep water in perpetual motion by means of various feedback systems. But as his understanding of the dissipation of energy matured, he realized the impossibility of such a task.

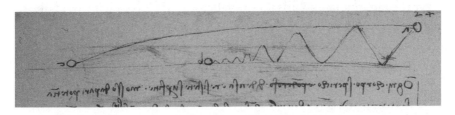

FIG. 5-11. Comparison between the trajectories
of a freely thrown and a bouncing ball.
Ms. A, folio 24r (detail).

"Descending water," he concluded, "will never raise from the place where it comes to rest to the height from where it started an amount of water equal to its weight."[94] In the end, Leonardo scoffed at attempts to build perpetual motion machines: "I have found among the excessive and impossible delusions of men the search for continuous motion, which is called by some the perpetual wheel."[95]

In the nineteenth century, the second law of thermodynamics was first formulated in terms of the dissipation of energy in thermal engines, but was soon recognized to be of much broader significance.[96] It introduced into physics the idea of irreversible processes, of an "arrow of time," as it came to be called. According to the second law, there is a certain trend in physical phenomena. As mechanical energy is dissipated and cannot be recovered, physical processes proceed in a certain direction—from or-der to disorder. To express this direction mathematically, physicists in-troduced a new quantity, called "entropy," which measures the degree of disorder, and hence the degree of evolution of a physical system. In its most general formulation, the second law of thermodynamics states that any isolated physical system will proceed spontaneously in the direction of ever-increasing entropy, or disorder.

Leonardo not only had a clear understanding of energy dissipation but also intuited its broader significance. He always paid special attention to the "consumption" of forms under the influence of physical forces over long periods of time. For example, his detailed description of the erosion of rocks carried by rivers and streams—from sharply angled fragments to smaller and rounder stones that eventually turn into gravel and fine sand—is a perfect illustration of an "entropic" sequence (as we would say today) toward ever increasing disorder (see p. 70). The entire passage would not look out of place in a modern textbook on thermodynamics. The same is true for Leonardo's vivid description of rock weathering: "Water wears away the mountains and fills up the valleys, and if it could, it would like to reduce the Earth to a perfect sphere" (see p. 70).[97]

Astonishingly, Leonardo associated these irreversible processes with a specific conception of time, as the founders of thermodynamics would do 350 years later. Physicists in the nineteenth and twentieth centuries dis-cussed the idea of a direction of time, manifest in the evolution of physi-cal processes from order to disorder. They called it the "arrow of time" to distinguish it from the reversible time coordinate in Newtonian physics.

Leonardo, using only slightly different language, introduced the notion of a physical quality of time. On a folio in the Codex Arundel, he jotted a brief reminder to himself: "Write of the quality of time as distinct from its geometry."[98] The quality of time he had in mind was that of "the consumer of all things"; in other words, time's irreversibility in the physical processes of transformation and decay.

The conception of time as the consumer of all things can be found already in Leonardo's early writings. A folio in the Codex Atlanticus, dating from around 1480, contains an evocative passage inspired by Ovid's *Metamorphoses*, in which Leonardo imagines the beautiful Helen of Troy as an old woman, her face ravaged by the passing of time:

> O time, consumer of all things! O envious old age, you destroy all things and consume all things with the hard teeth of the years, little by little, in slow death. Helen, when she looked in her mirror and saw the withered wrinkles that old age had made in her face, wept and wondered to herself why she had twice been carried away.[99]

Leonardo applied his qualitative conception of time and, accordingly, the conception of "movements of consumption," to three major domains: the transformations of the human body in the course of its life, those of the body of the Earth in the course of geological time, and to the consumption by attrition of the moving parts of machines.[100] His vision of the transformation and consumption of forms in these three domains as different manifestations of one universal process anticipated evolutionary thought in physics by more than three centuries and must be ranked as one of his greatest scientific achievements.

Weight, Force, and Motion

Weight was the second of Leonardo's four powers of nature. The relationships between weight, force, and motion were at the center of his attention when he began his theoretical studies of mechanics with a long series of empirical investigations of the medieval "science of weights." The main instrument of these investigations was the balance, and the basic theoretical framework was the classical Archimedean law of the lever (see p. 164).

Between 1490 and 1500, while he was painting the *Last Supper* in Milan, Leonardo made detailed studies of all the parts of a balance, experimenting with different kinds of suspensions or supports of the beams, different cords and weights, and so on. He systematically altered each variable in turn so as to get a clear understanding of the underlying principles. Not only that, he discussed possible errors arising from the differences between the mathematical treatment of a balance with the actual physical construction:

> The science of weights is led into error by its practice, which in many instances is not in agreement with this science, nor is it possible to bring it into agreement. This arises from the axes of the balances through which science is made from such weights. These axes, according to the ancient philosophers, were treated as having the nature of mathematical lines, and in some places as mathematical points. These points and lines are incorporeal, whereas practice treats them as corporeal, because this is what necessity demands for supporting the weight of these balances together with the weights on them that are to be judged, . . . These errors I set down here below.[101]

Leonardo then proceeded to list several possible errors. For example, he observed that the central line of the beams can run below, through, or above the fulcrum of a balance. "Only the one through the middle is perfect," he explained. "The one above is the worst; that below is less bad."[102]

These meticulous studies of the balance not only allowed Leonardo to calculate the forces and weights needed to establish equilibria in numerous compound systems, including balances, levers, and pulleys (see pp. 164–65), but also made him realize that the weight of a body is identical to the force of gravity acting on it. "The force is always equal to the weight that produces it," he noted in the Codex Atlanticus.[103] In view of his general acceptance of the Aristotelian view of gravity, which did not include the concept of an actual force, Leonardo's correct association of weight with a gravitational force, arrived at empirically, is rather remarkable.

The relationship between gravity and motion was of great interest to Leonardo in his anatomical studies because he wanted to understand the exact sequences of bodily movements in walking, running, jumping, and other activities (see pp. 213ff.). In preparation for detailed analyses of such movements Leonardo carried out many calculations to locate the center of gravity (called "center of mass" in modern mechanics) in a variety of geometric figures.[104]

The rules and principles for the exact determination of the center of gravity of triangles, squares, and other geometric figures had been established in antiquity by Archimedes in his treatise *On the Equilibrium of Planes*. Leonardo followed Archimedes closely in his presentation of proposals and proofs on centers of gravity, citing several passages from the great classic. Like Archimedes, he frequently divided geometric figures into triangles and used the law of the lever to show that pairs of those triangles balanced about a certain point. He may even have made physical models of his triangles and put equal weights at the angles to determine their centers of gravity.

Leonardo did not limit his investigations to plane figures, as Archimedes did, but also determined the centers of gravity of several solids. Most important, he discovered the exact location of the center of gravity of the tetrahedron, the regular solid composed of four equilateral triangles. The argument of Leonardo's proof is based on the tetrahedron's symmetry and sounds very modern to us.[105] He draws the height, which he calls an "axis," from the center of the figure's base to the opposite vertex (fig. 5-12) and reasons that, since the base is an equilateral triangle, the distribution of mass around this axis is symmetrical in all directions. Therefore, the center of gravity must lie somewhere on the axis. He then notes that, because of the tetrahedron's symmetry as a regular solid, the same argument can be made for any of its four axes. Hence, the center of gravity must lie at the

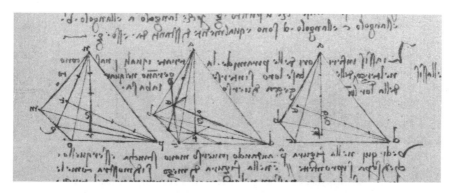

FIG. 5-12. Construction of the center of gravity of the tetrahedron. Codex Arundel, folio 218v (detail).

point where they intersect. Finally, he constructs that point of intersection and finds that it lies at a distance of one fourth of the height's length from the base.

On another folio in the same Codex Arundel, Leonardo writes down a concise summary of his result:

> The center of gravity of the body of four triangular bases is located at the intersection of its axes and will be in the fourth part of their length.[106]

The proof of Leonardo's theorem is concise and elegant, certainly one of his most significant discoveries in geometry.

Falling Bodies

During the last years of his first period in Milan, Leonardo not only immersed himself in detailed studies of statics—the "science of weights"—but also examined the motion of falling bodies. According to the Aristotelian view of gravity, this was an example of the natural motion of "weights" toward the Earth, and Leonardo did not fail to notice that free fall under the influence of gravity is an accelerated motion. He explained this fact in terms of the medieval impetus theory. In most cases, the concept of impetus had been applied to "violent" motion, that is, motion in which an object is forced to move against its natural tendency. Indeed, Leonardo himself defined impetus as "a power impressed by the motor in the moved object" (see p. 180). But since he had already established the identity of force and weight, it was natural for him to apply the concept of impetus also to motion under the influence of gravity, viewing the acceleration of a falling body as the continuous impression of impetus by the body's weight. "Impetus arises equally from weight as from force," he explained.[107]

Leonardo believed during most of his life that "pyramidal," or linear, relationships were universal in nature, as I have discussed (see p. 175),[108] and so it is not surprising that he asserted that the velocity of a falling body increases in direct proportion to time:

> The natural motion of heavy things, at each degree of its descent acquires a degree of velocity. And for this reason, such motion, as it acquires power, is represented by the figure of a pyramid.[109]

We know that the phrase "each degree of its descent" refers to units of time, because on an earlier page of the same Notebook he writes, "A weight that descends freely in every degree of time acquires . . . a degree of velocity."[110] In other words, Leonardo is affirming the mathematical rule that for freely falling bodies there is a linear relationship between velocity and time.

Leonardo's statements are entirely correct. In today's mathematical language, we say that the velocity of a falling body is a linear function of time, and we write it symbolically as $v = gt$, where g denotes the constant gravitational acceleration. This language was not available to Leonardo. The concept of a function as a relation between variables was developed only in the late seventeenth century. Even Galileo described the functional relationship between velocity and time for a falling body in words and in the language of proportion, as did Leonardo 140 years before him.[111]

Leonardo anticipated another discovery for which Galileo is famous. Instead of trying to verify his assertion about gravitational acceleration experimentally with falling bodies, which would have been almost impossible with the primitive clocks of his time, he had the same brilliant idea that Galileo had a century later—that a ball rolling down an inclined plane would accelerate in the same way as a freely falling object, only more slowly, which would allow one to measure the accelerated motion with reasonable accuracy even with simple instruments. "Although the motion is oblique," he reasoned, "it observes in each of its degrees an increase in motion and in velocity in arithmetic progression."[112]

Leonardo's sketch next to this statement is most ingenious (fig. 5-13). He has drawn the inclined plane as an isosceles triangle (*ebc*) that is meant, at

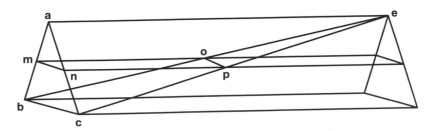

FIG. 5-13. Accelerated motion on an inclined plane.
Ms. M, folio 42v (as reconstructed by Clagett,
"Leonardo da Vinci: Mechanics").

the same time, to represent the arithmetic progression of the velocity with time. The triangle forms one face of an irregular pyramid whose opposite face is a vertical triangle (identical to the triangle *abc*), representing the linear relationship between velocity and time for the corresponding vertical fall. The correct relationships between the relevant variables can easily be recognized in Leonardo's diagram. The velocities at the end of both the vertical and the oblique descents are equal (both represented by *bc*), while the times involved in acquiring the velocities (represented by the edges of the two triangles) are clearly seen to differ. In addition, Leonardo marked the midway velocities (*mn* = *op*) to show that the same relationships hold for all intermediate velocities along the vertical and oblique descents.

Leonardo's diagram of the inclined plane is an impressive example of his great capacity for analyzing the elements of complex phenomena and presenting them visually with great clarity. However, it is doubtful that he actually experimented with balls rolling down inclined planes, as Galileo did. Had he done so, Leonardo would certainly have observed that the distances of falling bodies increase with the squares of the times and not linearly. Instead, he maintained erroneously that "in each doubled quantity of time, the length of the descent is doubled."[113]

As mentioned above, Galileo made the same error in his early work, but corrected it after experimenting with inclined planes (see pp. 175–76). The fact that Leonardo held on to his belief in a linear relationship between the distances traversed by falling bodies and the times elapsed seems to indicate that he never carried out the experiments he had designed so brilliantly. Marking off the distances covered by balls rolling down inclined planes during successive time intervals would have been relatively easy, whereas measuring their ever-increasing velocities during the same time intervals would have been a considerable challenge. Indeed, Galileo attained an understanding of the linear relationship between instantaneous speeds and times only with great effort and several years after he realized that the distances traversed by falling bodies increase with the squares of the times.[114]

In addition to the inclined plane, Galileo used the pendulum as a major tool for analyzing the effects of gravity on motion. Leonardo, too, studied the motion of the pendulum and used it as a regulator in clocks.[115] But he failed to recognize one of its most important properties: the period of the oscillation is independent of the mass of the bob. Galileo famously used this property of the pendulum to demonstrate that in a vacuum all objects will fall with the same acceleration and therefore will reach the ground at the same time, regardless of their mass.

Still, Leonardo did make some significant discoveries about pendulum motion. He described it as the interplay between natural motion (the downswing) and accidental motion (the upswing), and he realized that, because of the inevitable energy loss through friction, "the accidental motion will always be shorter than the natural."[116] This statement, on a folio of the Codex Madrid I, is illustrated with a simple sketch (fig. 5-14) that clearly shows the loss of energy on the upswing. It is another of Leonardo's many illustrations of the dissipation of energy.

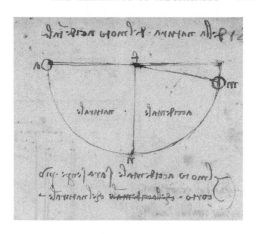

FIG. 5-14. Energy loss in pendulum motion. Codex Madrid I, folio 147r (detail).

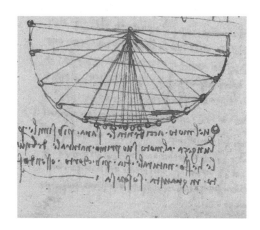

FIG. 5-15. Approximate regularity of pendulum swings for small arcs. Codex Madrid I, folio 147r (detail).

On the same folio, Leonardo analyzes the rate at which the arc of the pendulum diminishes under the influence of friction and notes that for small arcs the oscillations will be more uniform: "The smaller the natural motion of a suspended weight, the more the following accidental motion will be equal in length." Again, the observation is clearly illustrated in a small drawing (fig. 5-15). With this discovery of the increased regularity of pendulum swings for small arcs, and hence of the approximate regularity of their beats, Leonardo anticipated the theoretical formulations of Galileo and the practical applications to the development of pendulum clocks by about a century and a half.

Finally, still on the same folio of the Codex Madrid, Leonardo juxtaposes the trajectory of a pendulum and its interplay of natural and accidental motion with the trajectory of a stone thrown in an arc. Even though his detailed comparison of the two trajectories contains some errors, his observation that in both cases there is a similar interplay between gravity and impetus is prescient. In the seventeenth century, Galileo would make the same juxtaposition in his celebrated *Dialogue on the Two Chief Systems of the World.*

Ballistic Trajectories

The study of ballistic trajectories was of special interest to Leonardo in his work as a military engineer. In 1502, when he was fifty and had acquired great fame as an artist and engineer, he was hired by the papacy to travel throughout central Italy to inspect ramparts, canals, and other fortifications and make suggestions for their improvement.[117] He designed ingenious new fortifications. Instead of castles with high vertical walls he envisioned low bastioned fortresses arranged in a series of concentric curves so as to minimize the impact of cannonballs.[118]

To develop these effective military designs, Leonardo needed to have accurate knowledge of the trajectories of projectiles—a subject that was poorly understood and replete with erroneous assumptions at the time. Throughout the Renaissance and up to the late seventeenth century, the trajectory of a cannonball was pictured in all military treatises dealing with artillery as rising at an angle along a straight line, followed by a short curved section, and then falling in a perfectly vertical line.[119] Several variations of this picture were proposed by mathematicians in the sixteenth and seventeenth centuries. All of them—even Galileo in his early work— pictured the rising portion of trajectories as a straight line (fig. 5-16).

Leonardo, by contrast, did not fail to recognize (without naming it) the parabolic nature of ballistic trajectories. On the folio in Codex Madrid I cited above, his juxtaposition of the trajectory of a pendulum with that of a stone thrown in an arc is accompanied by a beautiful sketch of a series of ballistic trajectories for different launch angles (fig. 5-17).[120] The parabolic shapes of these trajectories are clearly visible.

Leonardo did not have the mathematical tools to calculate these shapes, and with the experimental equipment available at the time, he could not have determined them by monitoring the paths of actual cannonballs. But he used his powers of scientific analysis and his systemic way of thinking to find an ingenious solution to the problem. Instead of studying the trajectories of stones or cannonballs, he studied jets of water, where the trajectories can actually be seen. He realized that these jets were composed of water particles subject to the same forces of natural and accidental motion (of gravity and inertia, as we would say today) as the stones and cannonballs.

In Manuscript C, written shortly before Codex Madrid I, Leonardo describes how he systematically studied the trajectories of water jets: "Test in order to make a rule of these motions. Make the test with a leather bag full of water with many small pipes of the same inside diameter, installed along one line."[121] The accompanying drawing shows such a bag with water pouring out from four small spouts arranged at different angles, including one in a vertical direction (fig. 5-18). It is evident from this drawing that Leonardo's sharp eye perceived not only the correct parabolic shapes of the water jets but also, impressively, their slight distortions due to air resistance. The characteristic flattening of their ascending portion and the

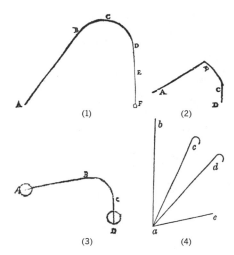

FIG. 5-16. Ballistic trajectories according to (1) Niccolò Tartaglia, 1537; (2) Girolamo Cardano, 1550; (3) Bernardini Baldi, 1621; and (4) Galileo Galilei, 1592.
From Bertoloni-Meli, *Thinking with Objects*.

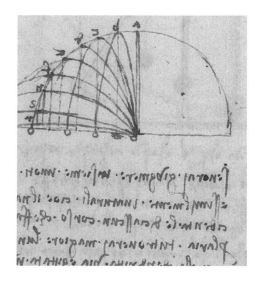

FIG. 5-17. Parabolic trajectories for different launch angles. As usual in Leonardo's diagrams, the direction of motion is from right to left.
Codex Madrid I, folio 147r (detail).

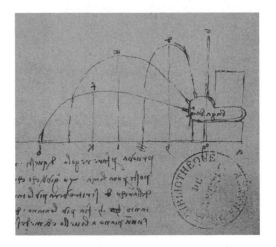

FIG. 5-18. Parabolic water jets with air resistance. Ms. C, folio 7r (detail).

steepening of their descending portion are clearly visible in the drawing.

The accuracy of Leonardo's drawing of these trajectories is truly astonishing. After Leonardo, the parabolic form of ballistic trajectories was observed by Galileo in 1609 and was proven mathematically by his most famous disciple, Evangelista Torricelli, in 1644. Torricelli also rediscovered, 150 years after Leonardo, the distortion of ballistic trajectories by air resistance, while Galileo failed to take this effect into account. The calculation of the exact ballistic curve with air resistance had to wait for Newton, who published it in his *Principia* in 1687.

The Origin of Physical Forces

Having established the identity of a body's weight with the force of gravity acting on it, it was natural for Leonardo to consider force as his third power of nature, after motion and weight.

The origin of physical forces was one of the most persistent and perplexing questions in the development of classical mechanics. Galileo did not address the problem, limiting his investigations to the motion of material bodies under the influence of various forces. He was criticized for failing to do so by Descartes, who vigorously promoted a thoroughly mechanistic view of the world in which both living and nonliving phenomena were reduced to the motions and mutual contacts of small material particles. The force of gravity, in particular, was explained by Descartes in terms of a series of impacts of tiny particles contained in subtle material fluids that permeated all space.[122]

Descartes's theory was highly influential throughout most of the seventeenth century, until Newton replaced it with his conception of gravity as a fundamental force of attraction between all matter, acting at a distance and diminishing with the square of that distance. Newton's conception, in turn, was criticized by many of his contemporaries, who were shocked by the idea that a force of attraction should act at a distance without being transmitted by any medium. The famous architect and mathematician Christopher Wren, for example, was reported to have "smiled at Mr. Newton's belief that [gravity] does not occur by mechanical means, but was introduced originally by the Creator."[123] The definitive solution of this vexing problem had to wait until the development of the field concept by Michael Faraday and James Clerk Maxwell in the nineteenth century and of Albert Einstein's theory of gravity (his general theory of relativity) in the twentieth.

Leonardo, who remained largely within the Aristotelian framework and was unencumbered by the fundamental division between mind and matter to be introduced a century later by Descartes, looked at the question of the origin of physical forces very differently. Since the movements of falling bodies, flowing water, rising air, and blowing wind were thought to be caused by the natural tendencies of these elements to move toward their proper places, the problem was reduced to explaining the origin of the "accidental" or "violent" forces that disturbed the balance of the elements.

For Leonardo, the principal accidental force came from the muscle power of animals and humans, which was indeed one of the main sources of energy in his time. The origin of this muscle power (for both humans and animals) was in the soul, from where it was transmitted to the body's muscles by invisible, nonmaterial nervous impulses that traveled through the sensory and motor nerves in the form of waves.[124] In other words, the origin of force was nonmaterial. "Weight is corporeal," he explained, "and force is incorporeal; weight is material and force is spiritual."[125] As I have mentioned, Leonardo often used the word "spiritual" in the sense of being immaterial and invisible, and this is how he described the ultimate nature of accidental forces:

> Force is nothing but a spiritual power, an invisible potency, which is created and infused, through violence from without, by sentient bodies in non-sentient ones.[126]

Similar clear and articulate definitions of accidental force appear repeatedly throughout Leonardo's Notebooks. But because of his use of the term "spiritual," there has been confusion about his conception of force among many historians. From their mechanistic perspectives, Leonardo's statements seem to reveal a spiritual, or even esoteric, dimension of his thought. Such a dimension may be discerned in some of his philosophical statements but not in his conception of physical forces, in my opinion. Leonardo's definition of force is unambiguous, derived from empirical evidence, and consistent with the overall framework of his scientific thought.

Force, Motion, and Work

In his work in mechanical engineering, Leonardo had ample opportunity to observe the effects of various forces in machines like the pulley and the lever. In particular, he paid special attention to the transmission of power and motion from one plane to another (see p. 162). It is typical of his scientific mind that he not only used this empirical knowledge to improve existing machines but also tried to derive general principles of mechanics from his observations.

One of those principles was the conservation of work, a special case of the conservation of energy. Several years after his comments on the conservation of energy in the Codex Madrid I, Leonardo stated in the pocket-sized Notebook known as Manuscript F:

> If a power moves a body a certain distance in a certain time, the same power will move half of this body in the same time twice that distance . . . [or] the whole distance in half that time.[127]

This statement of the conservation of work is in complete agreement with our modern definitions of work as "force times distance" and of power as "work over time."

Leonardo also discovered a principle that would become known as Newton's third law of motion two hundred years later. It states that "for every action, there is an equal and opposite reaction." In other words, physical forces always come in equal and opposite pairs. Leonardo did not formulate this observation as a general principle, but he clearly stated it in terms of many concrete examples.

One of his earliest Notebooks, Manuscript A, written around 1490 during his first years in Milan, contains a discussion of the rebounds of a small glass ball on a smooth polished stone. In his analysis, Leonardo stated that the force of the ball's rebound is equal to the force of its impact.[128] Around the same time, in 1485, he studied the flight of birds and developed his first designs of flying machines. Observing the wing movements of an eagle, he noted: "As much force is exerted by an object against the air, as the air exerts against the object."[129] Some twenty years later, he recognized the same principle in relation to the force of water on an oar: "The amount of movement made by an oar against still water equals the amount of movement made by water against a motionless oar."[130]

Finally, there is a principle of classical mechanics that is much more abstract and more general, which Leonardo anticipated by several centuries. Manuscript G contains a concise formulation, derived from observations of falling bodies: "Every natural action is made by the shortest way."[131] A more elaborate and effusive formulation is found on a folio of the the Codex Atlanticus where Leonardo discusses experiments with light rays:

> O marvelous necessity, with supreme reason you compel all effects to be linked to their causes, and by supreme and irrevocable law every natural action obeys you by the shortest operation."[132]

Leonardo's enthusiasm about this principle was fully justified. Nearly two centuries later it would turn out to be one of the most important principles of classical mechanics. It was formulated in the seventeenth century by the great mathematician Pierre de Fermat, who observed it in connection with geometrical optics, as Leonardo had done before him. Known today as Fermat's principle, or the "principle of least time," it states that light always follows the path of least time. In the nineteenth century, the principle was reformulated by the physicist and mathematician William Rowan Hamilton in much more abstract mathematical language. Hamilton's principle, also known as "principle of least action," is valid for all physical systems.*

Some of Leonardo's statements seem to imply that the path of least time for a natural process is also the shortest path, which is not always the case. Strictly speaking, therefore, Leonardo's principle cannot be viewed as an early formulation of either Fermat's or Hamilton's principle. But what it shares with both is the idea of a certain efficiency in natural phenom-

---

* Hamilton's principle states that any mechanical system can be represented by a certain mathematical quantity, called "action," which remains stationary under small variations of the system's variables. The mathematical formulation of the principle can be shown to be equivalent to Newton's equations of motion.

ena, called "necessity" by Leonardo, which can be measured by observing that the value of some quantity, or variable, becomes a minimum. The level of abstraction of this idea, which is clearly articulated in Leonardo's statements, is truly exceptional for his time.

Percussion—The Fourth Power of Nature

Leonardo's concept of "percussion," his fourth power of nature, corresponds to what we now call "impact" and refers to a broad range of phenomena. They include the striking of bells and tapping of vibrating plates; the impacts of hammers on nails and other surfaces; the impact of colliding balls, as well as the rebound of a ball from a firm surface; and even the destructive effects of mortars from the initial explosions to the impacts of their cannonballs and blast waves. In addition, Leonardo uses "percussion"—or, more frequently, "force of percussion" or "power of percussion"—to denote the energy transferred in the process of impact; in other words, as the equivalent of the modern concept of kinetic energy.

In all his studies of impact phenomena, Leonardo paid special attention to the transfer of energy from one body or medium to another, as well as to the eventual dissipation of that energy. He noted that energy transfer takes place in the process of impact when an initial motion comes to a sudden halt: "The blow is born of the death of motion."[133]

In a series of experiments, he struck various objects with a hammer and analyzed how much of the "force of percussion" is absorbed by the object and how likely the object is to break, depending on its supports and on the location of the blow. In one case, he showed how "the hand holding a stone that is beaten does not suffer as much as it would if it received the blow directly."[134] In another example of percussion with a hammer, he demonstrated how the kinetic energy of the impact is transformed into heat:

> If you beat a thick bar of iron between the anvil and hammer with frequent blows upon the same place, you will be able to light a match at the beaten place.[135]

For the extreme case of explosions, Leonardo gave vivid and detailed descriptions of the destructive effects of the blast wave, as in the following passage from the Codex Atlanticus.

If you discharge a small bombard in a courtyard surrounded by a convenient wall, any vessel that is there, or any window covered with cloth, will be instantly broken; and the roofs will be lifted off slightly from their supports; the walls and ground will shake as in a big earthquake; all the spiders' webs will fall down; small animals will perish, and every air-containing body nearby will suffer instant harm and some damage.[136]

Percussion also played an important role in Leonardo's studies of acoustics. As I discussed in my previous book, he observed from experiments with bells, drums, and other musical instruments that sound is always produced by "a blow on a resonant object," and he correctly deduced that this percussion causes an oscillating motion in the surrounding air.[137] Moreover, he described the phenomenon of resonance in detail:

The blow given to the bell will make another bell similar to it respond and move somewhat. And the string of a lute, as it sounds, produces response and movement in another similar string of similar tone in another lute. And this you will perceive by placing a straw on the string which is similar to that sounded.[138]

The observations of resonating bells and lute strings suggested to Leonardo the general mechanism for the propagation and perception of sound—from the initial percussion and the resulting waves in the air to the resonance of the eardrum. In Manuscript C, he illustrated his discovery with a charming little sketch (fig. 5-19) in which the generation, propagation, and perception of sound waves is represented symbolically by three little hammers striking a bell, a reflecting wall, and an ear.

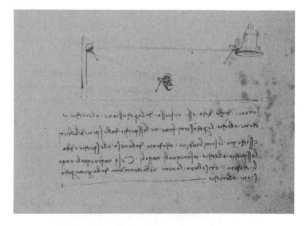

FIG. 5-19. Propagation of sound waves from a bell to the ear. Ms. C, folio 16r (detail)

Percussion also lies at the origin of Leonardo's experiments of dropping pebbles into the still water of a pond. His perceptive analysis of the subsequent phenomena includes detailed descriptions of the impact of the pebble, the generation of an up-and-down motion of the water particles, the spread of the "force of percussion" in circular waves, and its gradually diminishing "power" as the kinetic energy of the impact is dissipated by the water's viscosity.[139]

In his studies of percussion, Leonardo not only analyzed the transfers and transformations of kinetic energy in various impact phenomena but also tried to determine quantitative relations between the mass and velocity of a colliding object and the damage done by the "power of percussion" (kinetic energy) at impact. To do so, he dropped various weights from different heights on a slab of lead and measured the size of the dents produced by their impact.[140]

In another series of experiments, he shot arrows of different weights up to various heights and measured the penetrations of their shafts into soft soil, "the soil being of uniform resistance and the shafts of the same shape."[141] To achieve this uniformity, he produced arrows with identical hollow shafts and gave them different weights by placing stones inside the shafts. From all these experiments, Leonardo concluded correctly that the "power of percussion" is proportional to the weight of the falling object and to the height of its fall.

Collisions of billiard balls and similar objects were another major focus in Leonardo's investigations of percussion. He distinguished collisions between two moving objects from those between a ball and a firm wall or very heavy object (i.e., rebound phenomena):

> There are two kinds of percussion: when the object flees from the projectile that struck it, and when such a projectile rebounds back from the object struck.[142]

Leonardo studied both types of collision in great detail and in many variations. On a folio in Manuscript A (fig. 5-20), he sketched more than a dozen examples of both solid and breakable balls of different masses colliding at various speeds and angles of incidence. The brief verbal descriptions accompanying each sketch make it evident that these are records of systematic experiments with elastic and inelastic collisions, as they are called today.

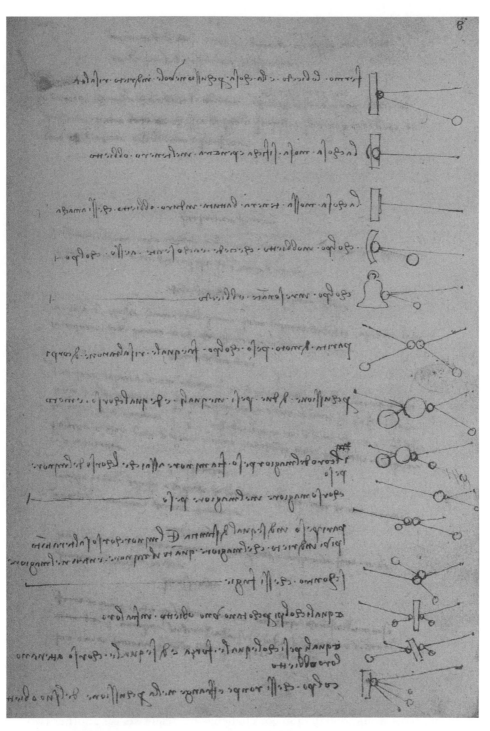

FIG. 5-20. Studies of elastic and inelastic collisions. Ms. A, folio 8r.

From these experiments, Leonardo tried to derive quantitative relations between the masses, velocities, and angles of the colliding balls. Sometimes he would pose collision problems in clear and concise language:

Ball *a* moves with three degrees of velocity, and ball *b* moves with four degrees of velocity. It is asked how such a percussion differs from one in which the ball [*b*] were to be at rest instead of approaching it [*a*] with the said four degrees of velocity.[143]

In other passages, Leonardo stated general rules about elastic collisions. A folio in the Codex Leicester, for example, contains the following statements.

If the percussor is equal and similar to the percussed, that percussor leaves its power completely in the percussed, which flees with fury from the site of the percussion, leaving its percussor there. But if the percussor—similar but not equal to the percussed—is greater, it will not lose its impetus completely after the percussion but there will remain the amount by which it exceeds the quantity of the percussed. And if the percussor is less than the percussed, it will rebound back through more distance than the percussed by the amount that the percussed exceeds the percussor.[144]

It is instructive to examine the three statements in this passage in some detail. The first is an accurate description of an elastic collision of two balls of equal mass, one moving and the other stationary, in which the "power" (kinetic energy) of the moving ball is completely transferred to the stationary target during the collision. In fact, Leonardo's description is the correct answer to the first problem posed in the preceding quotation.

The other two statements in the passage from the Codex Leicester describe two collisions between balls of unequal mass. In the first, the moving ball is heavier than the stationary target; in the second, it is lighter. Here Leonardo's descriptions are qualitatively correct, but he obviously struggles in his attempts to formulate precise relationships between quantities like mass, velocity, and momentum, which have different dimensions. This would remain a major challenge for scientists for another two centuries after Leonardo (see p. 173).

Indeed, the precise mathematical analysis of impact phenomena had to wait until the seventeenth century, when Huygens used the laws of both the conservation of energy and the conservation of momentum, formulated in terms of algebraic equations, to derive the exact rules for elastic collisions.

Rebound phenomena, in which an object collides with a firm wall, presented an easier problem to natural philosophers because they involve fewer variables. Central to Leonardo's descriptions of percussion and rebound was the rule that the angle of incidence equals the angle of reflection. From his early investigations of mechanics on, he repeatedly stated that fundamental rule, now known as the law of reflection. A succinct statement, recorded around 1500, can be found in the Codex Arundel:

> The angle made by the reflected motion of heavy bodies is equal to the angle made by the incident motion.[145]

In Manuscript A, written ten years earlier, we find a similar passage: "The line of percussion and that of its rebound are placed in the middle of equal angles."[146] The meaning of this statement is perhaps less evident, but it is clearly illustrated with a drawing (fig. 5-21) and accompanied by a more elaborate description:

> If the ball *b* is thrown to *c*, it will turn back through the line *cb*, necessarily making equal angles on the wall *fg*. And if you throw it through the line *bd*, it will turn back through the line *de*, and thus the line of percussion and the line of rebound will make an angle on the wall *fg* situated in the middle between two equal angles, as *d* appears in the middle between *m* and *n*.

The law of reflection was first formulated in optics by the great Arab mathematician Alhazen (Ibn al-Haitham) whose seven-volume work *Kitab al-Manazir* (Book of Optics) was available during the Renaissance in Latin translation and was discussed by several European philosophers. Leonardo became familiar with Alhazen's work through these authors and he used the law of reflection extensively in his explorations of spherical and parabolic mirrors, producing a series of precise and beautiful diagrams.[147]

But Leonardo went further. He was the first to recognize the broad

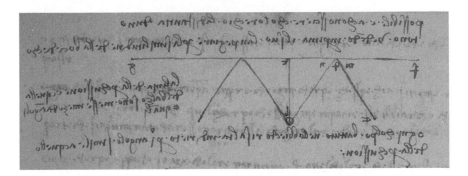

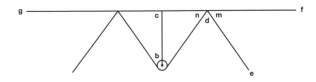

FIG. 5-21. Illustration of the law of reflection.
Ms. A, folio 19r (dcetail); reconstructed by Clagett,
"Leonardo da Vinci: Mechanics."

generality of the law of reflection, applying it not only to mechanics and optics but also to acoustics and hydrodynamics. On the same folio of Manuscript A where he discusses the rebound of a ball thrown against a wall, Leonardo adds a brief note: "The voice is similar to an object seen in a mirror."[148] In other words, the law of reflection holds equally for light and sound. Several years later, he applied the same reasoning to the rebound of a jet of water from a wall, noting, however, that some of the water peels off as an eddy after the reflection.[149]

As I have emphasized, this kind of systemic thinking is typical of Leonardo's scientific approach as a whole. In the preceding pages, I have discussed about a dozen of his discoveries and anticipations of abstract principles of mechanics that were centuries ahead of his time. They include his understanding of the relativity of motion, his intuitive grasp of the conservation of energy, and his emphasis on principles of conservation in general; his anticipation of the law of energy dissipation (the second law of thermodynamics), and his association of irreversible processes with a physical quality of time; his juxtaposition of the interplay between gravity and inertia in the swing of a pendulum with a similar interplay in the trajectory of a stone thrown in an arc; his experiments with an inclined plane

THE ELEMENTS OF MECHANICS

to study gravitational acceleration and with jets of water to study ballistic trajectories; his discovery of the principle now known as Newton's third law of motion, and his intuitive anticipation of Fermat's principle; and, last but not least, his derivation of the center of mass of the tetrahedron from symmetry arguments.

All these examples show a level of abstract thinking that makes so many of Leonardo's scientific statements sound utterly modern, in spite of the evident Aristotelian roots of his science. This level of abstraction is evident throughout Leonardo's "elements of mechanics," as well as in all other branches of his science.

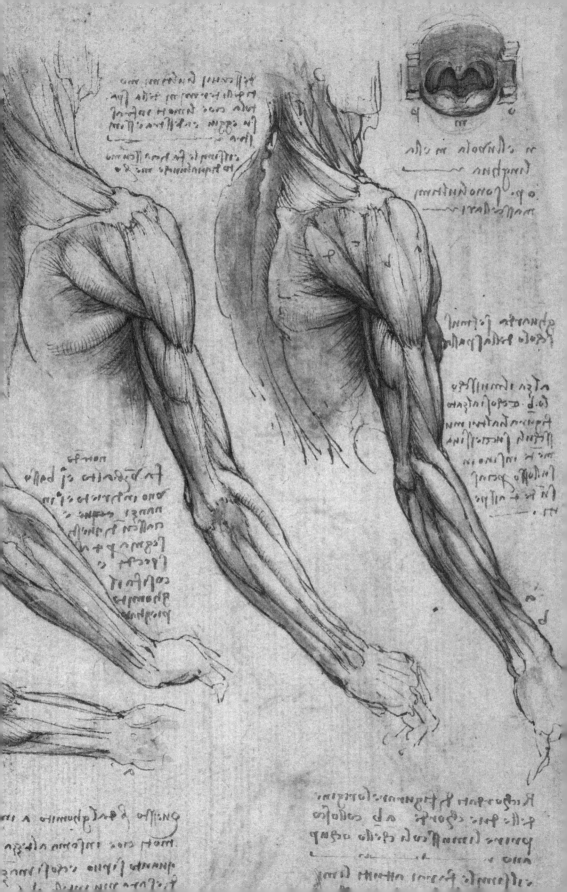

6

The Body in Motion

Leonardo perceived the living world as being in constant flux, its forms merely stages in continual processes of transformation; so it was natural for him to also understand the human body in terms of movement and development. He saw the body's continuous movements—epitomized in flowing gestures, curling hair, or floating draperies—as visible expressions of grace, and he was a master in portraying such graceful movements in his paintings. *The Madonna and Child with Saint Anne* (plate 7) is perhaps his finest demonstration of graceful gestures fused into a single, continuous flow.

The association of grace with smooth, flowing movements was common among artists in the Renaissance, but Leonardo was the only one who attempted to understand it within a scientific framework.[1] His analytic mind was fascinated by the autonomous, voluntary movements of the human body. Their investigation became a major theme in his anatomical work. In countless dissections and detailed anatomical drawings he explored the transmission of the forces underlying various bodily movements, from their origins in the center of the brain down the spinal cord, and through the peripheral motor nerves to the muscles, tendons, and bones.

Leonardo realized that in this propagation of the motor forces, their exact origins in the brain and transmissions through nerve impulses were invisible and hence inaccessible to further scientific analysis. He hypothesized that these pulses traveled through the nerves in the form of waves, and he called their movement "spiritual," by which he simply meant that

FACING Rotated views of the muscles
of the shoulder and arm, c. 1509–10
(detail, see plate 4).

it was immaterial and invisible.[2] Here is how he described the functional links between nerve impulses, muscles, tendons, and bones:

> Spiritual movement, flowing through the limbs of sentient animals, broadens their muscles. Thus broadened, these muscles become shortened and draw back the tendons that are connected to them. This is the origin of force in the human limbs. . . . Material movement arises from the immaterial.[3]

At various stages in his anatomical research, Leonardo investigated all the sections of this pathway of the body's motor forces. He began with detailed explorations of the brain, the cranial nerves, and the spinal cord,[4] and then proceeded, during the years 1506–8, with a series of anatomical drawings that are composite representations of several systems—nerves and muscles, muscles and bones, bones and blood vessels, and so forth—as well as representations of the entire body.[5]

FIG. 6-1. Measurements of muscular forces.
Ms. H, folio 43v (left) and folio 44r (right).

By that time, Leonardo had acquired sufficient understanding of the principles of mechanics to explore in detail how nature's "mechanical instruments"—the muscles, tendons, and bones—work together to move the body. In numerous drawings, he showed how joints operate like hinges, tendons like cords, and bones like levers; for example, in his exquisite demonstrations of the complex movements of the foot (see fig. 4-7), the arm (plate 3), and the hand (see fig. 4-2). He applied his knowledge of the rotational motion of axles to the movement of the "universal" ball-and-socket joints of the hip and shoulder.[6] In some anatomical studies, Leonardo drew cords or wires instead of muscles to better demonstrate the directions of their forces (see figs. 4-2 and 6-9). He also measured muscular forces by fastening cords to the hands and feet of a person and running them over pulleys with weights attached at their ends, not unlike modern weight-lifting machines (fig. 6-1).

Varieties of Bodily Movements

Leonardo used his knowledge of the mechanics of bodily movement to analyze a wide variety of actions of the human body down to their finest details. "After the demonstration of all the parts of the limbs of man and of the other animals," he wrote in a note to himself, "you will represent the proper operations of these limbs, that is in rising from lying down; in walking, running, and jumping in various ways; in lifting and carrying heavy weights; in throwing things to a distance, and in swimming; and thus in every action you will demonstrate which limbs and muscles cause the aforementioned operations."[7] Like the research projects in other branches of his science, Leonardo's program of anatomical research regarding bodily movements was comprehensive and very ambitious. It is also noteworthy that in this passage, as in many others, he considered the human body an animal body, as we do in biology today.

Leonardo distinguished between three types of bodily movements. "The movements of animals are of two kinds," he explained, "that is, motion in space (moto locale) and motion of action (moto azionale). Motion in space is when the animal moves from place to place; and motion of action is the movement which the animal makes within itself without change of place; . . . and the third is the composite movement (moto composto), combining action with locomotion."[8]

Movements of action, for Leonardo, included not only external bodily movements, such as pushing, pulling, and lifting, but also movements in

FIG. 6-2. Study of figures in action, c. 1506–8. Windsor Collection, *Figure Studies, Profiles, and Caricatures*, RL 12644r (detail).

the body's internal physiology, like breathing and digesting food. In the Windsor Collection, there are several folios with sketches of small naked figures performing a variety of actions (fig. 6-2), and the Notebooks also contain numerous detailed written descriptions of the combined operations of specific muscles and joints in carrying out those bodily movements.[9] In addition, Leonardo undertook elaborate studies of the actions of the heart, the lungs, and other internal organs (see chapter 8).

In his analysis of movements in space, or locomotion, Leonardo paid special attention to the positions of the body's center of gravity during the various phases of a particular movement. He observed that "motion is created by the destruction of equilibrium, that is, of equality [of weight]";[10] and furthermore, that "the locomotion of man or any other animal will be of as greater or less velocity as their center of gravity is further away from, or nearer to, the center of the supporting foot."[11] The Notebooks contain meticulous descriptions of successive shifts of positions in rising up from the ground, walking uphill (or against the wind) and downhill, running, and jumping.[12]

In particular, Leonardo was keenly aware of the undulating movements of the spine and body during walking and running. He noted the corresponding rise and fall of the shoulders and recognized accurately that these wave movements are possible because the spine is composed of many small bones (the vertebrae) strung together in a flexible column. On a sheet in the Windsor Collection (fig. 6-3), Leonardo's understanding of the undulating movements of the spine is vividly illustrated in a series of sketches of horses, cats, and—charmingly—dragons. The accompanying

FIG. 6-3. Studies of flexions of the spine in the
movements of horses, cats, and dragons, c. 1508.
Windsor Collection, *Horses and Other Animals*, folio 158.

text explains that the "serpentine winding" of the spine in the movements of these animals takes place longitudinally as well as laterally. During the same period, Leonardo produced a magnificent study of the vertebrae and vertebral column (see fig. 6-8), one of his most famous anatomical drawings.

In his instructions to painters on how to render human figures, Leonardo emphasized repeatedly that the figures' gestures and movements should portray the frame of mind—the thoughts, intentions, and emotions—that provoked them. "The most important thing that can be found in the discourses on painting," he wrote, "are the movements appropriate to the states of mind of each living creature, such as desire, contempt, anger, pity, and the like."[13] He admonished painters to "give your figures an attitude that is adequate to show what the figure has in mind."[14]

Leonardo saw the movements of the human body as the visible expressions of mental movements (moti mentali). Indeed, to portray the body's expressions of the human spirit was, in his view, the artist's highest aspiration. "That figure is not worthy of praise," he declared, "if it does not, as much as possible, express in gestures the passion of its spirit."[15] Leonardo himself excelled at this task. His celebrated Last Supper became famous throughout Europe immediately after its completion because of its novel composition and because of the eloquence and power of the protagonists' gestures and facial expressions, which displayed a wide range of intense emotions.[16] Similarly, the paintings of Leonardo's mature period, including the Saint Anne and the Mona Lisa, have always been considered masterpieces for the expression of their figures' inner lives.

Anatomy in the Renaissance

When we look at the large collection of Leonardo's anatomical drawings,* we marvel at their great accuracy and artistic beauty, but it is not easy to realize just how revolutionary they were in their time. Today we are used to seeing detailed wall charts of bones, muscles, and nerves in hospitals and doctors' offices, and we tend to forget that no even moderately accurate visualization of internal anatomical features was available to Leonardo's contemporaries. Even in the leading anatomical textbooks of the time, the body's internal organs and tissues were shown merely in diagrammatic

---

* Most of Leonardo's anatomical drawings (more than two hundred folios, almost always drawn on both sides) are contained in the Windsor Collection; see p. 355 for their scholarly edition in facsimile.

or symbolic forms. Leonardo, by contrast, represented the various parts of the body realistically, with their accurate shapes and relative positions, shown from several perspectives, and within the context of the body as a whole. "My configuration of the human body," he declared proudly, "will be demonstrated to you just as if you had the natural man before you."[17]

To fully appreciate the genius of Leonardo the anatomist, then, we need to juxtapose his drawings with the schematic illustrations that were typically found in the scholastic anatomical texts between the Middle Ages and the Renaissance. This has been done only recently. It is one of the novel features of a systematic revaluation of Leonardo's anatomical work by science historian Domenico Laurenza.[18] In his book *Leonardo: L'anatomia*, published in 2009, Laurenza carefully reconstructs how Leonardo moved among the leading physicians and anatomists of his time, how he studied the classical texts, was influenced by their anatomical visualizations, and then revolutionized them.

The integration of art and science—one of the defining characteristics of Leonardo's entire oeuvre—is apparent already in his earliest anatomical studies. According to Laurenza, the young Leonardo encountered two distinct anatomical schools that had been brought forth by the Italian Renaissance—the "anatomy of the artists" and the "anatomy of the doctors." The former was centered in Florence, the latter in Milan and Pavia. Leonardo maintained contacts with both schools, was influenced by both, and integrated their different approaches in his science and his art.

Many Florentine artists in the Renaissance had a keen interest in the anatomical study of muscles, which sometimes included dissections, so as to portray the gestures and movements of the human body—especially those of heroic, muscular figures—in realistic and expressive ways. Whereas Michelangelo was the outstanding artist of that school, a very influential early representative was the painter and sculptor Antonio del Pollaiolo, whose studies and paintings of muscular nudes were widely copied and became important models for other artists.[19] Pollaiolo's workshop was not far from the *bottega* of Andrea del Verrocchio where Leonardo received his training,[20] and it is possible that the young Leonardo observed dissections being carried out by Pollaiolo.[21]

Studies of the superficial anatomy of muscles were also an integral part of Leonardo's apprenticeship. In Verrocchio's *bottega*, plaster models of human limbs were made for that purpose, along with direct observation of the body's musculature. Leonardo demonstrated his considerable knowledge of the surface anatomy of muscles in one of his first paintings, the *Saint Jerome* (1480), in which the ascetic saint's pain and sorrow are powerfully expressed in the tense musculature of his neck and shoulder.[22]

The "anatomy of the artists" was largely limited to these superficial studies of muscles for portrayals of bodily gestures and movements. The necessity of anatomical studies for artists had already been emphasized by Leon Battista Alberti, a "universal man" and great idol of the young Leonardo.[23] In his book *De pictura* (On Painting), published in 1435, Alberti explained that bodily movement was the result of the concerted actions of nerves, muscles, tendons, and bones, and he encouraged artists to study all of these anatomical parts. However, before Leonardo none of the Florentine artists followed Alberti's exhortation, limiting their studies largely to external examinations of muscular or emaciated bodies.

Leonardo's contacts with the "anatomy of the doctors" also began during his youth in Florence. At the age of twenty, he had completed his apprenticeship, was recognized as a master painter, and was admitted to the guild of painters known as Compagnia di San Luca. Curiously, this guild also included physicians and apothecaries and was based in the hospital of Santa Maria Nuova. For Leonardo, this was the beginning of a long association with the hospital. For many years he used the guild as a bank for his savings, and his frequent visits to Santa Maria Nuova provided ample opportunities for him to mingle with some of Florence's leading physicians and anatomists.

It was customary at the time to perform autopsies in order to ascertain the cause of death in many cases.[24] The city of Florence was a leader in these practices, which were also occasions for new anatomical knowledge, and many of these post-mortems were carried out at Santa Maria Nuova. Leonardo very likely observed some of them at the hospital, and we know that he practiced dissections there himself when he was in his mid-fifties.

Leonardo's engagement with the anatomy of the doctors intensified after his move to Milan in 1482 at the age of thirty. At the Sforza court, he encountered a culture that was much more intellectual than artistic.[25] It was linked to the great universities of northern Italy, especially to the University of Pavia, which housed one of the leading medical schools of the time.

Leonardo adapted brilliantly to Milan's intellectual culture. Soon after his arrival, he embarked on an intense program of self-education involving systematic studies of the major fields of knowledge of his time. Contacts with scholars at the University of Pavia played an important role in this endeavor. Several medical scholars and anatomists lent him medical texts, provided him with explanations, and taught him how to dissect.

In Milan, Leonardo's anatomical studies went far beyond the superficial anatomies of the artists, but eventually he would integrate both approaches, verifying by means of dissection what he observed externally. The result was a series of "mixed" anatomical drawings (fig. 6-4), which he called images "between anatomy and the living."[26]

The Classical Medical Texts

During the Italian Renaissance, medical teaching at the great universities was based on the classical texts of Hippocrates, Galen, and Avicenna.[27] Most professors interpreted the classics without questioning them or comparing them with clinical experience. Practicing physicians, on the other hand, many of them without medical degrees, used their own eclectic combinations of therapies, including bloodletting and surgeries, which were often arranged according to the astrological calendar.

Leonardo carefully studied the classical medical texts, but he differed dramatically from most other Renaissance scholars by refusing to blindly accept the pronouncements of the classical authorities. In his anatomical studies, he generally began with summaries of their teachings, often transforming confusing written descriptions into far more intelligible visual forms. He would then proceed to verify the classical texts with his own dissections and, in doing so, did not hesitate to depart from the authorities by correcting them and adding new anatomical structures he had discovered.

The medical texts studied by Leonardo included the three classics that were most popular and most easily available during his time: *De usu partium* (On the Usefulness of the Parts) by Galen, the *Canon of Medicine* by Avicenna, and the *Anatomia* by Mondino (see pp. 144–45). He also owned a copy of the *Fasciculus medicinae* (Medical Collection), the first illustrated collection of medical texts, which included Mondino's *Anatomia* in its Italian edition. In addition, Leonardo borrowed other texts on medicine and anatomy from scholars known to him. All in all, he mentions about a dozen such works in his Notebooks.[28]

Before the publication of the *Fasciculus medicinae* in 1491, medical texts were not illustrated, but in some of them a separate plate was inserted at the end of the treatise, sometimes by a subsequent reader of the text. Some

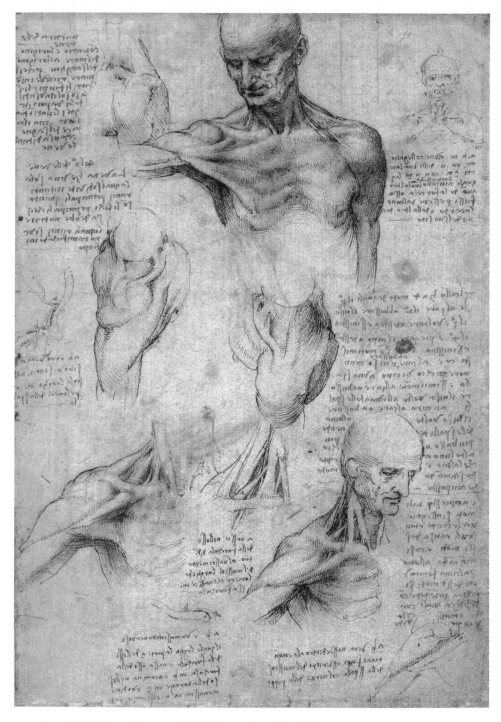

FIG. 6-4. Studies of the muscles of the neck and shoulder,
combining external observations and dissections, c. 1509–10.
Windsor Collection, *Anatomical Studies*, folio 137r.

of these plates of medical or anatomical illustrations were also available as loose leaflets, independent of the original treatise. Laurenza reproduces and discusses several of these early medical illustrations in his book. One of them, created by an anonymous Florentine artist in the 1470s and inserted at the end of a handwritten copy of Mondino's *Anatomia*, shows a human figure surrounded by lines of text (fig. 6-5). This plate exhibits several features that are typical of the medical illustrations of the time.

The text refers to the therapeutic practice of bloodletting, which was often prescribed by doctors but executed by barbers. The written lines surrounding the figure indicate which veins should be cut for specific illnesses. In addition, various parts of the body of the figure are labeled with the names of the signs of the zodiac, referring to the astrological context of surgical procedures.

From the way the notes are arranged around the figure or superimposed on it, it is evident that the text is more important. The figure serves merely as a schematic reference to aid the memory. Laurenza also points out that, even though the figure is rendered realistically and is not without artistic merit, it adds no information to the notes about bloodletting in the text. There is no representation at all of veins corresponding to the written instructions on how to cut them.

Leonardo revolutionized this relationship between text and figure in his anatomical drawings by giving priority to the visual image. Even a cursory look through the anatomical folios of the Windsor Collection makes it evident that his main focus is on the image. The accompanying text is secondary, often limited to explanatory captions, and is sometimes absent altogether. Indeed, he believed that his visual anatomical representations were more efficient than any written descriptions. He proudly asserted that they gave "true knowledge of [various] shapes, which is impossible for either ancient or modern writers . . . without an immense, tedious and confused amount of writing and time."[29] Leonardo's anatomical drawings were so radical in their conception that they remained unrivaled until the end of the eighteenth century, nearly three hundred years later.

A plate of the anatomical organs of a pregnant woman, created by an anonymous illustrator at the end of the fifteenth century (fig. 6-6), provides another instructive opportunity to highlight the dramatic differences with Leonardo's anatomical drawings. Again, the text invades the figure, whose schematic anatomical features cannot be understood without the explanatory captions superimposed on them. The contrast with

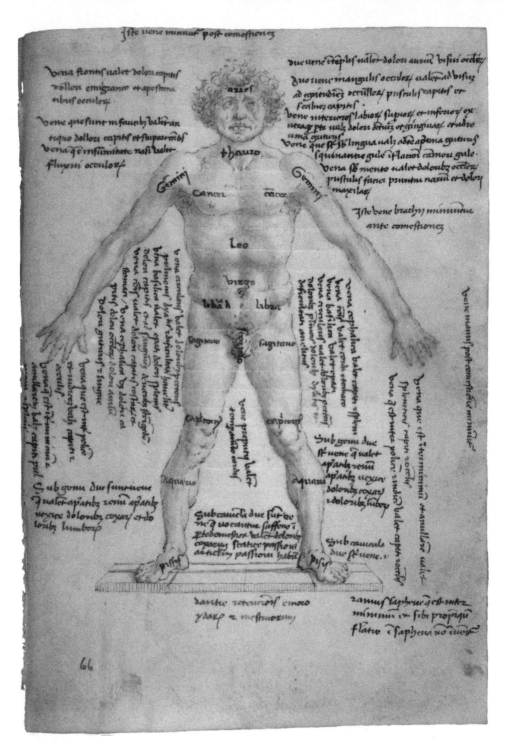

FIG. 6-5. Anatomical figure by anonymous
Florentine artist, fifteenth century.
Biblioteca Nazionale, Florence, Ms. CSP.X.42, folio 66r.

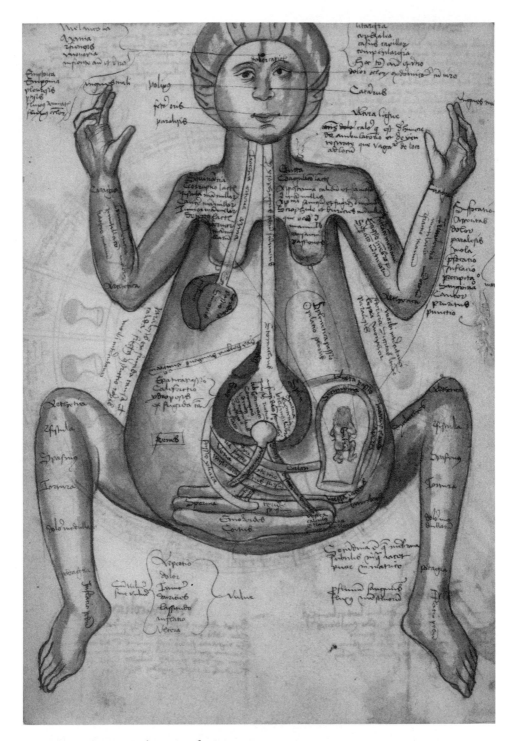

FIG. 6-6. Anatomical organs of a pregnant woman
by anonymous illustrator, end of fifteenth century.
Biblioteca Apostolica Vaticana, Ms. Pal. Lat. 1325, folio 349v.

Leonardo's famous representation of the same subject (see fig. 4-5), created around the same time, is stunning. So is the comparison between the symbolic representation of the fetus in fig. 6-6 with Leonardo's delicate and totally realistic drawings of the human fetus within the womb (plate 9).

A few years after the Latin edition of the *Fasciculus medicinae* was published, an Italian edition was printed in Venice with some significant changes to the illustrations. The figures are much more realistic and they are no longer invaded by the text. "All this seems to move in the direction taken by Leonardo," observes Laurenza, "except for the fundamental fact that Leonardo continues to work within the manuscript culture, innovating it but remaining its prisoner, while the [publishers of the *Fasciculus*] operate in a new world: that of the print culture."[30]

Leonardo was well aware of the great advantages of the printed image for exact replication and rapid dissemination, and throughout his life he was keenly interested in the technical details of the printing process.[31] He envisioned that his treatises would eventually be printed, and he insisted that his anatomical drawings should be printed from copper plates, which would be more expensive than woodcuts but much more effective in rendering the fine details of his work. "I beg you who come after me," he wrote on the sheet that contains his magnificent drawings of the vertebral column (see fig. 6-8), "not to let avarice constrain you to make the prints in [wood]."[32] However, the printing of his most finished anatomical drawings—with ultra-fine hatchings and in some cases with added watercolors—would have required highly sophisticated techniques of etching and other innovative techniques that were not available during his lifetime.[33]

Anatomical Drawings and Dissections

The reversal of the relationship between text and figure is not the only revolutionary aspect of Leonardo's anatomical drawings. In order to present the features of the human body accurately and realistically—"just as if you had the natural man before you" (see p. 133)—he introduced numerous innovations: drawing anatomical structures from several perspectives; presenting them in "exploded views" to illustrate how the parts of an ensemble (for example, the vertebrae of the spine) fit into one another; showing the removal of muscles in successive layers to expose the depth of an organ or anatomical feature, and so on. None of his predecessors or contemporaries came close to him in such anatomical detail, accuracy, and sophistication.

When he pictured muscles, tendons, and bones, Leonardo's main intention was always to show how they work together in the movements and activities of the entire body. To achieve this purpose, it was of paramount importance to him to demonstrate the three-dimensional relationships between various structures, so he repeatedly showed them from various perspectives. As he explained,

> If you want to know thoroughly the anatomical parts of man, you must either turn him, or your eye, in order to examine him from different aspects; from below, from above, and from the sides, turning him round and investigating the origin of each part; and by such a method your knowledge of natural anatomy is satisfied.[34]

This passage is part of a long sequence of paragraphs in which Leonardo sets out his ambitious plans for his magnum opus *De figura umana* (On the Human Figure; see p. 133). Having explained the need for presenting anatomical structures from several perspectives, he then describes how he would do that for each individual part:

> Therefore, through my plan you will come to know every part and every whole through the demonstration of three different aspects of each part. For when you have seen any part from the front with some nerves, tendons, and veins that arise from the side in front of you, the same part will be shown to you turned to its side or its back, just as though you had the very same part in your hand and went on turning it round from one side to another until you had obtained full knowledge of what you want to know.[35]

We do not know how much of this ambitious program Leonardo was able to carry out. But among the two hundred folios of anatomical drawings that have come down to us, there are many in which he demonstrates his technique of showing anatomical structures from different perspectives (for example, figs. 6-7, 6-8, 6-9, and 6-10). Sometimes these different perspectives form such a smooth sequence that they suggest continuous movement, almost like images on a strip of film (see plate 4).

Leonardo's sophisticated anatomical drawings were based on a large number of dissections of human and animal bodies, which he carried out

with the most delicate care and attention to detail, "taking away in its minutest particles all the flesh" to expose blood vessels, muscles, or bones
until the corpse's state of decay was too advanced to continue. "One single
body was not sufficient for enough time," he explained, "so it was necessary
to proceed little by little with as many bodies as would render the complete knowledge."[36] And even that long and difficult procedure was not
enough to satisfy Leonardo's high scientific standards. "This I repeated
twice in order to observe the differences," he tells us laconically.

On the same page, Leonardo also gives us a vivid account of the dread-
ful conditions under which he had to work. As there were no chemicals
to preserve the cadavers, they would begin to decompose before he had
time to examine and draw them properly. To avoid accusations of heresy,
he often worked at night, lighting his dissection room by candles, which
must have made the experience even more macabre. "You will perhaps be
impeded by your stomach," he writes, addressing an imaginary apprentice,
"and if this does not impede you, you will perhaps be impeded by the fear
of living through the night hours in the company of these corpses, quar-
tered and flayed and frightening to behold."[37]

It was Leonardo's passionate desire for scientific knowledge that gave
him the strength he needed to overcome his own aversion. His emotional
struggle is eerily reminiscent of a celebrated passage he wrote when he was
in his late twenties. It is one of the very few passages in the Notebooks
where he reveals his emotions, and its poetic language and symbolic nature have led some authors to speculate that it may refer to a childhood
memory or a dream:

> Drawn by my ardent desire, impatient to see the abundant variety
> of strange forms created by artistic nature, and having wandered
> for some time among the dark rocks, I reached the entrance of a
> great cavern, before which I remained for a moment, stupefied and
> unfamiliar with such a thing. I folded my loins into an arch, leaned
> my left hand on my knee, and with my right I sheltered my lowered,
> knitted brows; and I leaned repeatedly from one side to the other to
> see if I could discern anything inside. But the great darkness which
> reigned there did not let me. Having remained thus for some time,
> two things suddenly arose in me, fear and desire: fear of the menac-
> ing dark cave, and desire to see if there was some miraculous thing
> inside it.[38]

Leonardo's anatomical dissections, too, were driven by ardent desire. Overcoming his natural repugnance, he repeatedly entered the dark cave of the dissection room and returned with many "miraculous things" that would forever establish his fame as the greatest anatomist of his time.

In spite of the distasteful conditions under which Leonardo performed his anatomies, he never lost sight of the human dignity of the corpses he dissected. This is evident from his famous account of how, during one of his visits to the hospital of Santa Maria Nuova, he met an old man who then died in his presence:

> And this old man, a few hours before his death, told me that he was over a hundred years old and that he felt nothing wrong with his body other than weakness. And thus, while sitting on a bed in the hospital of Santa Maria Nuova in Florence, without any movement or other sign of any mishap, he passed out of this life.—And I made an anatomy of him in order to see the cause of so sweet a death.[39]

Leonardo's post mortem of the "centenarian" became a milestone in his work and led him to some of his most important medical discoveries (see p. 306). At the same time, the story is a moving testimony to the deep humanity of a great scientist.

In the long passage where he describes his techniques of dissection, Leonardo states that he had "dissected more than ten human bodies," and when he lived in Amboise in his old age, he told a visitor that he had "made anatomies of more than 30 bodies, male and female of all ages."[40] In addition, Leonardo dissected numerous bodies of animals, which were much easier to obtain, and he often transferred what he learned from these animal dissections to the human body. Laurenza points out that the resulting "hybrid anatomical representations" are often regarded by scholars as inaccuracies and signs of immaturity, but that a different interpretation may also be possible:

> These representations certainly reflect the difficulties, in those early years, of carrying out human dissections. However, Leonardo does not conceal the animal origin of his notions . . . thus emphasizing the animal aspects of the human anatomy.[41]

What emerges eventually from these conceptual developments is Leonardo's assured and accurate realization that the human body is an animal body.

The vast corpus of Leonardo's anatomical studies has been analyzed in impressive detail by several scholars of medical history. In addition to the recent reconstruction and revaluation by Laurenza, my main sources have been two standard volumes: *Leonardo da Vinci's Elements of the Science of Man* by Kenneth Keele and *Leonardo da Vinci on the Human Body* by C. D. O'Malley and J. B. Saunders.[42] In the following pages, I can only review the most outstanding of Leonardo's achievements in anatomy.

Bones and Joints

The human skeleton was, naturally, much easier to examine than the body's internal tissues and organs, which would have lost their form and structure soon after their dissection. Leonardo's renderings of bones and joints are among his most accurate, most finished, and most beautiful anatomical drawings. Indeed, they have been praised by Keele as extraordinary, even when compared with today's anatomical representations:

> Of all Leonardo's achievements those of his drawings of the human skeleton and muscles have perhaps been accepted as the most outstanding. This aspect of his exploration of the human body, supplemented by his unique capacity for presenting movements in visual perspectival form, make his drawings of bones live in a way seldom achieved by modern anatomical illustration.[43]

A magnificent folio in the Windsor Collection (fig. 6-7) shows Leonardo's nearest approach to drawing a complete skeleton. In a group of six drawings, several ensembles of bones and joints are displayed from the back, front, and side. The pelvis is shown for the first time in its accurate shape and place, correctly angled in relation to the spine and thigh-bones. The thorax, lumbar spine, and knee joints are shown with remarkable accuracy. According to Keele, "These drawings convey deep understanding of the statics [and] the transmission of weight in the human erect posture."[44]

Having demonstrated how the various parts of the skeleton fit together, Leonardo then proceeded to study them individually and in pairs on several other folios. These include studies of the vertebral column, the pelvis and legs, and the arms and hand, as well as the elbow and the knee and ankle joints.[45] All of these studies show the relevant bones and joints from several perspectives (front, side, and back) and in several positions: elbows bending and stretching, forearms rotating, legs in standing and kneeling positions—all of them precise analyses of the mechanics of bodily movement.

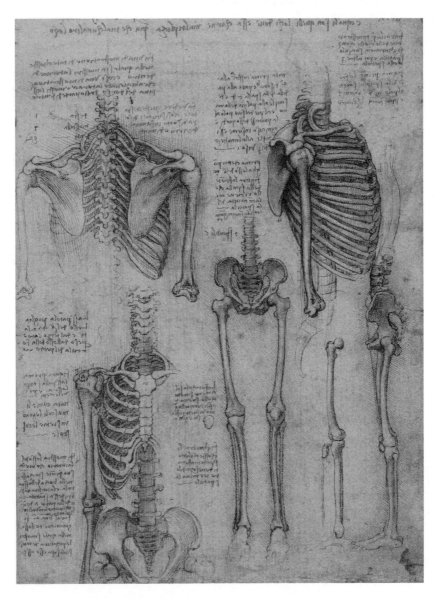

FIG. 6-7. The human skeleton, c. 1509–10.
Windsor Collection, *Anatomical Studies*, folio 142r.

On the celebrated folio of studies of the vertebral column (fig. 6-8), Leonardo demonstrates with wonderful accuracy how the vertebrae of the spine interlock to form a flexible column. The first drawing on the right (where Leonardo, writing and sketching from right to left, usually starts the page) shows the whole spine from the front. The five sets of vertebrae (cervical, thoracic, lumbar, sacral, and coccygeal in modern medical terminology) are clearly marked with letters, and in the text they are correctly enumerated in a little table, adding up to thirty-one in all. In addition,

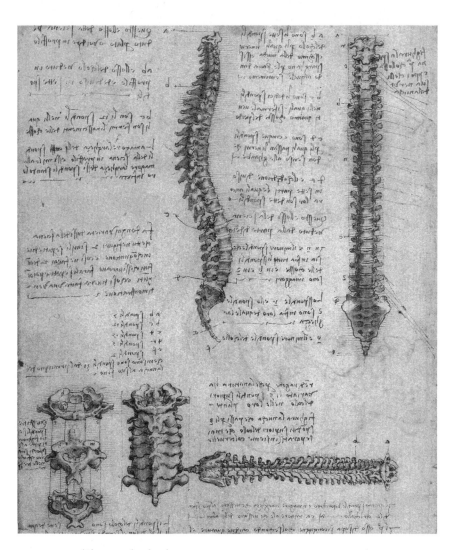

FIG. 6-8. The vertebral column, c. 1509–10.
Windsor Collection, *Anatomical Studies*, folio 139v.

Leonardo notes precisely where the sensory nerves going to the arms and legs emerge from the spinal cord.

The second drawing shows a lateral view of the vertebral column in which the spinal curvature is represented with complete accuracy—a feat unmatched by any other Renaissance anatomist. At the bottom of the page, the spine is depicted from behind, and to the left of this view the seven cervical vertebrae are shown with such precision that they can easily be identified by a modern medical practitioner. In addition, Leonardo depicts the first three cervical vertebrae in an "exploded view" in which the complex interlocking mechanisms of the joints, marked with letters, are clearly recognizable.

These drawings represent not only a tremendous scientific achievement but are also artistic masterpieces. Leonardo's precise renderings, together with his mastery of perspective and of light and shade, combine to convey a sense of complexity, elegance, and beauty unequalled in contemporary or modern scientific illustrations. No wonder he allowed himself a rare outburst of pride at the bottom of the page, stating that his presentations of the vertebrae "will give true knowledge of their shapes, knowledge that neither ancient writers nor the moderns would ever have been able to give without an immense, tiresome, and confused amount of writing and time."

In another beautiful set of drawings Leonardo presents his studies of the bones of the hand (fig. 6-9), employing delicate hatchings, two shades of brown ink, and wash over traces of black chalk. These studies were the first to show the bones of the wrist and hand accurately, displaying them from the front, the back, and from both sides. All bones are clearly labeled and enumerated in the accompanying text. In two sketches on the right margin, the bones of the fingers are covered by the flexor and extensor tendons, arteries, veins, and nerves—all labeled with letters and identified in the text below. The paragraph at the bottom of the page is devoted to a description of the movements of the hand at the wrist.

The text at the top of the page makes it clear that Leonardo intended this set of studies to be merely the first of a series of eight systematic demonstrations of the hand:

The first demonstration of the hand will be made of the bones alone. The second of the ligaments and the various chains of tendons which bind them together. The third will be of the muscles that arise from these bones; [the fourth and fifth] of the . . . cords

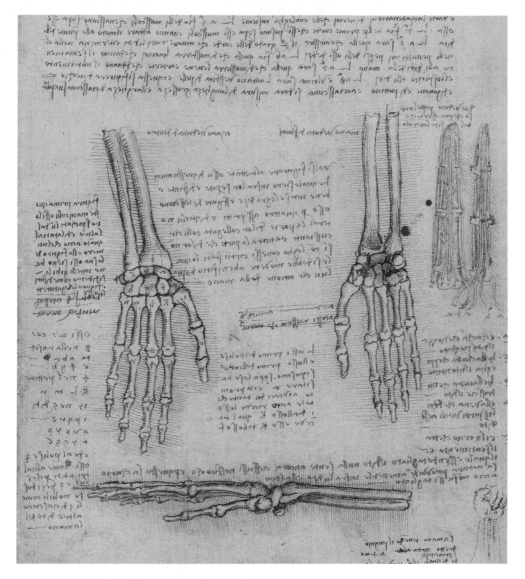

FIG. 6-9. Studies of the bones of the hand, c. 1509–10.
Windsor Collection, *Anatomical Studies*, folio 143v.

that move all the fingers. . . . The sixth will demonstrate the nerves
that give sensation to the fingers of the hand. The seventh will show
the veins and arteries that give nourishment and [vital] spirit to the
fingers. The eighth and last will be the hand clothed with skin, and
this will be illustrated for an old man, a young man, and a child.[46]

On another folio, he envisages an even larger set of forty demonstrations of the hand, concluding his long description with the ambitious statement:

> And thus in the chapter on the hand you will give forty demonstrations; and you should do the same with each limb. And in this way you will give full knowledge.[47]

We do not know how much of this grandiose project Leonardo was able to execute, since many of his anatomical drawings were lost. The Windsor Collection contains a few more detailed studies of the hand, as well as similar studies of the foot, again with all of its bones enumerated and shown in perfect anatomical shapes.[48] However, Leonardo himself acknowledged that it was unlikely that he would ever complete his anatomical studies in the systematic manner he envisaged. "In these I have been impeded," he wrote wistfully, "neither by avarice or negligence, but only by time."[49]

Muscles and Tendons

In Leonardo's scheme of systematic anatomical demonstrations "from within,"[50] the muscles and tendons are displayed in the second and third sets of dissections, after the demonstrations of the bones. In this endeavor, his goal was always to understand how muscles, tendons, and bones work together to bring about the movements of the body. As Keele puts it, "Leonardo was, one may say, dissecting the movements of the human body rather than a motionless cadaver."[51]

"There are six things that come together in the composition of movements," Leonardo explains in his *Anatomical Studies*, "that is, bone, cartilage, membrane, tendon, muscle, and nerve."[52] On another nearby page, he describes the nature and the functions of this sequence of anatomical structures in more detail:

> Tendons are mechanical instruments . . . which carry out as much work as is assigned to them [by muscles]. Membranes are joined to the flesh, being interposed between the flesh and the nerve. . . . Ligaments are joined to tendons and are a kind of membrane which bind together the joints of bones, and are converted into cartilage. They are as numerous in each joint as are the tendons which move

the joint and as the tendons opposite, which come to the same joint. And such ligaments are all joined and mixed together, aiding, strengthening, and balancing one another.[53]

These descriptions are remarkably accurate. The terms used by Leonardo are still used in medical science today.*

Implicit in the last two sentences of Leonardo's description is the recognition that pairs of antagonistic muscles keep the body in balance. On another folio, he describes these "pairs of muscles which are in opposition to one another" in more detail and discusses their importance for the maintenance of body posture.[54] Medical historian Sherwin Nuland has noted that the action of muscle groups acting reciprocally with antagonistic groups was not recognized again until the early twentieth century.[55]

In the long passage where Leonardo sets out his plans for multiple sets of dissections, he also discusses the tremendous difficulties encountered in dissecting the soft tissues of the body. He vividly describes "the very great confusion that results from the mix-up of membranes with veins, arteries, nerves, tendons, muscles, bones, and blood which itself dyes every part the same color; and the vessels emptied of this blood are not recognizable because of their diminished size; and the integrity of the membranes is broken in searching for those parts which are enclosed within them."[56] In view of the delicate and ephemeral nature of these tissues in the absence of preservatives, the clarity and accuracy of so many of Leonardo's anatomical drawings are all the more astounding.

To demonstrate the actions of muscles and tendons as "mechanical instruments," Leonardo developed an effective method for converting complex muscular patterns into simplified geometrical diagrams.[57] To begin with, he identifies the "central line" of action (known today as the "line of pull") for muscles of various shapes. Then he introduces a highly ingenious technical innovation: "Make a demonstration of the fine muscles by using rows of threads," he explains on the folio showing the anterior muscles of the leg (see fig. 4-7). "Thus you will be able to represent one upon the other, as nature has placed them; and thus you will be able to name

---

* The structures referred to as "membranes" by Leonardo are called "bursae" today. They are fluid-filled sacs, lined by so-called synovial membranes, which provide cushions between bones and tendons around a joint.

La sapienza è figliola della sperienza. (Codex Forster III, folio 14r)

Wisdom is the daughter of experience.

them according to the part they serve ... And having given such knowledge, you will draw alongside this the true shape, size, and position of each muscle."[58]

He makes a special point of drawing cords and not just lines: "When you have drawn the bones of the hand and wish to draw on this the muscles which are joined to these bones, make threads instead of muscles. I say threads and not lines in order that one should know which muscle goes below or above another muscle, which cannot be done with simple lines."[59] This note is found on the folio showing studies of the mechanisms of the hand that illustrate this technique (see fig. 4-2). The passage of the five flexor tendons under the transverse carpal ligament is clearly visible.

Finally, having replaced muscles and tendons by cords in some of his drawings, Leonardo proposes to construct a physical model of the resulting diagrams of muscular forces. He illustrates this last step of his technique with the muscles of the leg: "Make this leg in full rounded relief [that is, like a statue] and make the cords of tempered copper wires, and then bend them according to their natural form. After doing this you will be able to draw them from four sides, and to place them as they exist in nature and speak about their functions."[60] On the folio that contains these instructions (fig. 6-10), the resulting wire diagram of the leg muscles is shown from three sides.

Medical historians O'Malley and Saunders have identified a number of leg muscles in these cord diagrams, but they also point out that the drawings contain many inaccuracies.[61] Nevertheless, Leonardo's whole approach to the mechanical analysis of complex muscular patterns shows an unprecedented level of scientific and mathematical reasoning.

In his studies of the human muscular system, Leonardo concentrated his greatest efforts on the muscles of the shoulder and arm, the arm and hand, and the leg and foot. These are also the areas of myology (the study of muscles) in which he produced his most beautiful illustrations. The

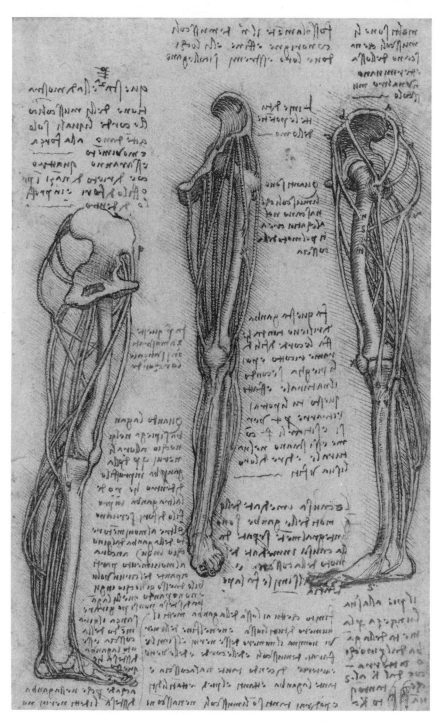

FIG. 6-10. Cord diagrams of the muscles of the leg, c. 1509–10.
Windsor Collection, *Anatomical Studies*, folio 152r (detail).

Windsor Collection contains about a dozen folios with studies of the shoulder region, among them the celebrated rotated views of the muscles of the shoulder and arm (see plate 4). This folio is preceded in the Collection by one showing another four sequential views.[62] The two folios together present a continuous series in which the body is turned through 180 degrees, from front to back (when viewed from right to left in the direction of Leonardo's writing and sketching). The result is an almost cinematographic display of muscle action during the rotation of the shoulder. In the drawings, various muscles are labeled with letters and described in the accompanying text, and even those without letters can easily be identified by a modern anatomist owing to their accurate representation.[63]

From a scientific point of view, Leonardo's studies of the deep structures of the shoulder are perhaps even more remarkable. In the four drawings shown in figure 6-11, for example, he employs several of his innovative graphic techniques. In the top drawing, the deltoid and trapezius muscles are lifted off to expose the underlying deep structures of the shoulder joint, all of which are so clearly rendered that they can easily be identified.[64] In the left drawing, the collar-bone is separated in an exploded view to show how it forms a joint with part of the shoulder blade. In the drawing at the bottom, all the muscles have been cut away so as to clearly expose the deep structures. The middle drawing on the right, finally, is a cord diagram of the shoulder muscles demonstrating their pull lines.

With his skillful use of these graphic techniques, Leonardo is able to demonstrate the spatial extensions and mutual functional relationships of the complex anatomical structures of the shoulder with compelling clarity. The lines of forces, parts labeled with letters, and other parts joined by guide lines make these drawings almost like a set of mathematical diagrams—geometric representations of functional anatomical relationships.

One of Leonardo's most finished studies of bone structures, drawn in the same exquisite style as his studies of the bones of the hand (see fig. 6-9), is his demonstration of the rotation of the arm (plate 3). The principal purpose of this series of five drawings is to demonstrate the mechanisms by which the palm of the hand is turned upward and downward, "toward the sky" and "toward the earth," as Leonardo puts it. It is one of Leonardo's many investigations of how bodily movements are generated by the actions of muscles and tendons.

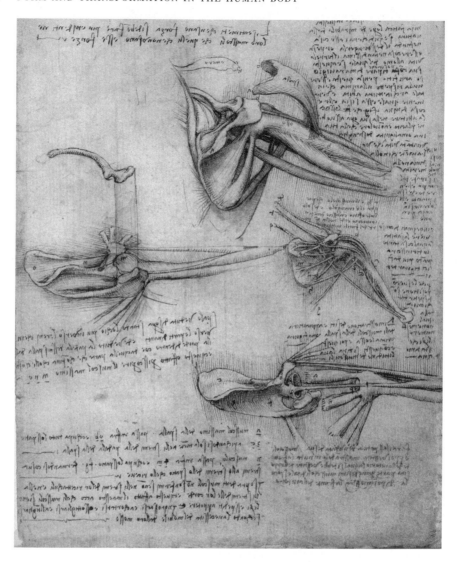

FIG. 6-11. Deep structures of the shoulder, c. 1509–10.
Windsor Collection, *Anatomical Studies*, folio 136r (detail).

The top drawing shows the outstretched arm with the correct proportional lengths of the bones. The biceps, the principal object of this study, is shown with its two heads joining halfway down the humerus (the bone of the upper arm). The muscle is depicted without its tendon, but the tendon's point of attachment to the radius (the smaller of the two forearm bones), just below the elbow, is clearly shown as a protruding band. The second drawing presents an exploded view in which the shape of the shoulder joint and the interlinking of the bones forming the elbow joint are demonstrated with great precision.

Leonardo discovered that the biceps not only bends the elbow but helps turn the palm upward by rotating the top of the radius. This is demonstrated in the third drawing. The two heads of the biceps are shown once more at their origin, and they are cut at point "b" ("d" in Leonardo's mirror writing) just before joining the tendon. The tendon's attachment to the radius is shown again, labeled with the letter "a" ("ʌ"). In this position, known as supination by modern anatomists, the two bones of the forearm, radius and ulna, lie parallel.

When the palm is turned downward in the position known as pronation, the rotated radius crosses the ulna. This is clearly demonstrated in the fourth drawing. Because of the crossing of the two bones, the forearm is slightly shortened, as Leonardo explains correctly in the accompanying text:

> The arm which has two bones interposed between the hand and the elbow will be somewhat shorter when the palm of the hand faces the earth than when it faces the sky, when a man stands on his feet with his arm extended. And this happens because these two bones, in turning the palm of the hand to the earth, become crossed.[65]

In the last drawing of this study, Leonardo analyzes the muscles that produce pronation. The biceps is now presented together with its tendon, and the so-called pronator muscle, which rotates the radius in the opposite direction to the biceps, is also shown. The pronator is accurately rendered as it originates at the end of the humerus (in front in the drawing) and runs like a strap to the top of the radius.*

It is worth noting that Leonardo accurately demonstrates the actions of the biceps and pronator muscles in these drawings but does not describe them in the accompanying text, confident that his visual demonstrations were more effective than any words could be.

On a double folio in the Windsor Collection we find another celebrated anatomical study, showing the anterior muscles of the leg and foot (see

---

* The full technical name of this muscle is pronator teres. The other pronator muscle, known as pronator quadratus, is depicted in the third drawing as a square band joining the two forearm bones close to the wrist (see Keele, *Leonardo da Vinci's Elements of the Science of Man*, p. 270).

fig. 4-7). Executed in three shades of brown ink and chalk with copious blocks of text arranged beautifully around it, the drawing shows a level of sophistication and elegance rarely, if ever, achieved in modern anatomical illustrations.

In the drawing, the muscle bellies of the short extensor muscles on the front of the foot (labeled a, b, c, d) are admirably rendered, and so are the long extensor muscles on the front and side of the shin (labeled f, n, r, s, t), which pull the foot and toes upward. All in all, about a dozen muscles can readily be identified.[66] In the text, Leonardo offers detailed descriptions of the synergistic actions of these muscles in producing the complex movements of the foot. In the lower right corner of the double page, almost hidden in the text, there is an exquisitely accurate drawing of the way the extensor tendon is inserted into the back of the big toe. According to O'Malley and Saunders, "nothing approaching such detail is to be found until comparatively recent times."[67]

Muscles and Nerves

In his studies of the body in motion, Leonardo traced various bodily movements back from the bones and joints, and the tendons attached to them, to the contraction of specific muscles; he also investigated the nerve impulses that trigger muscle contractions, following them through the motor nerves and the spinal cord into the brain, where he believed he had located the seat of the soul. His concise summary of this pathway of the body's motor forces is worth repeating (see p. 212):

> Spiritual movement, flowing through the limbs of sentient animals, broadens their muscles. Thus broadened, these muscles become shortened and draw back the tendons that are connected to them. This is the origin of force in the human limbs. . . . Material movement arises from the immaterial.[68]

The investigation of the nervous system was the last step in Leonardo's systematic demonstrations of the body in motion. Chronologically, however, it was where he began his anatomical work. I have discussed Leonardo's detailed explorations of the cranial nerves and his neurological theory of perception and knowledge in my previous book.[69] To summarize, he asserted that the sensory nerves carry all sense impressions to the brain, where they are selected and integrated before entering consciousness at the central cerebral ventricle (the "seat of the soul") to be judged by the intellect, influenced by the imagination, and then partly committed to memory.

Leonardo made a distinction between sensory and motor nerves, and he followed the nerve impulses for voluntary movement from the brain down the spinal cord and through the peripheral motor nerves to the muscles, tendons, and bones. His elaborate explorations of the anatomy of the nervous system began around 1489, when he was in his mid-thirties, and reached their peak about twenty years later, during the time he also produced the superb studies of muscles and bones I have just discussed.

Leonardo had an integrated view of the soul, seeing it both as the agent of perception and knowledge and as the force underlying the body's formation and movement.[70] The nervous system, accordingly, integrated all the movements and activities of the body. The sensory nerves carried sense impressions to the soul, and the motor nerves transmitted forces from the soul to the limbs, where the nerves, "having entered into the muscular fibers, command them to move."[71]

In a schematic drawing produced around 1508, Leonardo demonstrated how all these nerves are interconnected. It shows the entire nervous system: the brain and spinal cord, and the peripheral nerves to the trunk, the arms, and the legs (fig. 6-12). He presented this scheme from the front and from the left side. Between the two figures there is an explanatory note: "Tree of all nerves; and it is shown how all these have their origin from the top of the spinal cord, and the spinal cord from the brain."[72] A small sketch, also between the two figures, shows another diagrammatic outline of the nervous system. The note underneath reveals that the two figures shown on this folio were only the beginning of a much more ambitious project: "In each demonstration of the entire extent of the nerves, draw the external outlines, which denote the shape of the body."[73]

We do not know whether Leonardo ever carried out his plan, but since he produced an integral view of the skeleton (see fig. 6-7) and one of the internal organs (see fig. 4-5) around the same time, it is not unreasonable to assume that he also presented the entire nervous system on a single folio. At any rate, no such drawing has come down to us. However, the Windsor Collection contains numerous detailed studies of various parts of the nervous system.

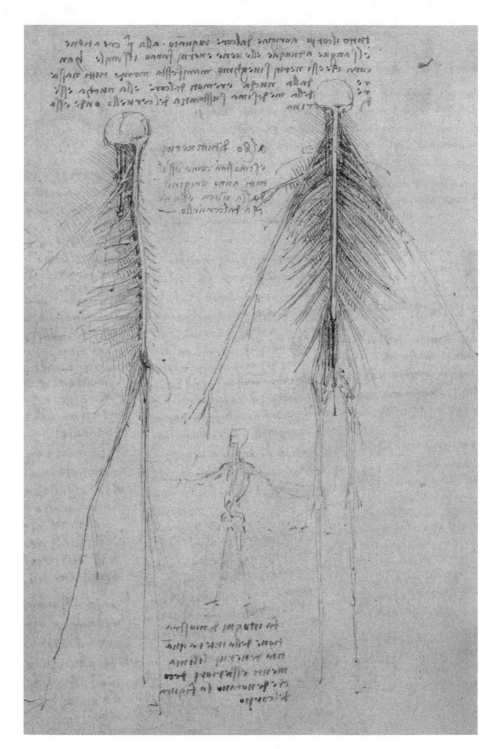

FIG. 6-12. "Tree of all nerves," c. 1506–8.
Windsor Collection, *Anatomical Studies*, folio 76v.

The early studies were largely based on dissections of animals, which were then projected onto human outlines. But during the period when he drew his "tree of nerves," Leonardo produced several accurate studies of the cranial and intercostal nerves, as well as the nerves of the shoulder, hand, and leg. Among them is a composite drawing (fig. 6-13) based partly on animal dissections, in which the curvature of the spine, the intercostal muscles, and the pelvic vessels and nerves are clearly and beautifully rendered.[74]

One area that fascinated Leonardo for two decades is the so-called brachial plexus, a complex bundle of nerves that forms the neural connections between the neck and the nerves of the arm. His early studies of this nerve complex, based on dissections of monkeys, date from 1487 and are among his very first anatomical drawings. The studies reached their climax more than twenty years later with a drawing that accurately represents the brachial plexus in all its complexity (fig. 6-14).

Above the main drawing on this folio, Leonardo wrote

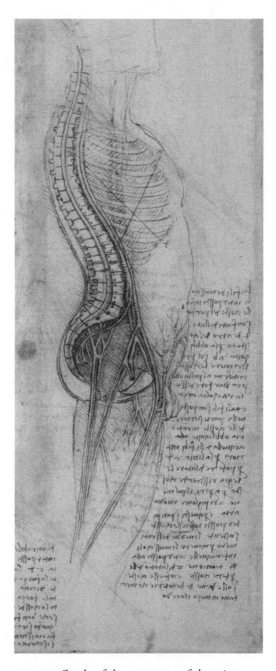

FIG. 6-13. Study of the curvature of the spine, the intercostal muscles, and the pelvic vessels and nerves, c. 1506–8. Windsor Collection, *Anatomical Studies*, folio 109v (detail).

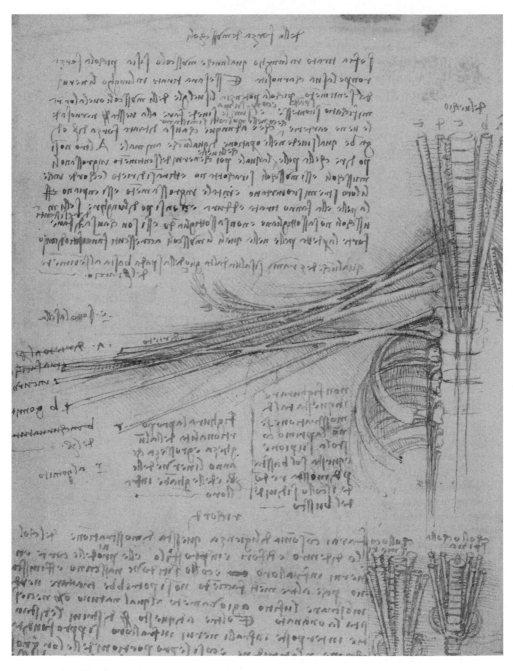

FIG. 6-14. The brachial plexus, c. 1508.
Windsor Collection, *Anatomical Studies*, folio 57v (detail).

"*del vecchio*," indicating that the dissection was performed on the "old man," the centenarian at Santa Maria Nuova in Florence. The study correctly depicts the brachial plexus as being formed by the lower four cervical nerves and the spinal nerve emerging from the first thoracic vertebra. Its extension from the lower part of the side of the neck to the underarm is clearly shown, and so is the complex pattern that arises as the five nerves repeatedly divide, merge with one another, and reunite. According to Kenneth Keele, "this whole map of the brachial plexus should be evaluated as a major discovery in Leonardo's [anatomical] explorations."[75]

After five hundred years, Leonardo's anatomical drawings have lost nothing of their splendor. We still admire their amazing accuracy and sophistication, and we are spellbound by their beauty. They are an enduring testimony to his genius, as both a scientist and an artist.

7

The Science of Flight

From the texts that accompany Leonardo's anatomical drawings we know that he considered the human body as an animal body, as biologists do today. He often transferred what he learned from numerous dissections of animals to the human body (see p. 227). But beyond these pragmatic aspects, Leonardo's anatomical studies of animals were grounded in a profound respect and compassion for all living creatures.[1] Thus it seemed natural for him, as Domenico Laurenza observes, "to give equal ontological and scientific dignity to humans and animals."[2]

Comparative Anatomy

Leonardo used his animal dissections to gain knowledge about human anatomical structures, but was also keenly interested in the many differences between the bodies of animals and human beings. "You will draw for this comparison," he wrote on a sheet showing the superficial muscles of a man's legs, "the legs of frogs which have a great similarity to the legs of man, in their bones as in their muscles. Then you will follow with the hind-legs of a hare which are very muscular and have agile muscles because they are not encumbered by fat."[3]

Leonardo's love of horses was well known to his contemporaries. He produced a wealth of magnificent studies of horses that are now assembled in a special volume of the Windsor Collection, and he is said to have written an entire treatise on the anatomy of the horse, now lost.[4] However, a folio in the Windsor Collection contains a superb study comparing the anatomy of the hind leg of the horse with that of the human leg (fig. 7-1). In both drawings, some of the hip muscles are represented by "cords" to show the exact lines of force. Moreover, and perhaps even more remarkably, the comparison between the bones of the lower leg and foot shows Leonardo's

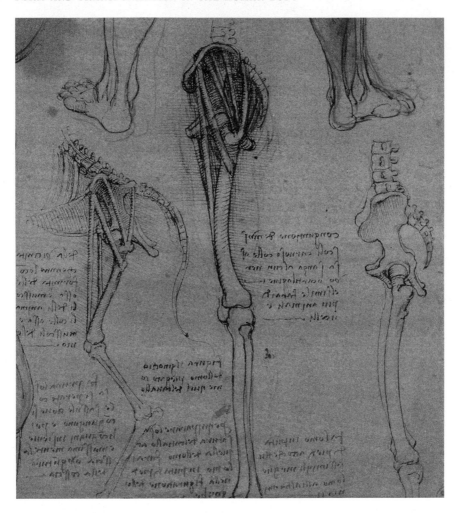

FIG. 7-1. Comparison of the human leg
with that of a horse, c. 1507. Windsor Collection,
Anatomical Studies, folio 95r (detail).

full appreciation of the fact that, compared with the human posture, the
horse stands on the tip of its toe.

In several of his comparative studies, Leonardo specifically contrasted
the limbs of various animals with those of the human body. In fact,
among his very first anatomical drawings there is an exquisite series of
four studies of a bear's foot in dissection.[5] Over the years, these comparative studies of animal limbs led Leonardo to the momentous conclusion
that their different structures should be seen as variations of a single
underlying theme:

All land animals resemble each other in their limbs, that is in muscles, sinews and bones, and these do not vary except in length and thickness.[6]

For Leonardo, the ability to recognize such anatomical similarities across a wide range of species constituted the very essence of "becoming universal" (*farsi universale*). This awareness of "universality" corresponds to what we call systemic thinking in contemporary science.[7]

Leonardo's "Evolutionary" Thought

Leonardo's statement about the similarities of the limbs of mammal species foreshadows a way of thinking that would re-emerge in biology in the eighteenth century and would lead, eventually, to the formulation of one of the cornerstones of modern biology: Charles Darwin's theory of evolution. The German school of Romantic biology, which included the great poet, dramatist, and scientist Johann Wolfgang von Goethe, recognized, as Leonardo had three hundred years earlier, a unity of patterns underlying the differences in animal shapes and sizes. Goethe and other biologists and philosophers of this school saw these patterns as manifestations of fundamental organic types, called "archetypes" (*Urtypen*).[8] This concept had a tremendous influence on biological thought in France and England during the nineteenth century. Darwin, in particular, acknowledged that archetype theory played a central role in his early conception of biological evolution.[9]

Another foundational idea for Darwin was the notion of gradual changes of anatomical structures over immense periods of time. As I have mentioned, that idea, too, can be found in Leonardo's writings; not in the context of gradually changing anatomical structures but of gradual changes in the strata and formations of rocks—the bones of the living Earth (see p. 74). Thus, Leonardo anticipated two key ideas that were important ingredients of Darwin's early conception of the origin of species.

Since Leonardo's science is utterly dynamic, it is perhaps not surprising that we can find an "evolutionary" flavor in many of his scientific writings. As I have mentioned, he perceived nature's forms—in mountains, rivers, plants, and in the human body—as being in ceaseless movement and transformation. The world he portrays, in both his art and his science, is a world in development and flux, in which all configurations and forms are merely stages in a continual process of transformation.

What is even more remarkable, however, is that Leonardo intuited both of the two contradictory theories of evolution that would dominate nineteenth-century scientific thought: the evolution of closed physical systems from order to disorder, described by the physicists who developed thermodynamics, and the evolution of life from disorder to ever increasing order and complexity, described by Darwin and other biologists.

I have discussed in detail how Leonardo anticipated both the first and second laws of thermodynamics, and how his thorough understanding of the dissipation of energy led him to deep insights about the nature of irreversible processes—the "consumption" of natural forms under the influence of physical forces over long periods of time (see pp. 184ff.).

Leonardo did not fail to notice the contradiction between this phenomenon of gradual transformations from order to disorder and life's continual creation of ever-increasing diversity, which he also observed. Within modern science, it took more than a hundred years to resolve the contradiction between the two theories of "evolution" developed in the nineteenth century.[10] For Leonardo, who never developed any kind of theory of biological evolution, the contradiction was not inherent in his science. He simply perceived it as two opposing trends in natural phenomena, and he asserted that the evolution toward ever-increasing diversity of living forms always outpaced the opposing trend of evolution toward increasing disorder. "Nature, capricious and taking pleasure in creating and producing a continuous succession of lives and forms," he wrote in the Codex Arundel, "is eager and quicker to create than time is to destroy."[11]

The Dream of Flying

The investigation of the body's voluntary movements was a major theme in Leonardo's anatomical studies; he also compared human movements with the movements of various animals. In particular, he analyzed the gait of horses and drew comparisons with the human manner of walking:

> The walking of men is always in the manner of the universal gait of four-footed animals; because just as they move their feet cross-wise, in the manner of the trot of the horse, so a man moves his four limbs cross-wise; that is, if he thrusts his right foot forward in walking, he thrusts the left arm forward with it, and so it always continues.[12]

In addition to the gaits of land-based animals, Leonardo studied the movements of fish in water; but what fascinated him more than any other animal movement was the flight of birds. It was the inspiration for one of the great passions in his life—the dream of flying.

The dream of flying like a bird is as old as humanity itself. But nobody pursued it with more ingenuity, perseverance, and commitment to meticulous research than Leonardo da Vinci. From his early years in Florence, he must have been well acquainted with the legendary flight of Daedalus and Icarus who, according to the Greek myth, escaped from Crete

FIG. 7-2. Andrea Pisano, *The Myth of Daedalus*, c. 1343. Sculpted panel on Giotto's campanile in Florence.

on wings of feathers and wax. Daedalus, the cunning craftsman, is depicted with his powerful wings in one of the sculpted panels that decorate the lower portion of Giotto's campanile (fig. 7-2). This image may well have been both inspiration and challenge to the young Leonardo, whose earliest drawings of flying birds, insects, and artificial wings date from around 1470, when he had just established himself in Florence as an independent artist.[13] During the same period, he painted his *Annunciation*, in which the angel's entirely realistic wings,* growing from the shoulder blades, were obviously modeled after the real wings of birds.[14]

However, Leonardo soon realized that flying like a bird would require more than finely crafted angels' wings. He would need to understand the subtle details of how birds sustain themselves in the air and be able to absorb that knowledge into the design principles of his flying machine. As his scientific mind matured after his move to Milan, he began to develop a comprehensive approach to this challenge that involved numerous disciplines—from aerodynamics to human anatomy, mechanics, the anat-

---

* The realistic wings painted by the young Leonardo were lengthened by a later hand in an unhappy alteration, apparently to make them look more canonical (see Bramly, *Leonardo*, p. 143).

omy of birds, and mechanical engineering. He diligently pursued these studies throughout most of his life, from his early years at the Sforza court in Milan to his old age in Rome.[15] No project is better suited to illustrate Leonardo's systemic, integrative approach to scientific research, its brilliant application to engineering, and his persistent endeavor to imitate nature than his lifelong quest for a "science of flight."

Air Pressure and Lift

Leonardo's first intense period of research on flying machines began in the early 1490s, about a decade after his arrival in Milan. By that time, he had been fully accepted as artist and "ducal engineer" at the Sforza court and had become Prince Ludovico's favorite court artist.[16] He had living quarters and a large space for his workshop in the Corte Vecchia, where Ludovico housed important guests, and was engaged in a flurry of intellectual and artistic activities. These included creating the molds for the *gran cavallo* (a large equestrian statue honoring Ludovico's father), painting the *Last Supper*, and carrying out experiments in optics and mechanics, as well as designing and testing his first flying machines.

From his early observations of birds in flight, Leonardo recognized the compression of the air under the bird's wing during the downward stroke as a critical element in the generation of lift. The detailed aerodynamics of a bird in flight is very complex and was understood by Leonardo only many years later (see pp. 261ff.). However, even his early explanations of aerodynamic lift in the 1490s contained some important insights into the physics of flight. As science historian Raffaele Giacomelli has noted, these partial insights are impressive in view of the fact that before Leonardo, no natural philosophers had bothered to wonder how birds sustain themselves in the air.[17] Following Aristotle, it was simply believed that birds were supported by air as ships are by water, their wings and tails being analogous to the ships' oars and rudders.

When Leonardo observed birds in flapping flight, he recognized in this process two phenomena of mechanics he had discovered in other circumstances. One was the fact that, unlike water, "air has the ability to compress and rarefy."[18] The other was the principle, now known as Newton's third law of motion, that for every physical force there is an equal and opposite reactive force (see p. 200). In his Notebooks, Leonardo mentioned both of these phenomena many times, and very early on he used their combined effect in his attempts to explain aerodynamic lift.

In his very first Notebook, the Codex Trivulzianus, begun in the late 1480s, we find a concise exposition of the basic idea, illustrated with two well-known examples from everyday life:

> When the force generates more velocity than the escape of the resisting air, that air is compressed in the same way as bed feathers when compressed and crushed by the weight of the sleeper. And the object that pressed on the air, meeting resistance in it, rebounds in the same way as a ball striking against a wall.[19]

In other words, aerodynamic lift is explained by the compression of air under the bird's wings during the downward strokes and the resulting upward rebound of the bird's body.*

In a passage in the Codex Atlanticus, written around the same time, Leonardo adds another important principle of mechanics to his explanation—the relativity of motion—which he also recorded many times in various manuscripts (see p. 177). He argues that, since the motion of an object against still air is equivalent to the motion of air against a fixed object, the force sustaining an eagle in the air during flapping flight is the same as the force of the wind pushing a sailing ship:

> As much force is made by the thing against the air as by the air against the thing. See how the wings striking against the air sustain the heavy eagle high up in the thin air. . . . See also how the air moving over the sea strikes against the swelling sails and makes the loaded and heavy ship run fast.[20]

From this observation, Leonardo derives a momentous conclusion. "Therefore, by these reasons, asserted and demonstrated," he continues in the same paragraph, "you will know that a man with his assembled and great wings, exerting force against the resisting air and conquering it, will be able to subjugate it and raise himself above it." What this means is that Leonardo's belief in the possibility of human flight was established during his earliest investigations. In his entire life, he never lost this belief. His firm conviction that, some day, human beings would be able to fly like birds was not based on hope, but was grounded in sound scientific principles.

---

* Leonardo's early explanation is only partly correct, as it disregards the lift generated by the low air pressure above the wings, which is the stronger effect (see p. 266).

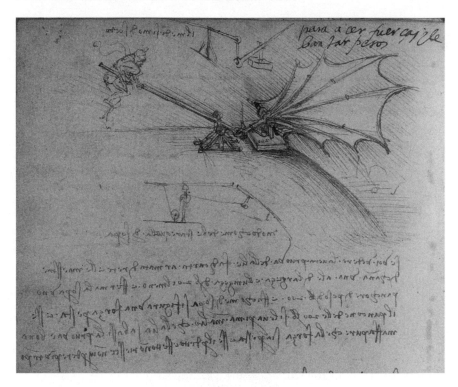

FIG. 7-3. Experiment to test the human capacity
to efficiently flap an artificial wing, c. 1487–90.
Ms. B, folio 88v (detail).

Having convinced himself that the critical challenge for human flight
was to flap artificial wings with enough force and velocity to compress the
air underneath and be lifted up, Leonardo set out to systematically test
this strategy. In a series of experiments at the Corte Vecchia that combined mechanics and human anatomy, he carefully measured the body's
capacity to generate various amounts of force in different bodily positions.[21] In addition, he designed a large membrane-covered wing (fig. 7-3)
to test the possibility of flapping it efficiently enough to lift a heavy plank
attached to its base. The aim of all these studies was to find out how a
human pilot might be able to lift a flying machine off the ground by flapping its mechanical wings.

Leonardo's first design of a "flying ship," based on these early experiments, is a rather strange contraption (fig. 7-4). It shows an upright craft
with four flapping wings, placed inside a vessel that is shaped like a bowl
and is accessible via a ladder and a hatch. The pilot, crouched down in the

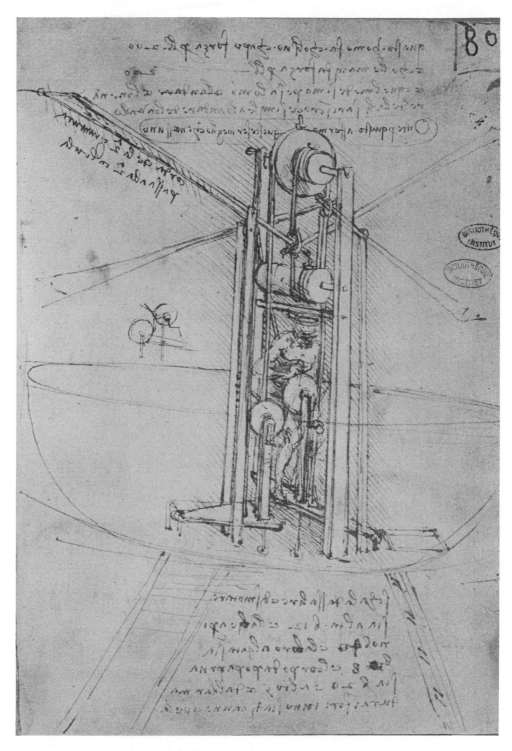

FIG. 7-4. Leonardo's "flying ship," c. 1487–90.
Ms. B, folio 80r.

FIG. 7-5. The "aerial screw," c. 1487–90.
Ms. B, folio 83v (detail).

center of the craft, generates the necessary force by pushing two pedals with his feet while simultaneously turning two handles with his hands. As Domenico Laurenza points out, "There is no note, no mention to be found . . . of how the pilot will steer the machine in flight. He becomes almost an automatic pilot: he simply has to generate the force to lift off the ground."[22]

Another flying machine designed by Leonardo around the same time is his famous "helicopter" or "aerial screw." It is based on the same idea of lift being achieved by means of proper compression of air, this time by a helical surface rotating rapidly through it. On a folio in Manuscript B, Leonardo drew a small sketch of such a device (fig. 7-5), and next to it he provided a succinct description of how it would work:

> Let the outer extremity of the screw be of steel wire as thick as a cord, and from the circumference to the center let it be 8 *braccia* [about 16 feet]. I find that, if this instrument, in the shape of a screw, is well made—that is, made of linen cloth with its pores stopped up with starch—and is turned swiftly, it will make a female screw in the air and will rise up high.[23]

In the same passage, Leonardo suggested trying out his aerial screw with "a small model made of paper, whose axis will be made of a fine steel blade, bent by force, and when released it will turn the screw." It is quite likely that Leonardo actually built a working model along those lines, which would have been similar to children's toys known in his time and still used today.[24] However, it is doubtful that the full-sized aerial screw could have been turned fast enough by human muscle power to provide sufficient lift. Be that as it may, we can easily recognize the aerodynamic principle by which Leonardo's craft was meant to rise into the air as the same principle underlying the functioning of the modern helicopter.

During the those years, Leonardo also designed a series of quite realistic machines for flapping flight in which the pilot is placed horizontally and controls a variety of subtle movements with his hands and feet. In addition, certain movements are achieved automatically by means of springs. Figure 7-6 shows an example from this series of designs. It is a highly finished technical drawing, accompanied by three sets of explanatory notes laid out neatly on the page. The plank on which the pilot is supposed to lie and the two foot pedals to operate the flapping of the wings are clearly visible. The pilot's hands and arms are used for maintaining balance and changing direction, not unlike in a modern hang glider.

Closer examination of Leonardo's drawing and text shows that during the downward stroke, the wings not only flap but also fold backward

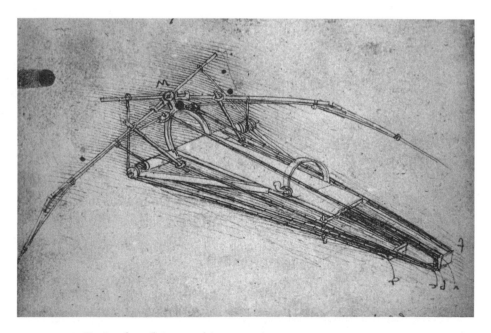

FIG. 7-6. Design for a flying machine, c. 1487–90.
Ms. B, folio 74v (detail).

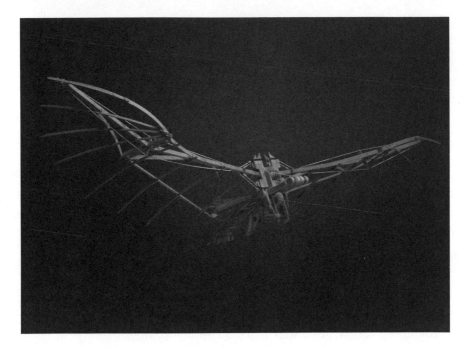

FIG. 7-7. Working model of the flying machine, wood, 1988.
Museum of the History of Science, Florence.

and inward, their tips moving toward the pilot's feet.[25] This elegant movement, imitating the actual wing motion of birds, is achieved by means of a
complex system of joints, pulleys, and springs—a masterpiece of delicate
mechanical engineering. The accompanying text includes the prudent reminder: "You will experiment with this machine over a lake and you will
wear as a belt a long wineskin, so that if you fall in, you will not drown."[26]

The drawing shown in figure 7-6, together with similar drawings from
the same period in Manuscript B and the Codex Atlanticus, represents
Leonardo's most sophisticated design of flying machines. These drawings
became the basis of several models built by modern engineers.[27] Figure 7-7
shows one of these models, built from materials that were available in the
Renaissance. The limitations of these materials—wooden struts, leather
joints and thongs, and skin of strong cloth—make it evident why Leonardo could not create a viable model of his flying machines, even though
they were based on sound aerodynamic principles. The combined weight
of the machine and its pilot was simply far too heavy to be lifted by human
muscle power.

Biomimicry

Eventually, Leonardo became aware that he could not achieve the required power-to-weight ratio for successful flapping flight. It would take him ten years to reach this conclusion, perhaps because those were years of frequent travels in central Italy with neither the sufficient time, nor the required ample workshop space, to test new designs of flying machines.[28] However, during his travels Leonardo continued to record his observations of birds in flight in two pocket-sized notebooks, now known as Manuscripts K and L.*

When he finally settled down in Florence, Leonardo intensified his ornithological studies, engaging in careful and methodical observations of birds in flight, down to the finest anatomical and aerodynamic details. He spent hours in the hills surrounding Florence, near Fiesole, where he could see eagles, swans, and other large birds in gliding and soaring flight. Leonardo intended these large birds to be his models for new designs of a flying machine that would imitate nature ever more closely, maneuvering with agility, keeping its balance in the wind, and moving its wings like a real bird.

He summarized his observations and analyses in a small Notebook called Codex on the Flight of Birds (Codex *Sul Volo*). Looking through the pages of this elegant manuscript, one almost has the impression that Leonardo wanted to become a bird himself. Not only does he call his flying machine *uccello* (bird) but he also uses anatomical terms for its parts, speaking, for example, of its "fingers" (wing tips) and the "tendons" (tie-rods) to move them. In some passages he shifts effortlessly back and forth between the third person (describing a bird in flight) and the second person (addressing himself or the pilot of his flying machine), for example, in the following series of instructions about how to keep one's equilibrium in the wind:

> If the bird should wish to turn quickly on one of its sides. . . . And if you wish to go west without flapping the wings. . . . That bird will rise up high . . . always turning on its right side or on its left side. . . . If in your straight rise the wind should be likely to upset you, then you are at liberty to bend by means of the right or left wing . . .[29]

---

* Manuscript K actually consists of three parts written at different times, the first of which, known as K¹, is the one with notes on the flight of birds.

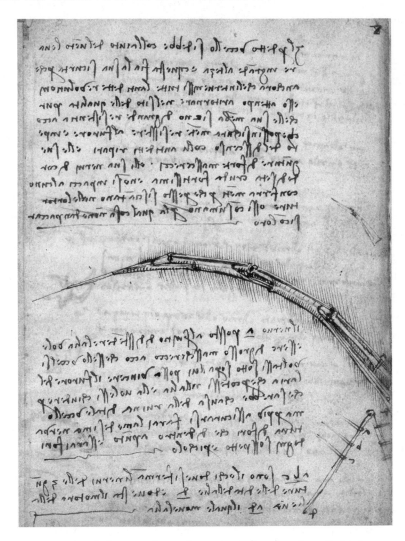

FIG. 7-8. Study for a mechanical wing imitating the wing of a bird, 1505. Codex *Sul Volo*, folio 7r.

This fusion of identities of the real and the mechanical bird in the text is matched, if not surpassed, by a similar fusion in Leonardo's drawings. His designs of mechanical wings sometimes mimic the anatomical structure of a bird's wing so accurately (for example, in fig. 7-8) that even experts find it hard to tell the difference. As I shall discuss in more detail, Leonardo's attitude of seeing nature as a model and mentor is now being rediscovered in the discipline of ecological design, and especially in the practice of biomimicry (see p. 324).

The Flight of Birds

To appreciate the significance of Leonardo's insights into the physics of flight, it is useful to review how modern scientists explain the flight of birds. Even a cursory glance at the relevant literature shows that this is an exceedingly complex subject that is still not fully understood. The science of flight is based on aerodynamics, which itself is part of the more general, and notoriously difficult, discipline of fluid dynamics

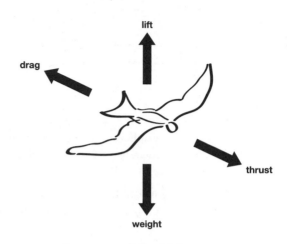

FIG. 7-9. Forces on a gliding bird.

(see pp. 33ff.). Even with powerful supercomputers at their disposal, scientists today are still not able to accurately model the turbulent flows of air around the surfaces of an airplane (or a bird).[30] In spite of these mathematical difficulties, however, the basic features of "animal aerodynamics" are now well known.[31]

To understand how birds fly, it helps to start with gliding flight, because a gliding bird is subject to the same forces as a gliding airplane with fixed wings. Two pairs of forces need to be balanced in steady flight: weight and lift, as well as thrust and drag (fig. 7-9). The direction of the lift is always perpendicular to the motion of the wing; it will be vertical only if the wing moves exactly horizontally (which flapping wings hardly ever do).

Both lift and drag are generated by the flow of air across the bird's wings. The principles are the same as for the flow around the wings of an airplane. Figure 7-10a shows a cross section of an airplane wing, also known as an "airfoil," with four streamlines indicating the flow of air across it. This could be a picture either of a wing moving through stationary air, or of air flowing over a stationary wing in a wind tunnel. The two points of view are entirely equivalent, as Leonardo da Vinci was the first to recognize (see pp. 265–66).

When the oncoming air hits the leading edge of the wing, it separates into two parts, one streaming above the wing and the other below. Because of the particular shape of the airfoil, this separation is not symmetrical. Above the wing, the streamlines are compressed; below the wing,

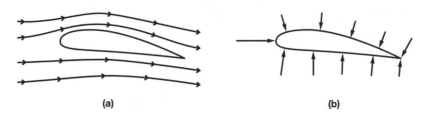

FIG. 7-10. Airflow over the cross section of a wing, showing
(a) streamlines and (b) distribution of pressures on the wing.
From Anderson, *A History of Aerodynamics.*

they expand, as can be seen in figure 7-10a. In terms of flow velocities, this
means that the velocity of the air flow increases above the wing while it
decreases below the wing.

Now there is a general theorem in fluid dynamics, known as Bernoulli's
theorem (after its discoverer, the eighteenth-century mathematician Daniel Bernoulli), according to which any increase in the velocity of a flowing
fluid is accompanied by a corresponding decrease in its pressure, and vice
versa.* Consequently, the distribution of pressures exerted on the surface
of the airfoil will be such that the pressures on the bottom surface (where
the air flow is slower) will be higher than the pressures on the top surface
(where the air flow is faster), as shown in figure 7-10b, in which the longer
arrows indicate areas of higher pressure. The net effect is an upward lifting
force on the wing.

This lifting force is always accompanied by a drag on the airplane (or
bird) due to air resistance. More precisely, the drag is caused by the friction between the body of the flying object and the streaming air, resulting
in shear stresses on the body surface. This "viscous drag" is not the only
type of drag on a flying object. There is also a "pressure drag" produced by
a turbulent wake behind the object, and the "induced drag" generated by
trailing vortices behind the tips of the wings. In practice, engineers usually
measure only the total drag.

The crucial quantity in wing design is the ratio of lift to drag. Since the
same thrust produces both lift and drag, the drag on the wings determines
an airplane's efficiency. A higher lift-to-drag ratio means that less thrust
(and hence less fuel) is needed to produce the necessary lift. In modern

* Bernoulli's theorem is simply a consequence of the conservation of energy. It was intuited and formulated correctly by Leonardo 250 years before Bernoulli for the specific case of jets spouting from
a tank at different heights (see pp. 183–84 above).

aerodynamics, it is relatively easy for scientists to calculate lift but very difficult to calculate drag, because of the critical role of air turbulence. In addition to the turbulent wake behind the flying

FIG. 7-II. Angle of attack, α, of a cambered airfoil.

object and the vortices behind the wing tips, the flow of air close to the body surface, in a region known as the boundary layer, is also highly turbulent. All these turbulences make calculations and mathematical modeling of the drag on an airplane extremely difficult.

The lift that can be achieved with a given amount of thrust depends not only on the shape of the airfoil but also on its angle relative to the oncoming air (see fig. 7-II). As the airfoil is tilted upward, the lift will increase with an increasing "angle of attack." Even a flat wing generates some lift when tilted upward. However, a convex, or "cambered," airfoil will produce more lift and less drag. We shall see that birds can adjust both the angle of attack and the camber of their wings to move effectively in various situations.

Naturally, a wing cannot be tilted upward indefinitely to obtain more and more lift. When the angle of attack reaches a certain critical value, a stall will occur: the airflow over the top surface of the wing will generate a strong turbulent wake, which causes a sudden loss of lift and increase of drag. Most airplanes will stop flying and will start to fall when their wings stall. Birds, amazingly, use the process of stalling very effectively in their landing maneuvers. Just before touching down on a perch, a bird will often sharply increase its angle of attack and, in a well-timed stall, will lose its speed and drop down safely on the perch.

Gliding is the simplest form of flight, and some birds glide most of the time they are airborne. In addition, eagles, hawks, and many other birds use a special form of gliding called soaring, which allows them to stay aloft for long periods of time without flapping their wings, by using rising air currents to increase their gliding time. Storks and albatrosses are masters of soaring. They may soar over the sea for days at a time, sometimes even crossing an entire ocean without flapping their wings.[32]

High lift-to-drag ratios are essential for soaring. Albatrosses and other large birds achieve these high ratios with their long and narrow wings. Smaller terrestrial gliders, like hawks and falcons, are able to adjust their wing area by bending the joints in their wings, reducing the area for fast gliding and expanding it for soaring. In addition, many soaring birds con-spicuously spread the large "primary" feathers at the tips of their wings. This reduces the trailing vortices at the wing tips and thus increases the lift-to-drag ratio.

With this understanding of the basic aerodynamics of gliding and soaring, we can now examine the much more complex actions of flap-ping flight. The most important, and somewhat counterintuitive, insight of modern animal aerodynamics has been the recognition that birds and other flying animals do not flap their wings to maintain themselves in the air, but do so only to produce thrust. Once they move forward against the air, their wings also generate lift owing to their aerodynamic shape, just as in gliding flight. This means that, in order to produce the required forward thrust, the flapping of a bird's wings must be much more complex than a simple up-and-down motion.

Indeed, slow-motion films of birds in flight have revealed that they do not flap their wings vertically but follow complex patterns that combine several types of motion.[33] The path of the wing tip is a curve, somewhat resembling the butterfly stroke of a swimmer, but the curve is tilted from the vertical by about 30 degrees—down and forward on the downstroke, up and backward on the upstroke. This has the effect that the lift pro-duced by the wings is tilted forward, so that the flapping generates both lift and thrust. For large birds like the albatross, the path of the wing tips is an oval; for some smaller birds it follows a figure eight, and flying insects trace out all kinds of complex loops.

In addition to moving their wings along a tilted curve, birds increase the angle of attack on the downstroke and decrease it on the upstroke; and many birds also change the wing area by folding the primary flight feath-ers on the outer part of the wing like a fan and pulling the wings closer to the body during the upstroke to reduce their effective surface area. All these movements combine to produce as much thrust and lift as possible on the downstroke and as little drag as possible on the upstroke.

As the bird's wing sweeps down and forward, the part near the wing's base (close to the bird's body) will experience a relative wind mostly from the forward motion, and hence the lift there will be more or less vertical. The wing tip, by contrast, will experience a flow of air that is caused by both the forward motion and the wing's flapping movement. It will be faster than at the base and will approach the wing from below. Consequently, the lift near the tip will be stronger and tilted forward. In flapping flight, the strength and direction of the air flow change gradually along the wingspan, producing most of the lift along the inner part of the wing and most of the thrust near its tip.

Understanding how birds fly involves many additional aspects. Just as terrestrial animals use different gaits for different speeds, so birds use different flapping patterns for slow and fast flights. The motions of their wings differ not only for different speeds but also from species to species. Then there are the numerous subtle movements birds use for takeoff and landing; for turning, climbing, and descending; for staying on course in the wind; and, in the case of small birds (as well as insects), for hovering in still air. With a keen eye, Leonardo observed the fine details of many of these movements, described them with great accuracy, and sketched them in lively and charming drawings. The critical question is: how much did he understand about the flight of birds?

Leonardo's Aerodynamics

Leonardo's first great achievement in formulating a proper science of flight was the recognition that such a science must be grounded in aerodynamics, to use our modern term. "To give the true science of the movement of birds in the air," he wrote late in his life, summarizing more than thirty years of research, "it is necessary first to give the science of the winds, and this we shall prove by means of the movements of water within itself."[34] In this passage, Leonardo asserts not only that the science of flight must be based on sound aerodynamics ("the science of the winds"), but also that the flows of air can be compared to flowing water, both being described by the same discipline of fluid dynamics, as we would say today.

Since the discovery of the basic principles of fluid dynamics was one of Leonardo's greatest scientific achievements (see p. 39), it is not surprising that his studies of the aerodynamics of flight led him to many pioneering insights. As mentioned above, he was the first to recognize and clearly formulate the principle of the relativity of motion, according to which a body moving through stationary air is equivalent to air flowing over a stationary body. "As it is to move the object against the motionless air," he wrote around 1505, "so it is to move the air against the motionless object."[35] Today, this is known as the principle of the wind tunnel, the most important experimental tool of modern aerodynamics.

Leonardo also realized that the relativity of motion implies that aerodynamic lift is generated by the same forces in both flapping and soaring flight. The passage in the Codex Atlanticus continues: "Therefore, the bird beating its heavy wings on the thin air causes it to compress and resist the bird's descent. And if air moves against the motionless wings, that air sustains the heaviness of the bird in the air." In the last part of this passage, Leonardo examines the three types of soaring flight—hovering in the wind, ascending, and descending:

> When the power of the motion of the air is equal to the power of
> the descent of the bird, that bird will stay in the air motionless. And
> if the motion of the air is more powerful, it will win and will raise
> the bird up. And if the power of the motion of the air is less than the
> weight of the bird, that bird will come down.

On another folio in the Codex Atlanticus, Leonardo reiterates the identity of the aerodynamic principles underlying soaring and flapping flight:

> When the bird finds itself within the wind, it can sustain itself on it
> without flapping its wings, because the function the wing performs
> against the air, when the air is motionless, is the same as that of the
> moving air against the wings when they are motionless.[36]

From that time on, Leonardo's notes on flight always treated flapping and soaring flight as equivalent. Both in his first set of notes in Manuscript K and in his more extensive records in the Codex on the Flight of Birds, the motion of birds is analyzed from these two perspectives.

In most of his analyses of flight, Leonardo reiterated his idea that birds are sustained in the air by the compression of air under their wings and the resulting upward rebound. This explanation is partially correct, but it is not the whole story. Today we know that aerodynamic lift is a consequence not just of the air pressure under the wing but also of the pressure *difference* between the air above and below the wing, and that the low pressure area above the wing actually generates most of the lift.

Late in his life, however, Leonardo realized the importance of the thin air above the wing. In fact, Manuscript E, composed around 1513–15 when

Leonardo was over sixty, contains an exact description of the air densities around the body of a flying bird:

> The air surrounding birds is above thinner than the usual thinness of the other air, as below it is thicker than the same; and it is thinner behind . . . and thicker in front of the bird.[37]

If we replace the terms "thin air" and "thick air" in this passage by "low pressure" and "high pressure," remembering that the concept of pressure was clearly defined only in the seventeenth century (see p. 169), the resulting description is very similar to how the pressure distribution around an airfoil is pictured in modern aerodynamics (see fig. 7-10b).

The flow of air around a bird's wings produces not only the upward lift but also a drag on the bird's forward motion due to air resistance in front of and turbulence behind the wings (see p. 261). When Leonardo examined these forces, he had already struggled for many years to understand the inertia and the dissipation of energy of bodies in motion, which he analyzed in terms of the medieval theory of impetus. By the time he composed the Codex on the Flight of Birds, he had convinced himself that drag was caused by the resistance of compressed air in front of the moving object, as well as by turbulence behind it (see p. 181).

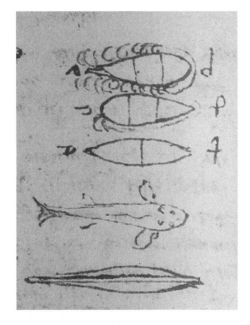

FIG. 7-12. Streamlined shapes of ships and fish. Ms. G, folio 50v (detail).

Leonardo investigated the effects of drag for motion in both in the air and water. A sketch in Manuscript G shows three ships of different shapes, as well as two kinds of fish (fig. 7-12). Both causes of drag—the resistance of the water in front of the ship and the turbulence on the side and in the back—are clearly visible in the sketch. Leonardo concludes that the drag on the ship shown on top will be the smallest because of its streamlined shape (as we would say today), and notes that "it resembles the shape of birds and fishes such as the mullet."[38]

Leonardo also tried to quantify the resistance encountered by a body moving through air.[39] He postulated correctly that it is proportional to the surface area of the body, and also to the body's velocity, which is incorrect (the air resistance is proportional to the square of the velocity). It is likely that the latter postulate was simply based on Leonardo's belief in the privileged role of linear relationships in nature, which was also held by Galileo in his early work (see pp. 175–76).

The Codex on the Flight of Birds

In the Codex on the Flight of Birds (*Codice sul volo degli uccelli*, or Codex *Sul Volo* for short), Leonardo summarized the observations and analyses of bird flight he made in Florence during a period of two years between 1503 and 1505. The elegant small Notebook is full of charming drawings of birds in flight (for example, plate 1), detailed descriptions of their turning maneuvers, their ability to maintain equilibrium in the wind, and various subtle features of active flight, as well as sketches of complex mechanisms he designed to mimic the birds' precise movements.

When Leonardo recorded his occasional observations of birds in flight during his travels in central Italy, he already had a treatise on this subject in mind. In Manuscript K, one of the Notebooks he carried with him during that time, he outlined a clear plan for such a work:

> Divide the treatise on birds into four books: the first on the flight maintained by beating the wings; the second on flight without beating wings, maintained by the wind; the third about flight common to birds, bats, fish, animals, insects; the last about instrumental motion.[40]

In the Codex *Sul Volo*, Leonardo more or less followed this plan. For some reason, he composed the eighteen folios of this Notebook in reverse order, so that the conceptual sequence runs from back to front.[41] The first part (folios with high numbers) deals mainly with flapping flight, while the second part (folios with low numbers) contains notes on how birds glide, soar, and maneuver in the wind. In both parts, the notes on bird flight are followed by sketches of mechanisms designed to imitate these natural movements with a mechanical "bird."

On the opening folio of the Codex *Sul Volo* (folio 18r in the reverse order), Leonardo summarized the main points of his analysis of flapping flight in six short paragraphs. A striking feature of these notes is that he quite naturally identifies the bones and joints of the bird's wing as the "elbow," "hand," "fingers," and so on. It is an eloquent testimony to the maturity of his studies in comparative anatomy. Even today, avian anatomists speak of the wing's elbow and wrist joints, of the hand (technically known as the carpometacarpus), and of its thumb (alula) and two fingers.[42] In an anatomical study of a bird's wing in the Windsor Collection (fig. 7-13), produced a few years later, these bones and joints, together with their

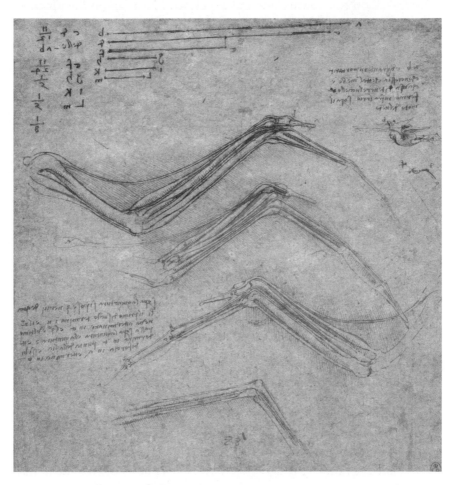

FIG. 7-13. Anatomy of a bird's wing, c. 1513.
Windsor Collection, *Anatomical Studies*, folio 187v.

tendons and ligaments, are pictured with great accuracy. The correspondences with Leonardo's famous study of the bones of the human arm and hand (plate 3) are quite evident.

Before evaluating Leonardo's description of flapping flight, I shall summarize once more the main characteristics of the bird's wing motion that have been identified by modern ornithologists (see p. 264). The wingbeat is tilted from the vertical, moving forward on the downstroke and backward on the upstroke. In addition, it is curved so that the wing tip describes an oval or more complex curve. The angle of attack is increased on the downstroke and decreased on the upstroke. The primary feathers are fully spread on the downstroke and are closed like a fan on the upstroke when the bird pulls the wings closer to the body. As in gliding flight, the pressure difference between the top and the bottom of the airfoil results in a net force on the wing, and because of the wingbeat's curved path this force is tilted forward, generating both lift and thrust. In this complex motion, most of the lift is produced along the inner part of the wing and most of the thrust near its tip.

In view of the fact that the subtle details of the dynamics of flapping flight have been revealed only recently with the help of high-speed photography and slow-motion filming, it is truly astonishing how many of its basic characteristics Leonardo identified correctly with his sharp eye and his great capacity of visualization. In his description of the wingbeat, he notes the wing's down-and-forward motion and accurately describes how its angle of attack is raised on the downstroke (by lowering the elbows):

> The lowering of the elbows at the time the bird is putting its wings forward, somewhat edgewise on the wind, guided by the impetus already acquired, is the reason why the wind strikes under that elbow and becomes a wedge on which the bird, with the aforesaid impetus and without beating its wings, comes to rise.[43]

The lift is attributed correctly to the relative airflow created by the thrust ("impetus") when the angle of attack is raised ("somewhat edgewise on the wind") but as in other passages, Leonardo does not recognize the important role of the low pressure above the wing, attributing the entire lift to the high pressure ("wedge") below it.

On the same folio, Leonardo also notes that during the wingbeat the path of the wing tips is not straight but describes an oval. "The course of the finger tips is not the same in going as in returning," he observed, "but is by a higher line, and the return is below it; and the figure made by the upper and lower lines is a long and narrow oval."

Even more remarkably, still on folio 18r, he sketches the changes in the angle of attack along this oval in a small drawing (fig. 7-14) together with the note:

> The palm of the hand goes from *a* to *b* [downstroke] always at about the same angle, pressing down the air, and at *b* it immediately turns edgewise and turns back, rising along the line *cd* [upstroke], and arriving at *d* it immediately turns full face and sinks along the line *ab*, and in turning it always turns around the center of its breadth.

As ornithologists do today, Leonardo compared this circular motion to that made by a swimmer's hand. In Manuscript F, composed a few years after the Codex *Sul Volo*, he noted:

> As the hand of the swimmer acts when it strikes and presses against the water and makes his body glide forward in a contrary motion, so acts the wing of the bird in the air.[44]

And finally, Leonardo did not fail to notice that the bird produces thrust and lift with two different parts of its wing. The thrust ("impetus") is produced by the outer wing (the "hand"), he explains, and the lift by the raised angle of attack of the inner wing (the "elbow"):

> The hand of the bird is what causes the impetus, and then its elbow . . . assumes a slanting position, and the air on which it rests becomes slanting, as if in the form of a wedge on which the wing raises itself up.[45]

Apparently, Leonardo realized the different functions of the inner and outer wing even before his methodical observations of bird flight in Florence. During his travels a couple of years earlier, he jotted down a quick note in his pocket book: "In beating the wings to remain up high and to go forward, from the hand back causes to stay up, and the hand causes it go forward."[46]

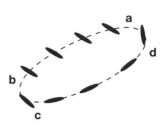

FIG. 7-14. Changes of angle of attack along oval wingbeat.
Codex *Sul Volo*, folio 18r (detail), and reconstruction.

In Leonardo's notes on birds in flight, there are also some contradictions and erroneous statements, which is not surprising given the great
complexity of the subject. However, the Codex on the Flight of Birds
leaves no doubt that he fully understood the essential features of both
soaring and flapping flight. Thus aeronautical engineer John Anderson
concludes in his *History of Aerodynamics*: "It is clear that Leonardo was
the first person to understand the mechanics of bird flight."[47]

In the Codex *Sul Volo*, Leonardo's summary of his understanding of
flapping flight is followed immediately by two folios with designs of mechanical wings. The first drawing (fig. 7-15) shows the left wing of the machine from the front.

The wing is connected to two foot pedals via a system of cords and
pulleys, which allows the pilot to raise and lower it with his feet. In addition, the wing can be rotated to change its angle of attack by means of a
handlebar, to be operated by the pilot's hands. It is evident that Leonardo
attempted here to imitate, as closely as possible, the complex motions of
flapping flight he described on the preceding folio.

While he studied the motion of birds in flight and designed mechanical wings to imitate it, Leonardo turned once more to the problem of
measuring human muscle power, which had occupied him so intensely ten
years earlier (see p. 254). But this time he approached the problem from
the perspective of comparative anatomy. "You will make the anatomy of
the wings of a bird together with the muscles of the breast, the movers of
those wings," he wrote in a note to himself, "and you will do the same for
man in order to show that there is the possibility in man to sustain himself
in the air by the flapping of wings."[48]

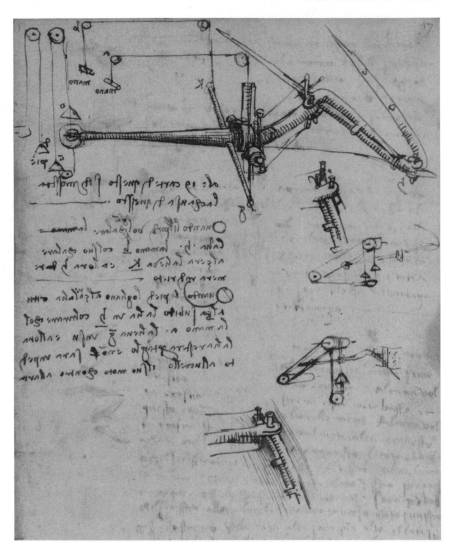

FIG. 7-15. Design for mechanical wing imitating
the flapping flight of birds. Codex *Sul Volo*, folio 17r.

Leonardo carried out these comparative anatomical studies with various birds and recorded them in the Codex Atlanticus on several folios.[49] The results convinced him that flying by beating mechanical wings might not be possible because of the limitations of our anatomy. "The sinews and muscles of a bird [are] incomparably more powerful than those of man," he explained in the Codex *Sul Volo*, "because all the fleshiness of the big muscles and fleshy parts of the breast goes to facilitate and increase the power of the wings' motion, and with that bone of the breast of one piece, which provides the bird's very great power."[50] This power is so strong, he observed, that it enables big raptors to sustain themselves in the air even while carrying heavy loads:

> All this strength is provided to enable [the bird], over and above the ordinary sustaining action of the wings, to double and triple its motion in order to escape from its predator or to pursue its prey. For this purpose, it has to double or triple its strength and, moreover, carry in its claws as much weight through the air as it weighs itself. Thus we see a falcon carry off a duck and an eagle a hare.[51]

However, Leonardo noted in the same passage that birds "need little power to keep themselves up in the air and balance on their wings and flap them on the currents of air and steer along their paths, a little movement of the wings sufficing; and the larger the bird, the slower [the wing motion]." He concluded from this observation that, even though human muscle power was too weak to lift a flying machine off the ground by flapping mechanical wings, a machine for soaring and gliding flight might be feasible, since this would require much less force.

For Leonardo, this was a momentous insight. After ten years of observations, studies, and design projects, he now saw a concrete way in which his unwavering belief in the possibility of human flight could be turned into reality. During the same year in which he compiled the Codex on the Flight of Birds, he summed up his new insight in a long and carefully worded passage in the Codex Atlanticus:

> The [mechanical] bird is an instrument working according to mathematical law, an instrument which it is within the power of man to make with all its motions, but not with such power [as the natural bird], its power extending only to the balancing movements. We may say therefore that such an instrument designed by man lacks only the soul of the bird, which must be counterfeited with the soul of the man.[52]

Leonardo was well aware that, even if he succeeded in designing a machine that could fly like a bird, he would never be able to completely match the bird's instinctive capacity to maneuver in the wind, keeping its equilibrium by responding to changes in air currents with subtle movements of its wings and tail. "The soul of the bird," he continued his summary, "will certainly respond better to the needs of its limbs than would the soul of the man, separated from them and especially from their almost imperceptible balancing movements."* However, he concluded that imitating the clearly perceptible balancing movements of birds should be sufficient to prevent the flying machine from crashing:

> The varieties of perceptible motions that we observe in birds . . . can be learned by man, and . . . he will to a great extent be able to prevent the destruction of the instrument for which he is the soul and driver.

The enthusiasm Leonardo must have felt when he reached these conclusions is reflected in his famous prophecy:

> The large bird will take its first flight from the back of the great Swan [Monte Ceceri, near Florence], filling the universe with amazement, filling all writings with its fame and [bringing] eternal glory to the nest where it was born.[53]

Leonardo placed this exuberant declaration on the inside back cover (that is, at the beginning) of the Codex on the Flight of Birds, and it seems that he intended it as an epigraph for his treatise. Apparently not quite satisfied with the wording, he composed a shorter version on the subsequent page:

> From the mountain that takes its name from the great bird, the famous bird will take its flight, which will fill the world with its great fame.[54]

Having convinced himself that a flying machine could be designed to imitate the soaring and gliding flights of birds, Leonardo used the last part of the Codex *Sul Volo* (pages with low numbers) to study how a gliding

---

* See p. 156 for my discussion of Leonardo's concept of the soul.

bird keeps its balance in the wind. For example, he devotes an entire folio to the analysis of how the bird, when pushed by a lateral gust of wind, balances itself by spreading or folding one or the other wing to various degrees, because "the forces of the wind striking the two wings will be of the same proportion as their extensions."[55] On the subsequent folios, Leonardo describes in detail how the bird's tail supports the balancing actions of the wings in steering and controlling the flight.[56]

These copious notes on the balancing actions of birds are followed, once again, by design sketches that attempt to imitate the birds' natural movements. This time, however, Leonardo's focus is no longer on the wingbeat but rather on the flexions and extensions of the wings that are critical to the balancing maneuvers in gliding flight (see fig. 7-8).

In Manuscript G, composed a few years later, Leonardo discusses a variety of techniques used by birds for takeoff, among them one that is especially relevant to his new design ideas for flying machines:

> The second method employed by birds at the beginning of their flight is when they descend from a height. They merely throw themselves forward and at the same time spread out their wings upward and forward, and in the course of their leap they lower their wings downward and backward, and thus, rowing, they proceed on their slanting descent.[57]

It is evident that this comes very close to describing the takeoff maneuvers of a modern hang glider.

Throughout the entire Codex *Sul Volo*, Leonardo shows supreme confidence in the feasibility of human flight. At times, he sounds as if mechanical flight had already become so routine for him that he could sprinkle his treatise with pieces of practical advice for would-be pilots. Thus he recommends, without any apparent sense of irony:

> The movement of the [mechanical] bird should always be above the clouds so that the wing does not get wet, and to survey more country, and to avoid the dangers of swirls of winds within the mountain passes, which are always full of gusts and eddies of wind.[58]

The Science of the Winds

In the years after the completion of the Codex *Sul Volo*, during his second period in Milan, Leonardo only rarely recorded observations on the flight of birds, and no further designs of flying machines have come down to us. It was not until eight years later that he returned to the subject of flight. He was then living in Rome, over sixty and rather lonely and depressed, his reputation as a painter having been eclipsed by younger rivals like Michelangelo and Raphael.[59] In spite of his somber state of mind, however, he continued his studies with great diligence.

During those years, 1513–15, Leonardo collected his scientific thoughts —many of a general, reflective nature—in a small Notebook now known as Manuscript E. This is the Notebook that contains the famous passage on his empirical method,[60] in addition to various notes on many of the grand themes he had pursued during his life: the "science of weights," geometry, motion, the flows of water, and especially the science of flight.

In the opening passage of the section on flight in this Notebook, which I have already quoted in part, Leonardo defines the proper theoretical framework for such a science:

> To give the true science of the movement of birds in the air, it is necessary first to give the science of the winds, and this we shall prove by means of the movements of water within itself. And this science, accessible to the senses, will serve as a ladder to arrive at the knowledge of things flying in the air and the wind.[61]

This passage is remarkable for several reasons. As I have discussed, Leonardo declares here that his science of flight is grounded in aerodynamics and, more generally, in fluid dynamics (to use modern scientific terms). To gain knowledge about the "science of the winds," he explains, he will study turbulent flows of water ("the movements of water within itself"), knowing from his lifelong observations that the principles of flow are the same for water and air (see p. 33). But unlike the movements of air, those of water are visible ("accessible to the senses") and hence can serve as a model ("a ladder") to gain knowledge about aerodynamics and about flight. Once again we encounter here an aspect of Leonardo's scientific thought that puts him centuries ahead of his time—the recognition of flow as a universal phenomenon of liquids and gases, and his use of the former as models of the latter.

Most of Leonardo's notes on flight in Manuscript E are concerned with his theoretical studies of the "science of the winds." Indeed, fluid dynamics was very much on his mind at that time, both in his science and in his art. Those were the years when he created his celebrated "deluge drawings"— violent and disturbing images that represent a visual catalogue of different types of turbulences in water and air (see p. 61).

Having defined the "science of the winds" as the proper framework for the study of the flight of birds, Leonardo then restates three of his most important discoveries in aerodynamics. The first is the fact that air, unlike water, is compressible. "Air can be compressed and rarefied almost infinitely," he notes, and he adds that, because of the thin air at high altitudes,* only large birds with great wingspans are able to fly there.[62]

Leonardo's second important discovery in aerodynamics is the principle of the wind tunnel, that is, the relativity of motion between a solid object and the surrounding air. Its formulation in Manuscript E is virtually identical to the one given ten years earlier in the Codex Atlanticus: "As it is to move the air against the motionless thing, so it is to move the thing against the motionless air."[63] His third discovery, finally, is that of the pressure distribution in the flow of air around a bird's wing—higher pressure on the bottom surface and lower pressure on the top surface— which is described correctly for the first time in Manuscript E, as I have discussed (see p. 267).

When Leonardo recorded these notes, he apparently no longer had the energy to review his previous observations on the flight of birds in the light of his late insight into the density distribution around the wings. Still, his full understanding of the origin of aerodynamic lift, together with his concise formulation of the principle of the wind tunnel, establishes Leonardo da Vinci as one of the great pioneers of aerodynamics. Indeed, in the opinion of aeronautical engineer John Anderson, "[Leonardo's] aerodynamic concepts were amazingly advanced and would have constituted a quantum jump in the state of the art of aerodynamics if they had been widely disseminated."[64]

Leonardo's mechanical "birds" with flapping wings were not destined to fly, even though their designs were based on sound aerodynamic prin-

* Leonardo's observation that birds cannot fly beyond certain altitudes because of the thinning of the air is correct in principle. In practice, however, the decrease of oxygen at high altitudes is a more severe limiting factor.

ciples (see p. 258). Nevertheless, the models built from those designs in recent years are extraordinary testimonies to his genius as a scientist and engineer. In the words of art historian Martin Kemp:

> Using mechanical systems, the wings flap with much of the sinuous and menacing grace of a gigantic bird of prey ... [Leonardo's] designs retain their conceptual power as archetypal expressions of man's desire to emulate the birds, and remain capable of inspiring a sense of wonder even in a modern audience, for whom the sight of tons of metal flying through the air has become a matter of routine.[65]

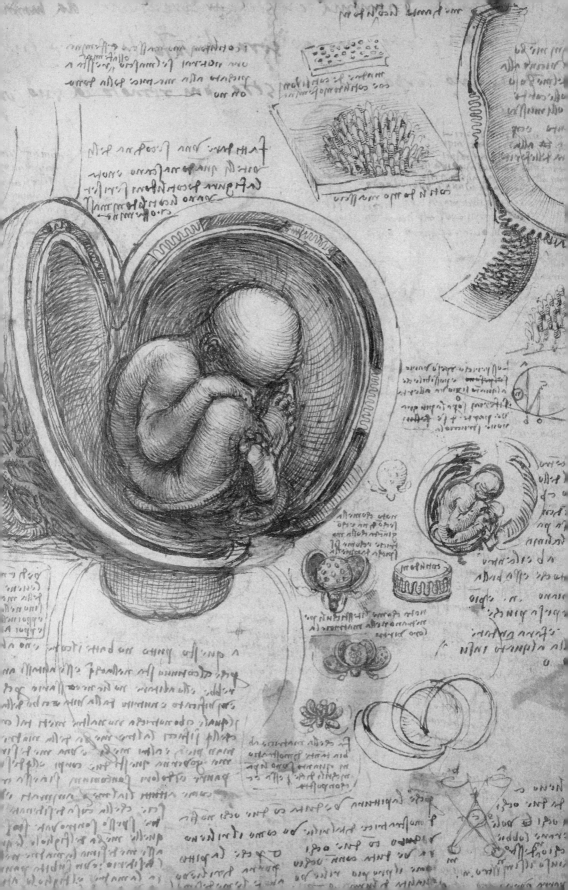

8

The Mystery of Life

The grand unifying theme of Leonardo's explorations of the macro- and microcosm was his persistent quest to understand the nature of life. Over the years, as he studied, drew, and painted the flows of water and air, the rocks and sediments of the Earth, the growth patterns of plants, and the anatomy of the human body, he correctly identified several of life's key biological characteristics.

Early on, he recognized the fundamental role of water as life's medium and vital fluid, the matrix of all organic forms (see p. 18). "It is the expansion and humor of all living bodies," he wrote in one of his earliest Notebooks, Manuscript C. "Without it nothing retains its original form."[1] He associated the fluidity of water with the fluid and dynamic nature of living forms. He was especially fascinated by water vortices and other forms of turbulence, recognizing them intuitively, as I have argued, as symbols of life—stable and yet continually changing (see p. 22).

Nature as a whole was alive for Leonardo, and he saw similar patterns and processes in both the macrocosm of the living Earth and the microcosm of an individual organism. In view of this systemic approach—seeking to understand a natural phenomenon by linking it to other phenomena through a similarity of patterns—it is not surprising that Leonardo developed a conception of life that was deeply ecological. This is evident throughout his manuscripts, as, for example, when he describes the continual processes of growth and renewal that are common to all life on Earth (see p. 67):

> Feathers grow on birds and change every year; hairs grow on animals and every year they change. . . . Grass grows in the fields and leaves on the trees, and every year they largely renew themselves.[2]

FACING The fetus in the womb, c. 1510–12 (detail, see plate 9).

Leonardo understood that these cycles of growth, decay, and renewal are linked to the cycles of life and death of individual organisms:

> Our life is made by the death of others. In dead matter insensible life remains, which, reunited to the stomachs of living beings, resumes sensual and intellectual life. . . . Man and the animals are really the passage and conduit of food.[3]

With these statements, he anticipated the concepts of food chains and food cycles that would become the central focus of ecologists more than four hundred years later.[4] It is also noteworthy that the phrase "life . . . sensual and intellectual" in this passage shows, like many other passages in the Notebooks, that Leonardo's concept of life included its cognitive as well as its biological dimensions.

Leonardo recognized that the energy driving the ecological cycles of growth and renewal, of life and death, flows from the sun. In his studies of plant growth he noted: "The sun gives spirit and life to the plants, and the earth nourishes them with moisture" (see p. 119).[5]

Finally, Leonardo understood that both plants and animals need the surrounding air to sustain themselves. In Manuscript G he noted that the branches of trees "take in the air which nourishes them" (see p. 120);[6] and in the Codex Atlanticus we find the observation:

> Where the air is not in the right proportion to accommodate the flame, there no flame can live, nor any terrestrial or airborne animal. . . . Where the flame does not live, no breathing animal can live.[7]

The critical role of water as the matrix and nourishing fluid of living tissues, the life-sustaining role of air (or its oxygen, as we know today), the continual growth and renewal of all organic forms, the cycles of life and death in the natural world, and the life-giving power of the sun were the fundamental characteristics of life that Leonardo observed and analyzed. He explored them in the macrocosm in his studies of fluid dynamics, geology, and botany; and late in his life he began to examine the same patterns and processes in the microcosm of the human body, recognizing them as components of a system of metabolic processes that are understood in modern science as key characteristics of biological life.

In his explorations of the nature and origin of life in the human body,

Leonardo focused on three interdependent processes. The first was the generation and transportation of the body's "vital spirits" which, according to the ancient philosophers, arose from a mixture of blood and air (identified in modern biochemistry as oxygenated blood). The second process was the ebb and flow of breath in the lungs, and the third was the digestion of food and the transport of nutrients to the bodily tissues by the blood.

Leonardo recognized that these interdependent processes are essential for sustaining life. He also realized that at their very core was the human heart, and he hoped that detailed investigations of the nature and actions of this mysterious organ would bring him closer to understanding the mystery of life.

The Human Heart

Throughout the ages, the heart has been the bodily organ that has served as the foremost symbol of human existence and emotional life. We metaphorically associate the heart with a variety of emotions. We "hold someone in our heart" (love) and speak of "a kind-hearted person" (compassion); we "take something to heart" (seriousness); we thank someone "from the bottom of our heart" (sincerity); we don't "have the heart" for a certain action (courage); and we make decisions "light-heartedly," "with a heavy heart," or after "a change of heart" (emotional depth).

In Leonardo's time, the associations of the human heart with life, consciousness, and emotion were much more than just metaphors. From antiquity to the Middle Ages, the heart had been considered a unique organ that generated the body's vital spirits (a mysterious "cardiac vapor"), as well as being the source of the body's heat.[8] For Aristotle (who was not aware of the central nervous system) the heart was not only the body's center of vitality but the very seat of the soul, that is, of intelligence, motion, and sensation. Galen, the leading medical authority in antiquity, emphasized the "noble nature" of the heart and maintained that, even though it might look like a muscle, it was something entirely different. It circulated the vital spirits throughout the body together with its "innate heat"; its expansion and contraction, for Galen, were signs of its role as an intelligent organ. Avicenna, the great physician and philosopher, attempting to integrate Aristotle's anatomy with Galen's physiology, saw the heart as the body's central and most important organ, but he stated that, being intelligent, it could delegate certain functions to other organs, especially to the brain.

Qui non estima la vita non la merita. (Ms. I, folio 15r)

One who does not respect life does not deserve it.

Leonardo's principal medical authority was Mondino, through whom he became acquainted with the works of Galen and Avicenna (see pp. 144–45). He accepted many of their concepts but readily departed from them when his observations taught him otherwise. Faced with this bewildering array of ideas about the heart inherited from antiquity, Leonardo concentrated on the twin problems, as he saw them, of how the actions of the heart maintained the blood at body temperature and how they produced the vital spirits that keep us alive. He adopted the ancient notion that these life-giving vapors arise from a mixture of blood and air—which is essentially correct, if we identify them with oxygenated blood—and he developed an ingenious theory to solve both problems.

As he did so often in his scientific investigations, Leonardo developed several theoretical models to explain the generation of body heat, discarding each model in turn when he found it unsatisfactory. His earliest attempt arose from the comparison of the flow of blood with the flow of water in "veins" inside the Earth and the flow of sap in plants. As I have discussed, Leonardo assumed that these three processes were all maintained by the same external power—the life-giving heat of the sun (see p. 18).

The sun, Leonardo thought, raises the "humors" (vital fluids) inside the three bodies: the water veins nourishing the Earth's vegetation, the sap nourishing the plant tissues, and the blood nourishing the tissues of the human body. Having been elevated to heights where they cool and condense, the fluids fall down again, only to be raised anew in continual circulation. After a few years, Leonardo realized that his analogy between the blood vessels inside the human body and water veins inside the Earth was too narrow, and eventually he reached a full understanding of the water cycle (see p. 29). As far as the movement of blood was concerned, he proceeded to develop a second model in which the heart acts like a stove, housing a central fire.

The idea of a cardiac "hearth" that generates the heart's innate heat had already been proposed by Aristotle. Leonardo linked this idea with a corresponding model of the water cycle in which water was supposed to be raised in special caverns inside the Earth in a process of distillation, fueled by the Earth's internal heat (see p. 28).

In the Codex Arundel, Leonardo made a small sketch showing the heart as a kind of furnace with inlet and outlet valves opening into separate chimneys, which represented the lungs. On a sheet in the Windsor Collection, he outlined a two-chambered heart with passages connecting the two chambers to the lungs, a clear analogy to his model of the oven.[9]

However, Leonardo soon became dissatisfied with the view of the heart as merely an oven containing the fire of bodily heat. As he made more detailed dissections of various parts of the heart, he became aware of their functions in regulating the flow of blood. Thus he embarked on creating his third, and far more sophisticated, model of the heart, using his understanding of turbulent flows of water and air and the role of friction to explain the origin of both the blood-air mixture of the vital spirits and the body's temperature. Although this model has serious flaws from the point of view of modern cardiology, it includes meticulous and accurate descriptions and drawings of many subtle features of the structure and actions of the heart and of the flow of blood—pioneering achievements in human anatomy.

To assess the significance of Leonardo's cardiac anatomy, it is useful to briefly review the modern understanding of the anatomy and physiology of the heart and blood circulation. The human heart is a pear-shaped structure made of a special muscle tissue and enclosed in a membranous sac. A wall of muscle (the septum) divides the heart into two cavities, each consisting of an upper chamber (atrium) and a lower chamber (ventricle), as pictured in figure 8-1.

Two types of blood vessels connect to the heart's four chambers: arteries, which carry blood away from the heart, and veins, which return blood to the heart. The main arterial vessel, the aorta, branches into smaller arteries, which in turn branch into still smaller vessels carrying blood to all parts of the body. Within the body tissues, the vessels are microscopic capillaries through which gas and nutrient exchanges occur.

After these exchanges, the blood converges from the capillaries into a network of minuscule veins, which in turn form larger veins that converge into the vena cava, the body's principal vein. The inferior vena cava is supplied by veins from the legs, the liver, and the kidneys. The superior vena cava receives blood from the head and neck.

In order to prevent the flow of blood from backing up, the heart is equipped with a series of valves at various openings: the tricuspid valve between the right atrium and right ventricle, the mitral valve between the left atrium and left ventricle, and several semilunar valves in the aorta and the pulmonary artery. During the cardiac cycle, these valves open and close in a precise rhythm.

The system of blood circulation consists of two distinct parts. The systemic circulation serves the entire body except for the lungs, while the pulmonary circulation carries blood to and from the lungs. In the systemic circulation, oxygenated blood from the lungs enters the left atrium via two pairs of pulmonary veins (one pair from each lung). When it is filled, the atrium contracts, sending the blood into the left ventricle. A large percentage of blood also enters the ventricle passively, without atrial contraction.

The powerful left ventricle then contracts, forcing the blood under great pressure into the aortic arch and on into the aorta. Three major arteries originate from the aortic arch, supplying blood to the head, neck, and arms. Other major arteries from the aorta supply blood to the kidneys, the spleen and liver, and the thighs and legs. At the periphery of the body, oxygen and nutrients diffuse into the tissue cells, while carbon dioxide (CO_2) and various metabolic waste products diffuse in the opposite direction, from the tissue cells into the capillaries, to be carried back to the heart by the veins. On this pathway of systemic circulation, part of the blood passes through the small intestine, where it absorbs nutrients from digested food, and proceeds to the liver for further digestion and regulation of various substances needed by the body. Another portion of blood goes through the kidneys, where the metabolic waste products are filtered out.

At the end of the systemic circulation, the blood, now low in oxygen and high in CO_2, begins the pulmonary circulation by entering the right atrium of the heart, from where it is pressed into the right ventricle. The right ventricle then contracts, forcing the blood into the lungs through the pulmonary arteries. In the lungs, the oxygen-poor and CO_2-rich blood flows through a vast network of capillaries surrounding the lungs' tiny air sacs. Oxygen from the inhaled air diffuses across the capillary membranes into the blood, where it binds to hemoglobin molecules in the red blood

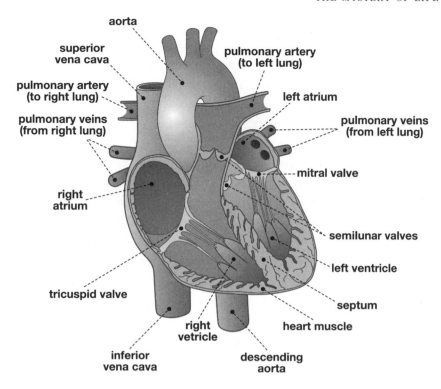

FIG. 8-1. The human heart and its blood vessels and valves.

cells, and CO_2 diffuses in the opposite direction to be exhaled. The oxygenated blood returns to the heart via the pulmonary veins, entering the left atrium to complete the cycle.

In addition, there is a separate system of so-called coronary vessels to nourish the muscle tissues of the heart itself (not shown in fig. 8-1). Two coronary arteries originate at the base of the aorta and carry oxygen-rich blood to all the tissues of the heart through a delicate system of branching vessels. A corresponding system of coronary veins collects the oxygen-poor blood and delivers it to the right atrium.

The blood circulation is precisely synchronized with the cardiac cycle, or heartbeat, which consists of three phases. In the first phase, the two atria contract in symmetry, emptying their contents into the ventricles. A fraction of a second later, the two ventricles contract simultaneously, forcing blood into arteries that exit the heart. During the strong ventricular contractions, known as systole, the tricuspid and mitral valves snap shut, producing the familiar "lubb" sound, the first part of the heartbeat.

Both atria and ventricles then relax briefly before the cycle repeats. At the beginning of the relaxation phase, known as diastole, the aortic and pulmonary valves (semilunar valves) close up, producing the characteristic "dubb" sound, the second part of the heartbeat. During the diastole, blood fills the atria and begins to flow passively into the ventricles. Both sides of the heart contract, empty, relax, and fill simultaneously. Therefore, only one systole and one diastole are felt.

It is noteworthy that in the pulmonary circulation, the arteries carry oxygen-poor blood (away from the heart) and the veins carry oxygen-rich blood (to the heart), whereas in the systemic circulation the oxygenated blood is carried by the arteries and the oxygen-poor blood by the veins. The pulmonary circulation intersects with the cycle of respiration in the lungs, while the systemic circulation intersects with the processes of digestion and waste excretion in the gastrointestinal tract, the liver, and the kidneys.

The Heart and Flow of Blood According to Galen

What Leonardo read about the structure of the heart and the flow of blood in the classical texts was quite different from our modern understanding.[10] Galen conceived of the heart as primarily a respiratory organ, made of a special substance and endowed with unique properties. Through some mysterious processes, the heart produced both the innate heat of the body and the force of life, or "vital spirits," and its most important function was to draw cool air from the lungs into the left ventricle, where the body heat was produced, to keep it from overheating.

This conception of the heart, which seems rather strange to us today, was derived from the fact that Galen's entire physiology of the human body was based on the traditional idea of the soul as a fiery vital breath (*pneuma*), which took the form of several distinct "spirits" that were seen as the primary movers of all bodily functions.[11]

Galen postulated three of these life-giving forces. The "natural" spirits were created by the liver. They transformed digested food into blood and distributed it through the veins. The "vital" spirits were produced by the heart by transforming a portion of the air drawn from the lungs, and were then distributed through the arteries. And finally, the "animal" spirits, the body's motor forces, originated in the brain and were transported to the muscles through hollow nerves.

Galen recognized only two chambers of the heart (the ventricles), viewing the atria merely as the endings of the vena cava and the pulmonary veins. He mentioned the cardiac valves but called them "orifices" and ignored their functions. Galen adopted the view of earlier Greek philosophers who had treated the arteries and veins as two completely separate systems, the former carrying the vital spirits (that is, essentially air) that were produced in the left ventricle of the heart, and the latter transporting the blood that was produced by the liver. However, Galen maintained that the arteries also contained some blood, which had passed through tiny pores in the septum from the right to the left ventricle.

Galen's ideas about the movement of blood were rather confused. He believed that blood was made in the liver out of food, and that it carried nutrients to the bodily tissues through the veins (that is, in the opposite direction of the actual flow of blood in the veins). Some of that blood was sucked into the heart's right ventricle when it dilated, and there it was "subtilized" (made thin and light). When the right ventricle contracted, some of this subtle blood seeped through the septum into the left ventricle, and the remaining blood passed to the lungs through the pulmonary artery to be exhaled together with air. In the left ventricle, the subtle blood combined with the air drawn from the lungs to form the vital spirits, which were then transported to the bodily tissues through the system of arteries.

However, in Galen's theory the distribution of the vital spirits through the arteries, as well as that of the blood through the veins, was not by circulation but rather by fluctuations similar to the ebb and flow of the air in the lungs and the trachea. This must have been a natural assumption for him, since he thought of the heart as a respiratory organ and ignored the functions of the cardiac valves.

Leonardo's Anatomy of the Heart

Leonardo's sophisticated studies of the movements of the heart and blood, undertaken in Milan and Rome when he was in his early sixties, are the culmination of his anatomical work. He not only understood and pictured the heart like no one before him; he also observed subtleties in its actions and in the flow of blood that would elude medical researchers for centuries.

Leonardo illustrated his discoveries in a series of stunning drawings, now in the Windsor Collection. One of the most impressive is also one of his last anatomical drawings, dating from 1513. It is a magnificent double sheet showing the heart of an ox from several perspectives (plate 10).

Leonardo's main purpose in this study was to demonstrate the coronary vessels.[12] In the two figures on the left side of the sheet, the coronary arteries are seen to originate at the base of the aorta and to divide into several branches. The pulmonary artery has been cut away so as to display the roots of the aorta and vena cava. The three cusps of the pulmonary valve are clearly visible in the orifice created by the severance of the artery.

The two figures on the top right show coronary veins as well as arteries. Their delicate branching patterns, crossing one another repeatedly, are beautifully rendered. The small sketches at the center of the right margin illustrate how the coronary vessels crown the heart, which explains their modern name. Below them, in the bottom right corner, Leonardo sketched the pulmonary valve and the tricuspid valve viewed from above, showing the latter both closed and open. The entire sheet is an impressive testimony to Leonardo's understanding of many subtle features of cardiac anatomy.

The main part of Leonardo's complex anatomical studies of the heart dates from the years 1511–13. At the beginning of this period, he recorded two major discoveries in a relatively large pocket book, now known as Manuscript G. The first was his conclusion that the heart, contrary to Galen's view, is a muscle; the second was the observation that it had four cavities, not two, as all earlier medical authorities had believed.

On the very first folio of Manuscript G, next to several sketches of a dissected heart, Leonardo states categorically, "The heart is a principal muscle of force, and it is much more powerful than the other muscles."[13] After studying the classical medical texts in which the heart was said to be made of a special substance endowed with rather mysterious properties, and developing several theoretical models that failed to explain those properties, recognizing the heart as a muscle was a major breakthrough for Leonardo. Since he had already studied the nature and actions of muscles extensively, he realized that the heart, like any other muscle, had to be nourished by special blood vessels, and that its actions had to be triggered by special nerves. "The heart in itself is not the beginning of life," he noted in the *Anatomical Studies*, "but is a vessel made of dense muscle, vivified and nourished by the artery and vein as are the other muscles."[14]

During subsequent years, Leonardo explored the pathways and branching patterns of the coronary arteries and veins in great detail, summarizing the results of his investigations on the exquisite double sheet of the Windsor Collection (plate 10). At the same time, he looked for the

nerves that stimulate the heart muscle and located them correctly in the
large network known today as the vagi, or "wandering nerves."* Leonardo
called them "reversive nerves," probably because of their frequent changes
of direction, and in a note to himself emphasized the importance of their
exploration:

> Follow up the reversive nerves as far as the heart and see whether
> these nerves give movement to the heart, or whether the heart
> moves by itself. And if such movement comes from the reversive
> nerves, which have their origin in the brain, you will make it clear
> how the soul has its seat in the ventricles of the brain.[15]

It is evident from this passage that Leonardo correctly traced both the
external and internal movements of the body back to the motor nerves and
their origins in the brain.

Indeed, just as he represented the anatomy of the body's muscles, tendons, and bones in terms of their movements, Leonardo pictured the heart
in motion from the very beginning of his investigations. The sketches in
Manuscript G are still quite imperfect, but already the heart is shown in
action, with contracted or dilated chambers.

The fact that the heart has four cavities, rather than two, was the second major discovery of Leonardo's early cardiac anatomy. He called the
atria the "auricles of the heart"—a term still in use today—and he correctly identified their role as "the heart's antechambers."[16] Leonardo was
well aware of the importance of this discovery, repeating his assertion in
several places in the *Anatomical Studies*.

Soon after his first sketches in Manuscript G, he produced a more
elaborate drawing (fig. 8-2) in which the atria are clearly distinguished
from the ventricles. In the accompanying text, Leonardo provides a succinct description:

> The heart has four ventricles, that is two lower ones in the sub-
> stance of the heart and two upper ones outside the substance of the
> heart. And of these, two are on the right and two on the left, and the

* The *vagus* (also used in the plural, *vagi*) has a more extensive course and distribution than any other
cranial nerve, traversing the neck, thorax, and abdomen. It supplies the heart and several other
organs with sensory and motor fibers.

ones on the right are much larger than the ones on the left. And the upper ventricles are separated from the lower ones by certain little doors, or gateways of the heart.[17]

On another folio, composed a couple of years later, he recorded a shorter version of the same statement: "The heart has four ventricles, that is two upper ones called auricles of the heart, and two below them called the right and left ventricle."[18]

Leonardo fully understood the functioning of the cardiac valves, and he demonstrated with amazing accuracy their shapes in various stages of opening and closing.[19] The characteristic H-shaped cleft of the mitral valve and the Y-shaped closures of the other three valves are clearly visible in several drawings of the cardiac orifices. The tricuspid valve, in particular, is explored in great detail, with respect to both its structure and its movements in action.[20]

Medical historian and Leonardo scholar Kenneth Keele has juxtaposed two of Leonardo's drawings of the aortic valve in open and closed positions with two modern pictures of the same valve, obtained by Keele himself by means of high-speed photography (fig. 8-3). The result is stunning. The triangular shape of the orifice between the open cusps, the cusps' wavy edges, and the detailed shape of the closed valve, as drawn by Leonardo five hundred years ago, are virtually identical to the anatomical features shown in the modern photographs.

In addition to demonstrating the precise shapes of the tricuspid and mitral valves in various positions, Leonardo dissected the so-called papillary muscles and showed how they contribute to controlling the actions of these valves by means of thread-like tendons, known today as chordae tendineae ("fibrous cords"). The way in which these fibers are attached to the valve cusps is

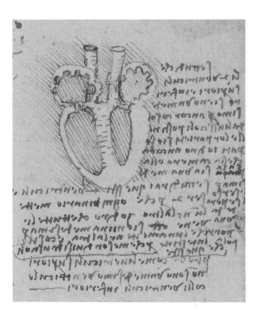

FIG. 8-2. The four chambers of the heart, c. 1511. Windsor Collection, *Anatomical Studies*, folio 155r (detail).

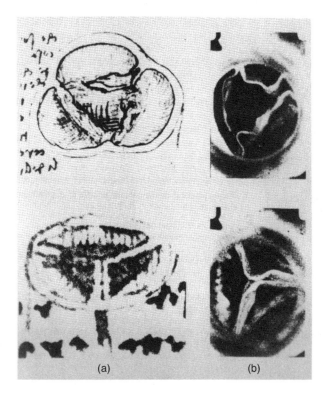

FIG. 8-3. Aortic valve cusps, open and closed,
as seen in Leonardo's drawings (a) and in modern
high-speed photography (b). From Keele,
Leonardo da Vinci's Elements of the Science of Man, p. 319.

demonstrated by Leonardo in great detail. It was not until the eighteenth
and nineteenth centuries that the papillary muscles and their tendons
were studied again with such meticulous care.[21]

Leonardo's precise representations of a variety of subtle anatomical
structures of the heart were matched by his accurate descriptions of many
cardiovascular functions. He correctly described the origin of the pulse as
being in the rhythmic contractions of the arteries, which help the heart
pump the blood and maintain a steady flow to the smaller vessels. He
clearly recognized the connection between pulse and heartbeat:

> The beating of the heart ... generates a wave of blood through all
> the vessels, which continually dilate and contract. ... And this we
> learn from the beating of the pulse when we touch the aforesaid
> vessels with the fingers in any part of the living body.[22]

Leonardo was also the first to appreciate that the heart shortens in systole (when it contracts) and lengthens in diastole (when it relaxes), which contradicted the traditional Galenic teachings and was confirmed 120 years later by the physician William Harvey. Even more remarkably, Leonardo offers the first correct interpretation of the cardiac impulse* against the chest wall:

> The time of the contraction of the heart and of the percussion by its apex against the rib cage, of the beating of the pulse, and of the entrance of the blood into the front gateway of the heart [the aortic orifice] is one and the same.[23]

The realization that the contraction of the ventricles, the thump of the heart's tip against the chest wall, the pulse, and the ejection of blood into the aorta all occur at the same time must be ranked as one of the greatest discoveries in Leonardo's anatomy of the heart. None of his predecessors and contemporaries were aware that these phenomena are all interrelated. Again, it was Harvey who rediscovered their connections after more than a century had passed.

"Flux and Reflux" of the Blood

In view of Leonardo's accurate visual demonstrations and verbal descriptions of so many subtle features of the cardiovascular system, it is difficult to believe that, unlike William Harvey, he did not recognize the circulation of the blood. Yet this is the case; so, why did Leonardo miss it and Harvey did not? Both of these brilliant scientists started from the same premise, the classical texts of Galen and Avicenna; both struggled with the same problems before the development of chemistry and the perfection of the microscope. Indeed, comparing Leonardo's cardiac anatomy and physiology with those of Harvey is as revealing as comparing his mechanics with that of Galileo, Harvey's contemporary (see p. 174).

One critical difference between the cardiac research of Leonardo and Harvey was that Leonardo refused to perform vivisections and therefore never saw the flow of blood through the heart and its vessels. He wrote

---

* The cardiac impulse is the result of the heart rotating, moving forward, and striking against the chest wall during systole, just as Leonardo described it.

about visiting an abattoir in Tuscany where he observed the slaughter
of pigs and drew conclusions from the ways in which the blood gushed
forth from their wounds, but performing vivisections himself was far too
repugnant for Leonardo. Harvey, by contrast, would open up dogs and
pigs, seemingly without any qualms, to observe and touch their "flagging"
hearts when the animals died.[24] In his accounts of these experiments,
which sound quite cruel today, Harvey explained calmly and in gruesome
detail how he obtained direct evidence about various aspects of the heart's
movements.

However, I believe that Leonardo might have recognized the circulation of blood even without evidence gained from vivisection. What
prevented him from doing so much more fundamentally was the way in
which he framed the whole issue of cardiac physiology. From his readings
of the classical texts, Leonardo distilled two central questions: how does
the heart generate the body heat that is characteristic of all mammals,
and how does it maintain the life force, or "vital spirit," that animates the
bodily tissues? He used his sophisticated understanding of turbulent flows
and of friction to develop a brilliant but erroneous model of blood flow,
involving continual currents swirling back and forth between the heart's
atria and ventricles, thereby producing both the body's innate heat and its
vital spirits.

Leonardo's outstanding discoveries in cardiac anatomy took him far
beyond the prevalent views of Galen. He recognized the heart as a muscle
exhibiting all the characteristics of muscular contraction; he identified
four chambers of the heart instead of two; and he emphasized that the
active movement of the heart was its contraction in systole, which expelled
blood from the ventricles into the main vessels, rather than an expansion
in diastole to draw air from the lungs into the heart, as Galen maintained.
In fact, Leonardo disproved this Galenic view by inflating the lungs of a
dead ox and observing that no air entered the pulmonary vein.[25]

In spite of all these advances, Leonardo clung to the fundamental Galenic idea that the blood moved in parallel in two separate vascular systems, and that this movement was one of ebb and flow—from the heart
out to the body's periphery and back to the heart—along both arteries and
veins. There is no suggestion in Leonardo's manuscripts of blood moving
through tissue from the arteries to the veins, and hence no indication of
any conception of circulation.

Leonardo maintained the Galenic conception of the ebb and flow of blood for several reasons. It was not contradicted by any of his observations, and it was consistent with his thorough studies of respiration—the ebb and flow of breath through the trachea and the lungs.[26] He saw the ebb and flow of blood as another manifestation of the cyclical movements so characteristic of human physiology.

Most important, perhaps, was the fact that Leonardo postulated a cyclical movement of blood right inside the heart to explain the generation of body heat. As the atria and ventricles contract and dilate in synchrony, the blood flows back and forth between them, and by the friction in this "flux and reflux" it is heated and "subtilized." This hydrodynamic model represented the very core of his cardiac anatomy, and it was fully consistent with the idea of cyclical movement of blood through the body. I think this was the main reason why Leonardo, with all his skillful anatomical dissections and great powers of observation, failed to recognize the blood's circulation.

When he discovered the atria, he immediately had the idea of explaining the body's heat as resulting from the friction of blood swirling through the four chambers of the heart. His extensive knowledge of turbulent flows allowed him to picture small vortices with great precision and to describe their motion accurately. On the very same folio of the *Anatomical Studies* that shows his first clear drawing of the atria and ventricles (see fig. 8-2), Leonardo provides a detailed description of his hydrodynamic model of heat generation:

> The upper ventricles continually make a flux and reflux of the blood which is continually pulled or pressed by the lower ventricles from [or into] the upper. . . . And so, by such flux and reflux, made with great rapidity, the blood is heated and subtilized, and is made so hot that, but for the help of the bellows called lungs, which, by being dilated draw in fresh air, pressing it into contact with the coats of the ramifications of the vessels, refreshing them, the blood would become so hot that it would suffocate the heart and deprive it of life.[27]

Leonardo attempts here to provide a scientific explanation of body heat, which was seen by Galen as an innate and rather mysterious property of the heart. He knows that the principles of flow are the same for blood as for any other liquid (see p. 33), that friction always generates heat (see p. 202), and that liquids expand (become "subtilized") when they are heated; and he assembles these observations from different areas of mechanics into a coherent (if incorrect) theoretical model.

Today we know that the body heat of mammals is the result of myriads of biochemical reactions throughout the bodily tissues, and that the body temperature is controlled by a heat-regulating center in the brain. Leonardo, living more than three hundred years before the development of biochemistry, could not have conceived of such an explanation. He erroneously assumed that the heat was generated by the swirling blood in the heart, but he was correct in his assumption that the blood is essential for distributing the heat throughout the body and maintaining a uniform body temperature. William Harvey, interestingly, did not attempt to explain the origin of body heat but simply assumed, without going into further details, that it was generated by the heart.[28]

Leonardo's description of the role of the lungs in the passage quoted above is very intriguing. He adhered to the Galenic view that the function of the lungs was to cool the blood, but he contradicted Galen by asserting correctly that no air passes from the lungs to the heart. Therefore, Leonardo concluded, the cooling of the blood takes place when the air in the lungs "is pressed into contact with the coatings of the ramifications of the vessels." This is a remarkably accurate description of the exchanges between the lungs' air sacs and the network of blood vessels surrounding them—though what is exchanged, in Leonardo's view, is not oxygen but merely heat.

On the subsequent folio, Leonardo continues his discussion of body heat with a detailed description of the turbulent flows of blood in the heart:

> And so, between revolving up and down successively it never ceases to flow through the cavernous recesses interposed between the muscles which contract the upper ventricle. And the whirling round of the blood in diverse eddies, and the friction it makes on the walls, and the percussions in the hollows, are the cause of the heating of the blood, and making it from thick and viscous to subtle and penetrative, suitable for flowing from the right to the left ventricle through the narrow porosities of the wall interposed between that right and left lower ventricle.[29]

In addition to generating the body temperature, the heating of the blood serves two further purposes, in Leonardo's view. On the right side of the heart, as he explains in the above passage, it is transformed from a "thick and viscous" liquid into one that is so "subtle and penetrating" that it can pass to the left ventricle through the invisible pores in the septum postulated by Galen. Harvey, in contrast to Leonardo, rejected the Galenic conception of invisible septal pores, postulating correctly that the blood moves from the right to the left side via the pulmonary circulation. However, Harvey was never quite comfortable with this argument, since it replaced the idea of blood seeping through invisible pores in the septum with that of its passage through equally invisible capillaries in the lungs.[30]

The Vital Spirits

Galen's physiology of the human body, as I have mentioned, was based on three types of life-giving forces, or "spirits," which were seen as different manifestations of the same vital breath (*pneuma*) and as the primary agents of all bodily functions (see p. 288). Leonardo, with his strictly empirical approach to scientific knowledge, which rejected all notions of supernatural forces,[31] eliminated two of Galen's spirits from his conception of human physiology. He questioned the ancient idea of animal spirits moving through hollow nerves as some kind of "psychic wind," and replaced it with the much more sophisticated concept of immaterial nervous impulses traveling through the sensory and motor nerves in the form of waves.[32] He also rejected the Galenic notion of natural spirits, the agents of digestion, as nonexisting entities for which he could find no evidence (see p. 309).

However, Leonardo retained Galen's concept of vital spirits as a fundamental force of life. Having identified the life-sustaining role of air in numerous observations of animal and plant life (see p. 282), he conceived of the idea that these spirits were some vital essence of air, which was isolated in the heart, intermingled with blood, and then transported to the periphery of the body in order to animate all bodily tissues. From our modern perspective we can see that Leonardo's intuition was absolutely right. Oxygen is the life-sustaining essence of the air, and oxygenated blood is the "mixture" of blood and air that animates the body's tissues.

Leonardo's theoretical model for the generation of the vital spirits is essentially the same hydrodynamic model he used to explain the generation of body heat, but the process is somewhat more complex and takes place predominantly in the left ventricle:

> [The blood] is more heated in the left ventricle where the walls are thick than in the right ventricle with the thin wall. And that heat subtilizes the blood and vaporizes it, and converts some of it into air. . . . The lung cannot send air into the heart, nor is it necessary, because as said, air is generated in the heart [and] mixed with heat and condensed moisture.[33]

According to Leonardo, the blood-air mixture is not obtained from air sucked in from the lungs, as Galen believed, but is produced by vaporizing part of the blood. In this way, the vital spirits are formed out of heat, humidity, and the mixture of blood and air. To generate the necessary heat for this process, the turbulence of the blood in the left ventricle must be much stronger than that in the right. For Leonardo, this is confirmed by the fact that the muscle walls of the left ventricle are thicker than those of the right ventricle. (We know now that this is due to the greater force needed to pump the blood through the systemic circulation.)

Like his explanation of the body temperature, Leonardo's account of the generation of the vital spirits integrates observations from several areas of his science into a coherent theoretical model: the relation between friction and heat, between air and the sustenance of life, and between heat and the body's living tissues—or between energy and metabolic processes, as we would say in modern scientific language. As Leonardo sums it up, "[without] flux and reflux the blood would not be heated, and consequently the vital spirits could not be generated, and therefore life would be destroyed."[34]

According to Leonardo, the vital spirits created in the left ventricle are "augmented and vivified" by further turbulences as the blood enters the base of the ascending aorta:

> The revolution of the blood in the anteroom of the heart at the base of the aorta serves two effects, of which the first is that this revolution, multiplied in many aspects, makes within itself great friction, which heats and subtilizes the blood, and augments and vivifies the vital spirits, which always maintain themselves in warmth

and humidity. The second effect of this revolution of the blood is to close again the opened gates [valve cusps] of the heart with its first reflected motion, with perfect closure.[35]

This passage summarizes what must be seen as the most sophisticated and most extraordinary piece of Leonardo's anatomy of the heart. To determine the precise shape of the turbulences in the aorta, he analyzed the flow patterns behind the aortic valve in incredible detail, picturing them repeatedly from various angles (for example, fig. 8-4) and describing them in long paragraphs on several folios of the *Anatomical Studies*.[36] He showed that the blood, as it passes through the valve's triangular orifice, forms three distinct eddies in a retrograde motion after impinging on the stationary blood already in the aorta. Leonardo demonstrated that these vortices are generated when the blood flows through three pouches in the wall of the aorta right behind the aortic valve. He called these hemispherical pouches "hemicycles." Today they are called the sinuses of Valsalva in honor of the anatomist Antonio Valsalva, who rediscovered them in the eighteenth century.

The most remarkable part of Leonardo's analysis of the three vortices is his discovery that they fill out the semilunar cusps of the aortic valve

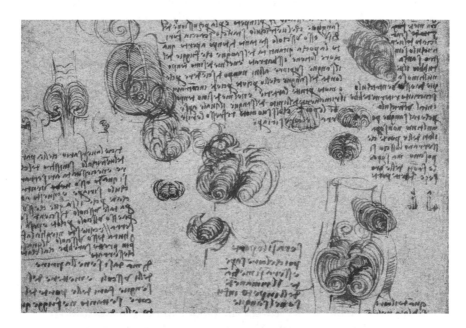

FIG. 8-4. Turbulent flows of blood at the base of the aorta, 1513.
Windsor Collection, *Anatomical Studies*, folio 172v (detail).

and begin to close it before the ventricle's contraction ends and the weight of the column of blood in the aorta shuts the valve completely. Leonardo finds a persuasive argument against the valve being closed by the weight of the blood above it alone: "The shape of the valve denies this, as it would quicker be crushed than shut," he writes next to a tiny sketch of a crushed valve cusp (see fig. 8-4, top left corner).[37]

Leonardo realized that determining the exact patterns of turbulence in the aorta was extremely difficult, and sometimes he was doubtful about whether his interpretation was correct. "Such doubts are subtle and difficult to prove and clarify," he mused.[38] But he did not leave things there. Incredible as it may seem, especially in view of his advanced age at the time, Leonardo planned to test his hypothesis by building a glass model of the aorta's base, including the aortic sinuses, with a valve taken from the heart of an ox. He would pour water into the model, with millet grains mixed in, to observe the turbulent flows.* "But first pour wax into the gate of the heart of an ox," he reminded himself, referring to a technique of dissection he had used many years earlier, "so that you may see the true shape of this gate."[39]

We do not know whether Leonardo ever built his glass model of the aorta. In any case, the experimental verification of his hypothesis (that the closure of the aortic valve is initiated by eddies of blood swirling through the aortic sinuses) had to wait for more than four hundred years.[40] This is certainly one of the most astonishing cases of a scientific discovery far ahead of its time. Here is how medical historian Sherwin Nuland tells the story.

> Until at least the early part of the twentieth century, it was assumed by all cardiac researchers that the valve between the heart and the aorta (the aortic valve) functions passively, like that of a standard water pump: When the heart contracts, it pushes blood out and forces the valve open so that the blood can be ejected upward into the aorta; when the pressure of contraction lessens, the valve is forced shut by the weight of the column of blood in the aorta, pressing down from above it. This seemed a perfectly straightforward explanation of the hydraulics of the system. . . .
>
> But in 1912 it was demonstrated that the dynamics are not quite as simple as had been thought. In fact, the process of valve closure

* For a description of Leonardo's techniques of flow visualization, see p. 38 above.

was shown to be somewhat more gradual than could be accounted for by an abrupt change in pressure. . . . Decades more had to pass before investigative technology had reached such an advanced state that the details could be satisfactorily explored and actually visualized. By the 1960s, dye and cineradiography methods had been sufficiently developed that it was possible to study flow patterns with extreme accuracy. It was demonstrated that some of the blood which is ejected into the aorta swirls into the [sinuses of Valsalva] and forms eddy currents that exert pressure on the upper surface of the valve, causing it to begin closing even before the ventricle has completed its contraction. This could not have been known without the new research methods.

Or so it was thought. Leonardo da Vinci had shown the same thing in the first decade of the sixteenth century. . . . Both his text and illustrations clearly show the correct mechanism of both opening and closure of the three leaflets that make up the aortic valve, including the fact that the initiation of the closure is due to eddy currents originating in the sinuses of Valsalva. He demonstrated repeatedly that the closure is gradual. Leonardo's observations are identical with those that would be made by groups of researchers in a series of studies beginning in 1969, and he drew the same conclusions from them as they would. . . . Of all the amazements that Leonardo left for the ages, this one would seem to be the most extraordinary.[41]

The Flow of Blood Through the Body

When Leonardo visualized the flow of blood through the chambers of the heart and through the arteries and veins, his main concern was to understand how the vital spirits are generated, how they "vivify" the bodily tissues, and how the blood nourishes these tissues. In other words, he wanted to understand the body's basic metabolic processes (as we would say today)—the very essence of biological life.

The precise mapping of the movement of blood through the heart and body was a secondary issue for Leonardo. As medical historians O'Malley and Saunders point out, his views on the subject varied considerably over time, and he never recorded them in a complete statement.[42] According to these scholars, Leonardo's theory of the movement of blood may be summarized as follows.

At the beginning of the cardiac cycle, the right atrium contracts, while the right ventricle dilates, which makes the blood rush into the ventricle, thus creating turbulent currents. Then the right ventricle contracts, while the atrium dilates, sending blood back to the atrium and also into the lungs through the pulmonary arteries. A small portion of "subtilized" blood seeps through the septal pores into the left ventricle. This process is facilitated by the dilation of the left ventricle during the contraction of the right. During its dilation, the left ventricle also receives blood from the contracting left atrium.

Then the left ventricle contracts, sending part of the blood back to the atrium and ejecting the other part through the aorta. In the forceful flux and reflux between the left atrium and ventricle, the blood is heated considerably and some of it is vaporized to generate the vital spirits, which are sent to the bodily tissues through the aorta together with the ejected blood. A small part of the vaporized blood enters the lungs through the pulmonary veins and escapes into the bronchi. On both sides, the blood is cooled in the lungs before returning to the atria and ventricles.

In Leonardo's cardiac cycle, the atria and ventricles contract and dilate alternatively, as in the modern understanding of blood circulation. However, the ventricles on the left and right do not contract in symmetry, so as to facilitate the passage of blood through the septal pores when the right ventricle contracts while the left expands. In other words, the right atrium and left ventricle contract in symmetry during the first phase of Leonardo's cardiac cycle; and the left atrium and right ventricle contract in symmetry during the second phase.

One problem of this scheme was to explain how the blood could pass back and forth between the atria and ventricles through the valves that separate the two chambers on both sides of the heart. Leonardo tackled this problem by proposing a complex sequence of synchronized actions of the two valves through rhythmic contractions and relaxations of their papillary muscles. As Keele commented, "It is a very neatly reasoned and consistent account of papillary muscle action. Its error lies basically in Leonardo's ignorance of the fact that the papillary muscles are part of the main mass of cardiac muscles, all of which contract in systole."[43]

As far as the flow of blood through the arteries and veins was concerned, Leonardo retained the fundamental Galenic idea of its ebb and flow in two separate vascular systems, as I have mentioned. This implied that once the blood reaches the body's periphery, it is used up by the tissues for their "vivification" and nourishment, which means that it has to be constantly replenished. Following Galen, Leonardo assumed that the blood was formed in the liver, which is not completely wrong, since some essential components of blood (including blood-clotting substances and other plasma proteins) are indeed manufactured there.

In his dissections of blood vessels, Leonardo's starting point was the works of Avicenna and Mondino (see pp. 144–45). In both of these texts, there is considerable confusion between arteries and veins, some of which is still present in Leonardo's work when he uses the term *vene* indiscriminately to describe either arteries or veins. Leonardo produced most of his drawings of blood vessels several years before his sophisticated anatomies of the heart. They reached their climax with his dissection of the centenarian around 1508, which he documented with a series of magnificent anatomical drawings (see p. 227).

One of the most accomplished of these anatomical studies shows the blood vessels of the liver, known today as the portal and hepatic vessels (fig. 8-5). The major branches of these arteries and veins, as well as the fractal structures of their ramifications in the body of the liver, are depicted by Leonardo with such clarity that modern medical scientists can easily identify them.[44]

Blood is carried to the liver via two large vessels. The hepatic artery carries oxygen-rich blood from the abdominal aorta, and the portal vein carries blood containing digested food from the gastrointestinal tract. These blood vessels subdivide repeatedly in the liver until they form minute capillaries through which the blood enters into clusters of hepatic cells, known as lobules. The liver tissue is composed of thousands of these lobules, in which the nutrients transported by the blood are further digested and various toxic substances are filtered out. Having thus been cleansed, the blood is collected by a corresponding network of minuscule veins that converge into progressively larger veins, eventually ending up in the single hepatic vein that drains the blood into the vena cava.

The drawings in figure 8-5 show that Leonardo has dissected the blood vessels out of the liver substance in order to clearly demonstrate their branching patterns. The figure on top of the page shows the networks of blood vessels formed in the liver by the hepatic artery and the portal vein. Leonardo shows how the artery arises from the abdominal aorta and how it gives off several branches that carry blood to the other digestive organs before it enters the liver.*

---

* In modern terminology, only the branch entering the liver is called the hepatic artery; see O'Malley and Saunders, *Leonardo da Vinci on the Human Body*, p. 304, for the names of the other branches.

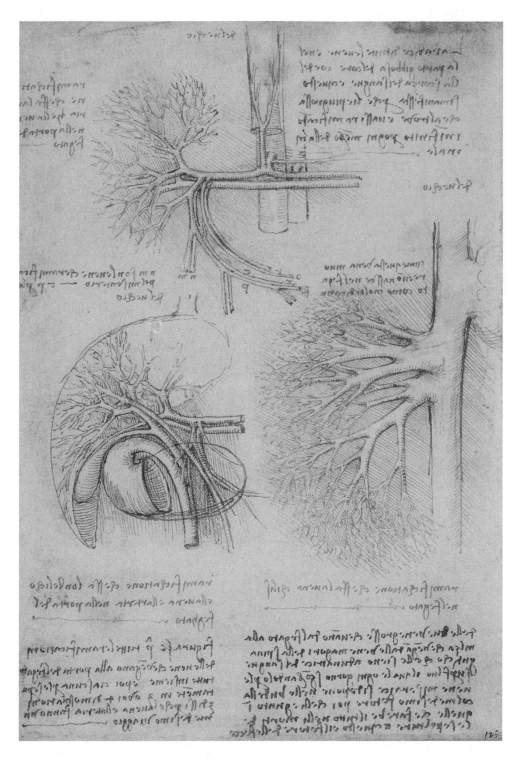

FIG. 8-5. The blood vessels of the liver, c. 1507–8.
Windsor Collection, *Anatomical Studies*, folio 60v.

Below the branches of the hepatic artery lie the portal vein and its ramifications (the hepatic veins). Both the hepatic artery and the portal vein are seen to divide into right and left branches in the liver before generating a sequence of smaller and smaller ramifications. The large network of hepatic veins with its numerous branches, all draining into the vena cava, is demonstrated in the lower right figure. In addition, Leonardo loosely sketched the opening of the vena cava into the right atrium of the heart.

The figure on the left is partly the same as that above. But here Leonardo has indicated the outlines of the liver and stomach, and he has added some of the vessels, or "ducts," that carry bile from the liver to other digestive organs. In the words of O'Malley and Saunders, "The gall bladder, cystic duct, hepatic ducts and common bile duct passing to the duodenum are amazingly portrayed."[45] The fact that medical scientists today have no problem identifying these anatomical details, which are bewildering for the lay person, is impressive testimony to Leonardo's anatomical skills, holistic memory, and mastery of pictorial demonstration.

Leonardo performed his anatomy of the centenarian, who died in his presence without having experienced any major illness, "in order to see the cause of so sweet a death" (see p. 227). His careful dissection of the old man's entire body resulted not only in a series of superb studies of the blood vessels and internal organs, but also in Leonardo's most famous and most amazing medical discovery.

He observed that the blood vessels of the old man had thickened with age and that their ramifications had become tortuous. This observation was reinforced when he had the opportunity, by coincidence, to dissect the smooth and straight vessels of a two-year-old child around the same time. Leonardo recorded his observations on a folio of the *Anatomical Studies* with two small sketches, one showing a straight set of branching vessels labeled "young," and the other a set of twisted ramifications labeled "old" (fig. 8-6). In a note next to the drawings, under the heading "Nature of the ves-

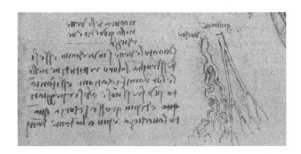

FIG. 8-6. Blood vessels of the young (right) and the old (left), c. 1507–8. Windsor Collection, *Anatomical Studies*, folio 69r (detail).

sels in youth and old age," Leonardo succinctly describes the condition now known as arteriosclerosis:

> In so far as the vessels become old, their straightness is destroyed in their ramifications, and they become so much more sinuous, or twisting, and of a thicker coat, as their age is fuller in years.[46]

On the verso of the same folio, Leonardo develops the theme of arteriosclerosis with age at some length, together with his brilliant and correct interpretation of the condition, which he observed to be particularly severe in the blood vessels supplying the centenarian's liver. Leonardo begins his analysis with a description of the thickening of the vascular walls, due to their prolonged contact with the nourishing blood, the consequent obstruction of the blood flow, and the effects of the diminished blood supply on the liver:

> The artery and the vein . . . acquire so thick a skin that it restricts the passage of the blood. . . . And these vessels, apart from the thickening of their skin, grow in length and twist like a snake, and the liver loses the nourishment of the blood which was carried there by that vein. Thus the liver becomes desiccated and like congealed bran both in color and in substance, so that when it is subjected to the slightest friction, its substance falls away in small particles like sawdust and leaves behind the veins and arteries.[47]

According to O'Malley and Saunders, this passage contains not only the first description of arteriosclerosis in medical history, but also the first vivid and clear account of cirrhosis of the liver.[48] A couple of paragraphs below, on the same page, we find Leonardo's moving story of his encounter with the old man at the hospital of Santa Maria Nuova, followed by his diagnosis of the centenarian's cause of death:

> And I made an anatomy of him in order to see the cause of so sweet a death, which I found to be weakness from lack of blood and deficiency of the artery [aorta] that nourishes the heart and other parts below it, which I found very dry, thin, and withered. . . . The other anatomy was that of a child of two years in which I found everything to be the opposite to that of the old man.

In the margin of the same page, Leonardo concludes his analysis by summarizing arteriosclerosis in general terms:

> The aged who enjoy good health die through lack of nourishment. This happens because . . . the passage of the meseraic [portal] veins is continually constricted by the thickening of the skin of these veins, progressing as far as the capillary vessels which are the first to close up entirely.

This passage is noteworthy also because it shows that Leonardo was the first to use the term "capillary vessels" (*vene capillari*) for the minute blood vessels, which were unknown in his time.

Finally, at the end of the page, Leonardo puts his discovery into the overall context of his science of living forms by observing that the aging process of blood vessels represents a general pattern of behavior of living tissues:

> [The] coat of the vessels acts in man as in oranges, in which, as the skin thickens, so the pulp diminishes the older they become.

After Leonardo's detailed description and correct interpretation of arteriosclerosis, it took another three hundred years for the condition to be rediscovered. In the early nineteenth century, the anatomist Antonio Scarpa accurately described and illustrated the thickening of the arterial walls, based on his dissections, but attributed the process merely to "internal unknown causes." Some thirty years later, the pathologist and surgeon Jean Lobstein coined the term "arteriosclerosis" and, like Leonardo, asserted that the condition was due to an "abnormal state of nutrition" of the tissues.[49] The role of cholesterol and other fatty substances in the thickening and hardening of the blood vessels was investigated only during the second half of the twentieth century and is still not fully understood today.

The Digestive System and the Body's Metabolic Processes

When Leonardo investigated the blood supply to the liver and reflected on the effects of the nutrients in the blood on the arterial walls, he had already carried out extensive studies of the entire digestive system.[50] During the three years preceding his dissection of the centenarian, 1506–8, he produced a series of splendid representations of the abdominal muscles, as well as the stomach, intestines, liver, and spleen. Moreover, he demonstrated the blood supply to the gall bladder and depicted the intricate intestinal blood vessels. Leonardo was the first to show the correct relative positions of the small and large intestines, and the first to recognize and clearly depict the appendix.

In his investigations of the physiology of digestion, Leonardo accurately described the act of swallowing and the passage of food through the esophagus to the stomach and the intestines. However, he was unaware of peristalsis, the waves of contractions of the intestine that pass along the food, having never observed it because of his ethical objection to vivisection. Leonardo attributed the movement of food through the digestive tract to pressure exerted by the diaphragm in its downward movement. A folio of the *Anatomical Studies* contains a detailed discussion of the diaphragm's dual function* as "the motor of food and air."[51]

Without access to biochemistry, Leonardo explained the process of digestion that we now associate with digestive enzymes by assuming that the food was broken down and liquefied in the stomach and intestine by bodily heat in a process of "coction." This was consistent with the Galenic teachings, but Leonardo disagreed with Galen's view that the heat required for digestion arose spontaneously in the intestine. Instead, he postulated that it was conveyed to the digestive tract by the blood in the intestinal arteries.

Leonardo correctly observed that the partially digested nutrients are carried to the liver in the portal vein. He assumed with Galen that in the liver they were transformed into blood, and he also recognized that various impurities were eliminated from the blood to be excreted with the urine. However, he erroneously saw the bile (which facilitates the digestion and absorption of fat) as a collection of waste products, stored in the gall bladder to be excreted.[52] Leonardo's errors are hardly surprising; any explanation without the help of chemistry was bound to be woefully inadequate.

Having analyzed the entire process of digestion from the mastication and swallowing of food to its transformation into blood and the excretion of waste products, Leonardo turned his attention to the metabolic processes at the body's periphery. Without a microscope, he could not observe the exchanges of nutrients, wastes, and "vital spirits" (oxygen) between the

* Leonardo was not completely wrong with his explanation. Acting as a partition between the cavities of the chest and the abdomen, the diaphragm not only is the chief muscle used in respiration, but also stimulates the stomach and liver during its downward movement, thus facilitating the digestive processes.

tissues and capillaries, but he showed remarkable intuition in postulating them. In fact, he intuited a fundamental aspect of tissue metabolism (to use the modern scientific term) that would be rediscovered only in the twentieth century.

Leonardo recognized three important functions of the blood: to "vivify" the tissues by supplying them with oxygen (the "vital spirits" generated in the heart), to nourish them with the nutrients absorbed from digested food, and to carry away broken-down tissues and wastes.[53] The vital spirits, according to Leonardo, dissolve in the capillaries; their life-giving essence (identified today with oxygen) is absorbed by the tissues, and the remaining hot moisture "evaporates through the terminations of the capillary vessels at the surface of the skin in the form of sweat."[54] Leonardo's chemistry is of course rather fuzzy here, and he did not realize that the blood, having delivered the oxygen to the tissues, continues its circulation by draining into a network of veins. Nevertheless, his intuitive understanding of the diffusion of oxygen from the capillaries to the tissues is remarkable.

Similarly, Leonardo correctly intuited the exchange of nutrients and waste products between the capillaries and tissues, and it was this intuitive understanding that led him to one of his most extraordinary discoveries. He had previously compared the life-sustaining role of air for animals to the way air sustains the "life" of a flame (see p. 282). Now he extended this analogy to the nourishment of the bodily tissues by the blood in a long passage, titled "how the body of an animal continually dies and is reborn":

> The body of anything that is nourished continually dies and is continually reborn, for nourishment cannot enter except into those places where past nourishment has been exhausted; and if it has been exhausted it no longer has life. If you do not supply nourishment equal to the nourishment that has departed, then life fails in its vigor; and if you take away this nourishment, life is totally destroyed. But if you supply just as much as is destroyed daily, then as much of the life is renewed as is consumed, just as the light of the candle is made from the nourishment it receives from the liquid of that candle.[55]

The brilliant intuitive insight expressed by Leonardo in this passage is that every living organism, like a burning flame, needs to feed on a continuous flow of nutrients to stay alive, and that these nutrients are continually absorbed and transformed by the body's living tissues while waste products are generated in the same process. Leonardo continues the passage by elaborating his analogy between the continual stream of death and renewal in living tissues and the process of combustion in the flame of the candle:

> And this light is also continually renewed with the swiftest assistance from below by as much as is consumed above in dying, and in dying the brilliant light is converted into murky smoke. This death is continuous as long as the smoke continues, and the continuity of the smoke is equal to the continuing nourishment; and in the same instant the whole light dies and is completely regenerated together with the motion of its nourishment.

With this insight, described precisely and in beautiful poetic language, Leonardo was centuries ahead of his time. The definition of living organisms as "open systems" that need to feed on a continual flux of energy and matter was proposed by the biologist Ludwig von Bertalanffy in the 1940s, more than four hundred years after Leonardo had clearly recognized it. "The organism is not a static system closed to the outside and always containing the identical components," Bertalanffy wrote. "It is an open system in a (quasi-) steady state . . . in which material continually enters from, and leaves into, the outside environment. . . . The fundamental phenomena of life can be considered as consequences of this fact."[56]

It took another thirty years before a mathematical theory of open systems was formulated. This was achieved by physicist and chemist Ilya Prigogine, who called those systems "dissipative structures" and was awarded the Nobel Prize for his theory.[57] Today, the flow of energy and matter through a dissipative structure is considered a defining characteristic of life, and the image of the flame of a candle is still used as a simple illustration.

The Nature of Life

Leonardo's sophisticated studies of the heart and the flow of blood, undertaken in old age, represented the climax of his anatomical research and, at the same time, led him to some of his deepest insights in his long-standing quest to understand the nature of life. He recorded meticulous descriptions of many subtle features of cardiac physiology—including the coordinated actions of the heart's four chambers (when his contemporaries knew

of only two), and the corresponding synchronized actions of its valves—
and he illustrated them in a series of superb drawings. According to Keele,

> Leonardo's success in cardiac anatomy [is] so great that there are as-
> pects of the work which are not yet equaled by modern anatomical
> illustration. . . . His consistent practice of illustration of the heart
> and its valves, both in systole and in diastole, with a comparison
> of the position of the parts, has rarely if ever been performed in
> any anatomical textbook. . . . His detailed treatment of the valves
> and their movement is such that it is difficult to find illustrations in
> modern books with which to compare them.[58]

Leonardo missed some crucial details of the mechanics of blood circulation because of his pursuit of the forbidding challenge to explain (without chemistry) the generation of the body's temperature and of the blood's
life-giving essence, identified today with oxygen. At the same time, this
focus on metabolic processes enabled him to recognize many fundamental
features of tissue metabolism that would be rediscovered only centuries
later. These include the insights that heat energy supports the metabolic
processes; that oxygen (the "vital spirits") sustains them; that there is a
constant flow of oxygen from the heart to the tissues at the body's periphery; that the blood returns from these tissues carrying metabolic waste
products; and, finally, what must be seen as Leonardo's crowning achievement, the insight that the continuous absorption and transformation of
nutrients, with concomitant generation of waste products, is a necessary
condition for sustaining life and hence a fundamental property of all living
tissues. It is evident from these achievements that Leonardo's theory of
the functioning of the heart and the flow of blood led him to understand
some of the defining characteristics of biological life.

The Origin and Early Development of Human Life

While Leonardo immersed himself in studying the subtleties of the movements of the heart and the flow of blood, he became intensely interested in
another dimension of the mystery of human life—its origin and unfolding
in the processes of reproduction and embryonic development. The main
body of Leonardo's embryological studies is represented on three Windsor folios (196, 197, and 198) dating from 1510 to 1512. These folios were
initiated during his last years in Milan, but in Rome he added several important notes and smaller sketches.[59]

In Leonardo's time, there was a lively debate among philosophers as to whether the human embryo was formed from two "seeds" (one from the male parent and the other from the female), or whether its characteristics derived from the male seed only. Aristotle taught that all inherited characteristics came from the father, with the mother providing merely the nutritive bed in which the embryo could grow. Galen, by contrast, believed

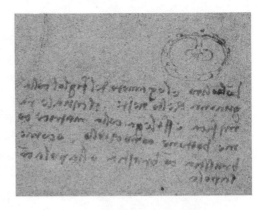

FIG. 8-7. Early embryo (1–2 months). Codex Atlanticus, folio 1064r (detail).

that both the father's and the mother's "seeds" contributed to the embryo's characteristics. After examining both theories, Leonardo unequivocally sided with the Galenic view, offering the following empirical support for it.

> If a black man impregnates a white woman, she will give birth to a grey child, and this shows that the seed of the mother has equal power in the embryo to the seed of the father.[60]

The way Leonardo exposed Aristotle's ideological bias with this simple observation is typical of his assured empirical approach to knowledge.

Although Leonardo performed at least one anatomy of a human fetus,[61] most of his embryological studies were based on dissections of cows and sheep. The earliest stage of embryonic development he recorded is in a small sketch in the Codex Atlanticus. It shows an oval embryonic body surrounded by amniotic fluid and attached to the uterine wall by a bodystalk (fig. 8-7). According to Keele, an embryo of that appearance, whether animal or human, can be dated as being one to two months old.[62]

Leonardo observed the same structure in the ovaries of flowering plants, where the seed, during its first stage of development, remains attached to the ovarian wall by a stalk. As I have discussed, he viewed this stalk as the equivalent to the umbilical cord, as botanists do today.* This observation, recorded on one of the folios of embryological studies in the

---

* See pp. 124–25 above. The equivalent of an umbilical cord in plants was rediscovered by Nathaniel Highmore two hundred years after Leonardo observed it (see Needham, *A History of Embryology*, p. 107).

FIG. 8-8. Study of an early embryo. Codex Atlanticus, folio 313r (detail).

Windsor Collection,[63] is an impressive testimony to Leonardo's ability of seeing similar patterns in different living systems.

Another rough sketch from around the same time shows a fetus in utero in the second month surrounded by a series of membranes (fig. 8-8). In an accompanying sketch below, Leonardo shows the appearance of the fetus after he has extracted it and has sliced the membranes, opening them like flower petals.

The most sophisticated and famous of Leonardo's embryological drawings depicts a five-month-old fetus as seen through the incised uterine wall (plate 9). The positions of its head and limbs are rendered with an accuracy never seen before and, one feels, with awe and reverence for the unfolding of human life.

Below and to the right of the main drawing, Leonardo shows the stages of his dissection, as he delicately slices and peels off the fetal membranes. Again, these images are strikingly reminiscent of the petals of a flower or the outer layers of a seed—moving visual expressions of Leonardo's deep conviction of the unity of life at all scales of nature.

At the center of the right margin we find an intriguing diagram of a ball rolling up an inclined plane. In the accompanying note, Leonardo explains that it can do so because a piece of lead weighing more than the rest of the ball is attached to its periphery at the uphill end. This diagram provides a concise mechanical explanation of how the weight of the head of the fetus enables its body to rotate within the uterine "ball of water" right before birth, so as to engage head first.

Although Leonardo's representation of the fetus in the womb is obviously based on the dissection of a pregnant woman after her unfortunate death, the uterine membranes and placenta, as shown in figure 8-9, are those of a cow. To experts in anatomy this is evident from the stylized rendering of the placenta as a series of individual patches attached to the fetal membrane. These placental patches, known as cotyledons, are typical of the uterus of cows.

In the sketches at the top of the page of plate 9, Leonardo studies in great detail how the protuberances of the fetal membranes, known today

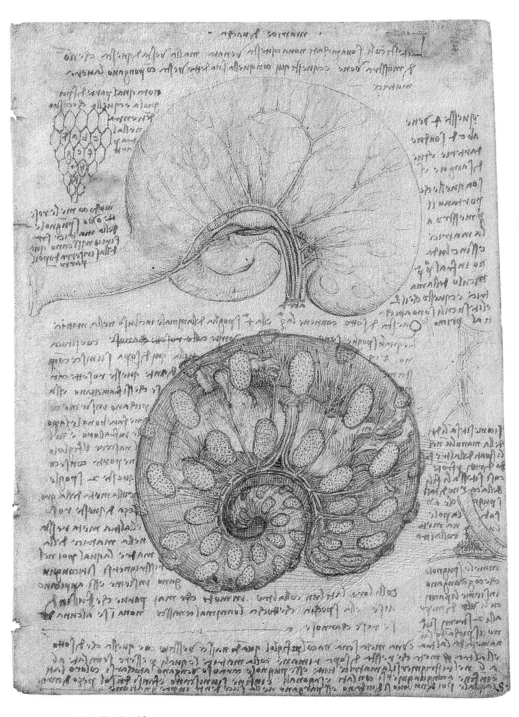

FIG. 8-9. The fetal calf in utero, c. 1506–8.
Windsor Collection, *Anatomical Studies*, folio 52r.

as villi, interlace with corresponding projections from the uterine wall. Here too, the interdigitation of the placenta and uterine wall is shown to occur in the distinct patches (cotyledons) typical of cows.

Leonardo dissected animals not only to find anatomical analogies to the human body but also to explore the animal body in its own right. Among his anatomical studies in the Windsor Collection there is a sheet with two superb illustrations of the pregnant uterus of a cow (see fig. 8-9). The upper drawing shows the crescent-shaped womb with the uterine and ovarian blood vessels. In the drawing below, as Keele explains, Leonardo has removed the uterine wall to expose the scattered cotyledons and reveal the fetus through the inner transparent membranes.[64] The fetal calf, lying in the womb upside down with its head to the left, its forelegs and cloven hoofs on top, and its body stretching across to the hind legs on the right, is rendered with the same delicacy and care as the human fetus. In fact, Leonardo's "transparency technique" is so subtle that the calf's body is often not recognized at first viewing.

Leonardo's embryological drawings are graceful and touching revelations of the mysteries surrounding the origins of life. They epitomize the artist's great care, sensitivity, and tremendous respect for life, both animal and human, and exude a tenderness that is deeply moving. In the words of physician Sherwin Nuland,

> [His] depiction of a five-month fetus in the womb is a thing of beauty. . . . It stands as a masterwork of art, and, considering the very little that was at the time understood of embryology, a masterwork of scientific perception as well.[65]

In addition to studying the anatomies of human and animal embryos in the womb, Leonardo tried to determine their rates of growth both in utero and after the baby's birth. "The child," he noted, "grows far more day by day when lying in the body of its mother than it does when it is outside her body."[66] Instead of doubling its size every nine months after its birth, he went on to observe, the child's growth rate diminishes progressively until the body reaches its adult height. Leonardo was the first embryologist to make such quantitative observations of fetal growth. The next time was in the late nineteenth century, when the anatomist Charles Minot recorded the same diminishing rates of growth that Leonardo had observed.[67]

As in his cardiovascular studies, Leonardo was keenly interested in the body's metabolic processes in his embryological research. On the famous folio showing the fetus in the womb (plate 9), he added a long note in the lower left corner describing its vital processes, beginning with a summary of its basic metabolism:

> The heart of this child does not beat, nor does it breathe, because it lies continually in water, and if it breathed it would drown; and breathing is not necessary because it is vivified and nourished by the life and food of the mother. This food nourishes the creature not otherwise than it does the other parts of the mother, that is, the hands, feet, and other parts.[68]

Leonardo was wrong about the fetal heartbeat. Nevertheless, this passage shows his remarkable insight into the exchanges of oxygen and nutrients between mother and fetus. On another folio, the connections between maternal and fetal metabolism are explored in astonishing detail. In several drawings, Leonardo studies the organs involved in these processes: the placenta, through which all metabolic exchanges are mediated; the umbilical vessels, which carry the nutrients; and the stomach, liver, and kidneys.[69] This is also the folio on which Leonardo recorded his quantitative observations on fetal growth. The accuracy and sophistication of his investigations are so impressive that the eminent embryologist and science historian Joseph Needham called Leonardo "the father of embryology regarded as an exact science."[70]

Even more amazing, perhaps, is the fact that Leonardo found it quite natural to include speculations about the embryo's mental life in his discussion of its vital processes. The passage quoted above continues without any break with these inspired thoughts:

> One and the same soul governs these two bodies; and the desires, fears and pains are common to this creature as to all other animated parts.[71]

On another folio, Leonardo continues his meditation on the relationship between the souls of mother and child, describing the gradual emancipation of the infant's own soul:

> The soul of the mother . . . first composes within the womb the human figure, and in due time awakens the soul which is to be its inhabitant. This at first remains asleep under the guardianship of the soul of the mother who nourishes and vivifies it through the umbilical vein.[72]

The conception of the human soul conveyed in these two passages is extraordinary. To Leonardo's contemporaries, it must have sounded heretical: it flatly contradicted the Church's doctrine of the soul's divine nature. Leonardo, of course, did not publish his views, but somehow the rumor that he entertained a radically different concept of the human soul reached Pope Leo X, who swiftly barred him from conducting further autopsies or human dissections.[73]

From the scientific perspective of the twenty-first century, on the other hand, Leonardo's contemplations seem to foreshadow the views of modern cognitive science to an amazing degree. I have argued elsewhere that his conception of the human soul as the force underlying the body's formation and movements, and the agent of perception and knowledge, corresponds rather precisely to the concept of cognition in contemporary cognitive science.[74]

In his embryological studies, Leonardo applies this concept to describe in beautiful poetic language the relationship between the soul of the mother and that of her unborn child. What he asserts (to put it into modern scientific language) is that the formation of the embryo's body and the organization of its metabolism is at first guided by the cognitive processes of the mother, and that gradually the self-organizing capabilities of the fetal organism emerge as its cognitive functions develop. The subtlety of Leonardo's description of the emergence and development of the embryo's mental life together with its body is truly astounding.

The Mystery of Life in Leonardo's Art

During the last decade of his life, Leonardo became fascinated by the mystery of the origin of life not only in his science but also in his art. Two of his most personal paintings, *Leda and the Swan* and the *Mona Lisa*, are devoted to this majestic theme. His *Leda* is his only female nude and his only painting inspired by a myth of antiquity—the seduction of the beautiful wife of the king of Sparta by Zeus in the guise of a swan. The painting was not initiated by a commission, and yet Leonardo worked on it for many years, evidently following a strong personal impulse.

Leonardo finished the painting in Rome and it accompanied him to his last abode in Amboise, together with the *Mona Lisa*, the *Saint Anne*, and the *Saint John the Baptist*.[75] After Leonardo's death, many copies of the *Leda* were made by Italian and Flemish painters but the original, unfortunately, was lost or destroyed sometime during the seventeenth century.

FIG. 8-10. After Leonardo, *Leda and the Swan*
("*Spiridon Leda*"), c. 1503–15. Uffizi Gallery, Florence.

The copy housed today in the Uffizi Gallery in Florence (fig. 8-10) may well be the one that comes closest to Leonardo's original.

The painter does not show us the sexual embrace of the woman and the swan, but presents the fruits of their union—two sets of twins tumbling out of their broken eggs. Thus Leonardo turns the ancient myth into an allegory of the mystery of female fertility and procreative power. The painting is erotically charged not only by the woman's naked beauty and the sinuous and unnaturally enlarged neck of the swan, but also by the abundant fertility of the landscape around them—thick grasses and phallic reeds bursting forth from moist, swampy soil.* By depicting the abundance of these generative forces in the realms of plants, animals, and humans, and by using the Greek myth to transcend species boundaries, Leonardo celebrates the universal mystery of life's inherent procreative power.

The *Mona Lisa* (plate 11), Leonardo's most famous painting, was originally a portrait of a young Florentine lady, Lisa del Giocondo. It was commissioned by her wealthy husband, but for some unknown reason the painting was never delivered. Leonardo kept it in his possession until he died and over the years transformed it into his personal meditation on the origin of life. The painting is different from his other portraits; indeed, it is different from *all* other portraits. The striking difference, in the words of Daniel Arasse, is "the strong contrast between the mellowness of the figure and the wild austerity of the archaic landscape that is its background."[76]

This background is the bare, craggy setting with its mythical rock formations, lakes, and streams that Leonardo painted throughout his life. This time, however, the landscape is not merely a distant backdrop but has become a major protagonist, as essential as the figure in the foreground. The forms of the Earth are portrayed in ceaseless movement and transformation, as the primordial waters cut through the rocks, carving out valleys and depositing masses of gravel and sand that, eventually, will become fertile soil. What we see here is the birth of the living Earth out of the waters of the primeval oceans.

As Martin Kemp has noted, the background and the foreground of the painting are aesthetically interlinked by a series of similarities, many

* These details are more apparent in Leonardo's studies than in the existing copies of the painting; see p. 100 above.

surface details of the landscape being echoed in the figure's hair and garments—the spiral folds of veil across her left breast, the delicate cascades of her hair, and the luminous highlights on both the drapery and the landscape.[77] These are symbolic reminders of the similarities between the living Earth and the living human body, and in particular between the origin of life in the macro- and microcosm. We know that the *Gioconda*, as she is called in Italy, was a young mother; and Kenneth Keele, examining the woman in the portrait with the eye of a physician, has argued that Leonardo's Mona Lisa was actually pregnant when she sat for the portrait.[78]

The evidence cited by Keele to support his hypothesis is persuasive—her erect posture, sitting well back in a comfortable armchair and slightly gripping the chair's arm with her left hand; the slight puffiness of her fingers, from which she has removed all her rings; the "matronly" shape of her entire body and fullness of her breasts; and the skillfully concealed outline of her abdomen. "Within her body," Keele writes, "is a new living world in the form of a babe growing out of the amniotic waters just as the great world grows out of the waters of the sea."[79] The Mona Lisa's mysterious, knowing smile, in this interpretation, is a subtle allusion to the mysterious secret in her womb.

Whether or not we accept Keele's suggestion of actual pregnancy, it is evident that the central theme of Leonardo's celebrated masterpiece is the procreative power of life, both in the female body and in the body of the living Earth. The *Leda* and the *Mona Lisa*, then, are both meditations on the origin of life, the theme that was foremost in Leonardo's mind during his old age.

Leonardo knew very well that, ultimately, the nature and origin of life would remain a mystery, no matter how brilliant his scientific mind. "Nature is full of infinite causes that have never occurred in experience,"[80] he declared in his late forties, and as he got older, his sense of mystery deepened. Nearly all the figures in his last paintings have that smile that expresses the ineffable, often combined with a pointing finger. "Mystery to Leonardo," wrote Kenneth Clark, "was a shadow, a smile and a finger pointing into darkness."[81]

CODA

Leonardo's Legacy

I argued in the Prologue to this book that Leonardo's greatest legacy to us may be his systemic thinking, together with his deep respect for nature and for life. In his mind, the two were closely connected. To gain knowledge about a natural phenomenon, for him, meant connecting it with other phenomena through a similarity of patterns; and such systemic knowledge he also saw as the basis for love. "For in truth," he asserted, "great love is born of great knowledge of the thing that is loved."[1]

Today, it is becoming increasingly evident that systemic thinking is critical to solve our major global problems; yet our sciences and technologies remain narrow in their focus, unable to understand systemic problems from an interdisciplinary perspective; and our business and political leaders are often incapable of "connecting the dots." This is exactly what we can learn to do from Leonardo da Vinci's unique synthesis of art, science, and design.

As we recognize that most of our sciences, technologies, and business activities are not life-enhancing but life-destroying, we urgently need a science that honors and respects the unity of all life, recognizes the fundamental interdependence of all natural phenomena, and reconnects us with the living Earth. Indeed, what we need today is the kind of science Leonardo da Vinci anticipated and outlined five hundred years ago.

Leonardo did not pursue science and engineering to dominate nature, as Francis Bacon would advocate a century later. Throughout this book I have shown that he had a deep respect for life, a special compassion for animals, and great awe and reverence for nature's complexity and abundance. As he put it succinctly: "One who does not respect life does not deserve it."[2]

FACING Leonardo's self-portrait (detail), c. 1512. Biblioteca Reale, Turin.

While a brilliant inventor and designer himself, Leonardo always thought that nature's ingenuity was vastly superior to human design. He felt that we would be wise to respect nature and learn from her. "Though human ingenuity in various inventions uses different instruments for the same end," he declared, "it will never discover an invention more beautiful, easier, or more economical than nature's, because in her inventions nothing is wanting and nothing is superfluous."[3]

In the designs of his flying machines, Leonardo tried to imitate the flight of birds so closely that he almost gives the impression of wanting to become a bird.[4] He called his flying machine *uccello* ("bird"), and his designs of its mechanical wings sometimes mimicked the anatomical structure of a bird's wing so accurately and, one almost feels, lovingly, that it is hard to tell the difference.

This attitude of seeing nature as a model and mentor is now being rediscovered in the practice of ecological design. Like Leonardo da Vinci five hundred years ago, ecodesigners today study the patterns and flows in the natural world and try to incorporate the underlying principles into their design processes.[5] When Leonardo designed villas and palaces, he paid special attention to the movements of people and goods through the buildings, applying the metaphor of metabolic processes to his architectural designs.[6] He applied the same principles to the design of cities, viewing a city as a kind of organism in which people, material goods, food, water, and waste need to flow with ease for the city to be healthy.

In his extensive projects of hydraulic engineering, Leonardo carefully studied the flow of rivers in order to gently modify their courses by inserting relatively small dams in the right places and at the optimal angles. "A river, to be diverted from one place to another, should be coaxed and not coerced with violence," he explained.[7] These examples of using natural processes as models for human design, and of working with nature rather than trying to dominate her, show clearly that, as a designer, Leonardo worked in the spirit that the ecodesign movement is advocating today.

Leonardo's deep respect for nature and for life, which is evident in his art, his science, and in his designs, is perhaps his most important legacy for our time. Our great challenge, as I have said, is to build and nurture sustainable communities—communities designed in such a way that their ways of life, businesses, economy, physical structures, and technologies respect, honor, and cooperate with nature's inherent ability to sustain life.

The first step in this endeavor is to become ecologically literate; the second step is to apply our ecological knowledge to the redesign of our technologies and social institutions, so as to bridge the current gap between human design and the ecologically sustainable systems of nature. In both of these endeavors—ecoliteracy and ecodesign—we can find great inspiration in the genius of Leonardo da Vinci.

CHRONOLOGY OF LEONARDO'S LIFE AND WORK

The following pages are based on the fairly detailed account of Leonardo's life in chapters 3 and 4 of my previous book, *The Science of Leonardo*.

Tuscany, 1452–81 (age 0–29)

Born in Vinci to mother Caterina and father Ser Piero da Vinci; later barred from university because of illegitimate birth; raised by grandparents and uncle Francesco

1452

Begins apprenticeship with Andrea del Verrocchio (perhaps earlier); "many flowers portrayed from nature" (Codex Atlanticus, folio 888r)

1467

Contributes to *Baptism of Christ* by Verrocchio (1470–75); Verrocchio installs ball on cathedral dome

1471

1464

Grandfather dies; Leonardo moves to Florence to live with his father

1469

Painting of a dragon on a wooden shield

1472

Master painter at age 20; becomes member of the guild of painters in Florence but remains in Verrocchio's workshop; cartoon for tapestry (lost); *Dreyfus Madonna*; first designs of costumes and sets; early inventions (1473–77): optical devices, vacuum to raise water, diving apparatus, olive press; attempts at perpetual water motion

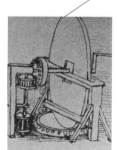

Annunciation
(1473–75);
drawing of
Val d'Arno

Leaves Verrocchio's
workshop;
establishes himself
as independent
artist but without
great diligence

Saint Jerome

1473

1477

1480

1476

1478

1481

Ginevra de' Benci
(1476–78)

Floods in Florence; Pazzi
conspiracy; begins "two
Virgin Marys," one of
them identified with the
Benois Madonna; first
drawings of mechanical
devices; first military
inventions (1478–81)

Slighted by Lorenzo
de' Medici; begins
Adoration of the Magi

First Period in Milan, 1482–99 (age 30–47)

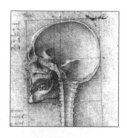

Moves to Milan, leaving *Adoration of the Magi* unfinished; writes letter to Ludovico Sforza, offering his services as engineer and artist, and receives no immediate response

Milan struck by plague (1484–85); Leonardo embarks on sustained self-education (1484–89); begins first Notebook (Codex Trivulzianus); begins to assemble personal library; first scientific studies: light, vision, perspective, sensory perception; geological and botanical studies for *Virgin of the Rocks*; early studies of flight of birds; urban designs in response to plague

Forms friendship with architect Donato Bramante; architectural designs for *tiburio* of Milan's cathedral

Begins work on equestrian statue *(il cavallo)*; first phase of anatomy (skulls with path of vision, nervous system); studies of muscles; mechanical inventions for transmission of movement and power; sets for Sforza wedding

1482   **1484**   **1487**   **1489**

1483   **1485**   **1488**

Begins *Virgin of the Rocks* (Louvre version, 1483–86)

Portrait of a Musician

Lady with an Ermine (1488–90)

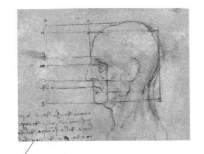

Receives full recognition at Sforza court; begins to write *Paragone* (comparison of painting with the other arts); establishes large workshop in Corte Vecchia; has 40 books in personal library; visits Pavia with Francesco di Giorgio; studies in Visconti Library; meets Fazio Cardano; begins to show strong interest in mathematics; revisions of earlier technical drawings with additions of theoretical comments; studies of human proportions; draws *Vitruvian Man*; designs of flying machines; designs and produces lavish festivals and spectacles for the Sforza court—for example, "Masque of the Planets"

Clay model of *cavallo*; stage designs; begins to study Latin

Meets mathematician Luca Pacioli and studies Euclid with him; drawings for Pacioli's *De divina proportione*; *La Belle Ferronière*

Milan occupied by the French; Leonardo leaves city with Pacioli

1490   **1493**   **1496**   **1499**

1491   **1495**   **1498**

Assembles notes about the *cavallo*; studies geometry and optics; studies mechanics; plans to write *Elements of Machines*

Begins *Last Supper*

Visits Genoa and Savoy; climbs Monte Rosa; completes *Last Supper*; decoration of Sala delle Asse; extensive studies of mechanics (1497–99)

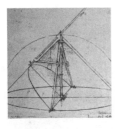

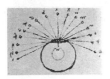

Venice, Florence, and Romagna, 1500–1506
(age 48–54)

Visits Mantua;
portrait of Isabella
d'Este; arrives
in Venice in
March; design
of submarines;
work on canals;
invents beveled lock gate; visits Friuli as military
engineer; plan of dam across Isonzo; leaves
Venice after one month; returns to Florence
to continue mathematical studies with Pacioli;
studies for *Madonna and Child with Saint Anne*

Travels extensively through Tuscany
and Romagna as military engineer
for Cesare Borgia; inspects
fortresses; meets Macchiavelli;
draws maps of regions in Tuscany
and Romagna, plus the beautiful
map of Imola; designs for canals
and for draining marshes

1500

1502

1501

Madonna with a Yarn-Winder
(1501–7)

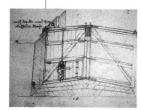

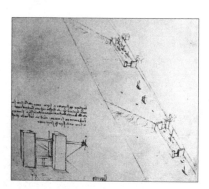

Leonardo's father dies;
Leonardo designs
fortifications (Piombino);
hydraulic works

1504

Leaves Florence,
abandoning *Battle
of Anghiari*

1506

1503

1505

Returns to Florence; begins
Battle of Anghiari; serves
as military engineer for the
Signoria; travels in Tuscany;
maps; extensive geological
studies; scheme to divert Arno;
design for "industrial" canal

Has 116 books in personal library;
begins *Mona Lisa* and *Leda*;
begins mathematical work *On
Transformation* (Codex Forster);
extensive studies of flight of birds;
designs of flying machines

Second Period in Milan, 1506–13
(age 54–61)

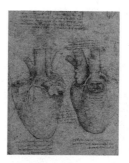

Arrives at French court in Milan; governor Charles d'Amboise is great admirer of Leonardo; has unprecedented freedom to pursue scientific studies; develops and refines *Saint Anne*, *Mona Lisa*, *Leda*; designs sets and costumes; designs villa and gardens for Charles; works on Lombard canals; extensive studies of hydraulics and fluid dynamics (1506–9); anatomical studies of superficial muscles, blood vessels, nerves, and internal organs (1506–8)

Divides his time between Florence and Milan; helps sculptor Giovanni Francesco Rustici with bronze statues; collaborates with Giovanni Ambrogio de Predis on *Virgin of the Rocks* (London version); reviews Notebooks and plans to organize them; maps out several comprehensive treatises; extensive anatomical studies of internal organs, especially of the heart and flow of blood (1508–13); dissects "centenarian" in Florence; begins botanical studies (1508–12)

1506

1508

1507

Meets Francesco Melzi, who will become his disciple and lifelong companion; uncle Francesco dies; visits Florence for six months to settle will

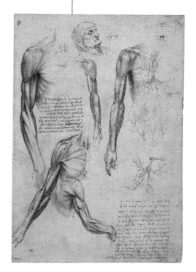

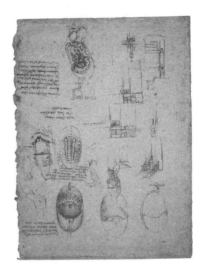

Continues work on *Mona Lisa* and *Leda*; designs equestrian statue for Trivulzio monument (unfinished); meets anatomist Marcantonio della Torre

1510

Spends two years at Melzi estate; studies of water turbulence; dissection of animals; botanical studies; design for enlargement of Villa Melzi

1512

1509

Assembles notes on painting into *Libro A* (lost); geological studies in Lombardy

1511

French expelled from Milan in December; Leonardo retreats to Melzi estate

1513

Leo X (Giovanni de' Medici) elected pope; Leonardo leaves for Rome, invited by the pope's brother Giuliano de' Medici

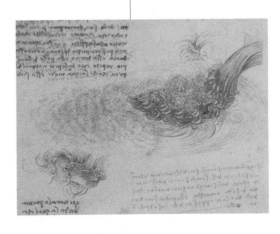

Rome and Amboise, 1513–19
(age 61–67)

Arrives in Rome in October; resides in Belvedere villa in Vatican; draws self-portrait

1513

François I captures Milan; Leonardo builds a mechanical lion for the French king; studies cardiology and embryology; clashes with the pope over the nature of the soul; is barred from conducting dissections; finishes *Saint Anne*, *Mona Lisa*, *Leda*; paints *Saint John the Baptist*

1515

1514

Is venerated but no longer in fashion as painter; lonely and depressed; deluge drawings; continues scientific studies with undiminished energy; extensive botanical studies; geometry of transformations; travels to Civitavecchia, Parma, Florence, Milan; plans to drain Pontine marshes; consults on architectural and engineering projects; mechanical and optical designs

Is visited by Cardinal of Aragon;
suffers paralysis in right arm;
travels to Romorantin with the king;
urban designs for a new capital

1517

Dies at Cloux;
leaves artistic and
intellectual legacy
to Francesco Melzi

1519

1516

Moves to Amboise, invited by François I;
resides at Château de Cloux; holds long
conversations with the young king;
reorganizes Notebooks, probably in view
of publication; plans several new treatises;
creates designs for rebuilding royal château

1518

Creates spectacular
performances; designs
costumes and emblems

NOTES

Citations of Leonardo's manuscripts refer to the scholarly editions listed in the Bibliography. Most of these editions are now available also online at e-Leo (www.leonardodigitale.com), published by the Biblioteca Leonardiana in Vinci. I have retranslated some of the passages by staying closer to the original Italian texts to preserve their Leonardesque flavor.

Prologue: Leonardo's Genius
1. See Murray, *Genius*.
2. See Steptoe, *Genius and the Mind*.
3. See David Lykken, "The Genetics of Genius," in Steptoe, *Genius and the Mind*.
4. Clark, *Civilisation*, p. 135.
5. A list of scholarly editions of Leonardo's Notebooks (facsimile editions together with transcribed, dated, and annotated versions of the original texts) is given on pp. 355–56 below.
6. Quoted by David Lykken in Steptoe, *Genius and the Mind*, p. 31.
7. Quoted by Bramly, *Leonardo*, p. 281.
8. Ms. E, folio 55r.
9. See Capra, *The Science of Leonardo*, pp. 158–59, for a more detailed discussion of the modern scientific method.
10. See Capra, *The Web of Life* and *The Hidden Connections*.
11. Orr, *Hope Is an Imperative*, pp. 13ff.
12. Quoted in Capra, *The Hidden Connections*, p. vii.
13. See ibid., pp. 116ff.
14. See ibid., pp. 121ff.; see also Wheatley, *Leadership and the New Science*.

Chapter 1. The Movements of Water
1. Ms. K, folio 2r.
2. For a brief account of the extent, history, and locations of Leonardo's Notebooks, see Capra, *The Science of Leonardo*, pp. 133ff. A list of current facsimile editions and scholarly transcriptions is given on pp. 355–56.
3. A copy of the original manuscript is now housed in the Collection Vittorio Emanuele of the Biblioteca Nazionale Centrale in Rome.

4. Ms. C, folio 26v.

5. Codex Arundel, folio 210r.

6. Ibid.

7. See also Capra, *The Science of Leonardo*, p. 104.

8. Codex Arundel, folio 57v.

9. Ibid.; see also Keele, *Leonardo da Vinci's Elements of the Science of Man*, p. 89.

10. See Capra, *The Science of Leonardo*. p. 77.

11. See Kemp, *Leonardo da Vinci*, pp. 94–96; see also Capra, *The Science of Leonardo*, pp. 47–49.

12. See Capra, *The Science of Leonardo*, pp. 106–8.

13. See ibid., p. 177.

14. Ms. G, folio 93r.

15. Codex Atlanticus, folio 171r-a.

16. Codex Arundel, folio 57v.

17. See Capra, *The Web of Life*, pp. 169ff.

18. See Capra, *The Science of Leonardo*, p. 111.

19. See Codex Atlanticus, folio 785r-b.

20. See Bramly, *Leonardo*, p. 335

21. Codex Atlanticus, folio 302.

22. See Capra, *The Science of Leonardo*, p. 115.

23. See ibid., pp. 163ff.

24. Ms. A, folio 26r.

25. Codex Arundel, folio 57v.

26. Ms. A, folio 56v.

27. See Capra, *The Web of Life*, pp. 5–6.

28. See Capra, *The Science of Leonardo*, p. 146.

29. See Capra, *The Web of Life*, pp. 100ff.

30. See Capra, *The Hidden Connections*, pp. 81–82.

31. Ms. A, folio 56v.

32. Codex Leicester, folio 33v.

33. Ms. H, folio 77r.

34. Codex Arundel, folio 235r.

35. Codex Atlanticus, folio 468.

36. See Kemp, "The Crisis of Received Wisdom."

37. Ms. A, folio 56r.

38. Ibid.

39. See Keele, *Leonardo da Vinci's Elements of the Science of Man*, p. 81.

40. Codex Leicester, folio 28r.

41. See Capra, *The Science of Leonardo*, p. 77.

42. Codex Leicester, folio 3v.

43. Codex Leicester, folio 34v.

44. Codex Leicester, folio 32v.

45. Windsor Collection, *Anatomical Studies*, folio 137r.

46. Ms. G, folio 48v.

47. See Capra, *The Science of Leonardo*, pp. 168–69.

48. See Emboden, *Leeonardo da Vinci on Plants and Gardens*, p. 24.

49. See Capra, *The Science of Leonardo*, pp. 83ff.

50. See Maccagni, "The Italian School of Hydraulics."

51. See Zammattio, "The Mechanics of Water and Stone."

52. Codex Leicester, folio 15v.

53. Ms. F, folio 2v.

54. Macagno, *Leonardian Fluid Mechanics*, pp. 4–5.

55. See Arasse, *Leonardo da Vinci*, p. 130; see also Capra, *The Science of Leonardo*, p. 91.

56. Ms. A, folio 61r; see also Capra, *The Science of Leonardo*, p. 230.

57. Codex Atlanticus, folio 506r.

58. Codex Atlanticus, folio 299r.

59. See Macagno, *Leonardian Fluid Mechanics*.

60. See Capra, *The Science of Leonardo*, pp. 161ff.

61. See, for example, Moin and Kim, "Tackling Turbulence with Supercomputers."

62. For a review of the basic concepts and techniques of nonlinear dynamics, see Capra, *The Science of Leonardo*, pp. 112ff.

63. See Davidson, *Turbulence*, pp. 23–24.

64. See Sabersky et al., *Fluid Flow*, p. 4.

65. Capra, *The Science of Leonardo*, pp. 229ff.

66. Ms. F, folio 87v.

67. See Capra, *The Science of Leonardo*, p. 231.

68. See Capra, ibid., p. 96.

69. See Arasse, *Leonardo da Vinci*, p. 380.

70. See Capra, *The Science of Leonardo*, p. 229.

71. See ibid., p. 109.

72. Ms. F, folio 13r.

73. Codex Leicester, folio 27r.

74. Ms. I, folio 115r.

75. Ms. F, folio 34v.

76. Ms. K, folio 1r.

77. See Merzkirch, *Flow Visualization*.

78. Ms. G, folio 10r.

79. Codex Atlanticus, folio 465.

80. I am grateful to Michael Nauenberg for this observation.

81. Codex Trivulzianus, folio 31r.

82. See Keele, *Leonardo da Vinci's Elements of the Science of Man*, pp. 135–36.

83. Codex Leicester, folio 26v.

84. See Capra, *The Science of Leonardo*, pp. 149–50.

85. Codex Leicester, folio 26v.

86. Ms. K, folio 101v.

87. Ms. M, folio 42v.

88. Ms. I, folio 68r.

89. Codex Leicester, folio 25r.

90. Ms. H, folio 55.

91. Ms. I, folio 106r.

92. See Fasso, "Birth of Hydraulics."

93. See Capra, *The Science of Leonardo*, p. 188.

94. See ibid., pp. 199ff.

95. See Fasso, "Birth of Hydraulics."

96. Ms. F, folio 16r.

97. Codex Atlanticus, folio 1018r.

98. Ms. A, folio 57r.

99. Ms. A, folio 57v.

100. See Capra, *The Science of Leonardo*, p. 158.

101. See Fasso, "Birth of Hydraulics."

102. Ms. A, folio 60r.

103. I am grateful to Prof. Ugo Piomelli, fluid dynamicist at the University of Maryland, for this and many other comments on Leonardo's drawings and descriptions of turbulent flows.

104. Codex Atlanticus, folio 813.

105. Ms. F, folio 13v.

106. Windsor Collection, *Landscapes, Plants, and Water Studies*, folio 46r.

107. Ms. F, folio 2r.

108. Ugo Piomelli, personal communication, February 2005.

109. Ms. F, folio 3r.

110. See Davidson, *Turbulence*, pp. 17ff.

111. Windsor Collection, *Landscapes, Plants, and Water Studies*, folio 48r.

112. In the early 1990s, Leonardo's statement caught the eye of fluid dynamicist Ugo Piomelli, who pointed out its significance to several colleagues at that time; Piomelli, personal communication, 2005.

113. Ms. I, folio 78r.

114. Windsor Collection, *Landscapes, Plants, and Water Studies*, folio 42r.

115. See Capra, *The Web of Life*, pp. 133–34.

116. Ms. E, folio 54r.

117. See Giacomelli, "La scienza dei venti di Leonardo da Vinci."

118. See Zöllner and Nathan, *Leonardo da Vinci*, pp. 526–35.

119. See Capra, *The Science of Leonardo*, pp. 125–26.

120. Arasse, *Leonardo da Vinci*, p. 419.

Chapter 2. The Living Earth

1. See Gould, "The Upwardly Mobile Fossils of Leonardo's Living Earth."
2. Ms. A, folio 55v.
3. Ibid.
4. Codex Leicester, folio 34r.
5. Codex Leicester, 34r.
6. See Capra, *The Web of Life*, pp. 100ff.
7. Lovelock, *Healing Gaia*, pp. 31ff.
8. Kemp, "Body of Earth and Body of Man," p. 105.
9. Wicander and Monroe, *Essentials of Geology*, p. 7.
10. Ms. E, folio 4v.
11. Ms. F, folio 11v.
12. Codex Atlanticus, folio 901r.
13. Codex Leicester, folio 10r.
14. Codex Leicester, folio 6v.
15. Gould, "The Upwardly Mobile Fossils of Leonardo's Living Earth."
16. Codex Atlanticus, folio 508v.
17. See Capra, *The Science of Leonardo*, p. 163.
18. See ibid., p. 111.
19. See ibid., pp. 209–10.
20. See ibid., pp. 109–10.
21. Arasse, *Leonardo da Vinci*, p. 217.
22. Codex Leicester, folio 10r.
23. Codex Arundel, folio 138r.
24. Codex Leicester, folio 31r.
25. See Gohau, *A History of Geology*, pp. 139ff.
26. Codex Atlanticus, folio 1040v.
27. Codex Arundel, folio 30v.
28. See Gohau, *A History of Geology*, p. 139.
29. Martin Kemp in Kemp and Roberts, eds., *Leonardo da Vinci*, p. 108.
30. See Capra, *The Science of Leonardo*, pp. 85–86.
31. See Kemp, *Leonardo da Vinci*, pp. 94–96.
32. Pizzoruso, "Leonardo's Geology."
33. See Wicander and Monroe, *Essentials of Geology*, pp. 57ff.
34. Pizzorusso, "Leonardo's Geology."
35. See Capra, *The Science of Leonardo*, p. 85.
36. Pizzorussso, "Leonardo's Geology."
37. Emboden, *Leonardo da Vinci on Plants and Gardens*, pp. 125ff.
38. Codex Leicester, folio 10r.
39. See Kemp, *Leonardo da Vinci*, p. 312.
40. Codex Leicester, folio 9v.
41. Ms. F, folio 80v.
42. Codex Leicester, folio 10r.

43. Ibid.
44. Codex Leicester, folio 8v.
45. Codex Leicester, folio 9v.
46. Codex Leicester, folio 10r.
47. Codex Leicester, folio 8v.
48. Codex Leicester, folio 9r.
49. Codex Leicester, folio 10r.
50. Gould, "The Upwardly Mobile Fossils of Leonardo's Living Earth."
51. Ellenberger, *History of Geology*, vol. 1, p. 101.
52. Kemp, "Analogy and Observation in the Codex Hammer."
53. Ms. F, folio 80r.
54. Ms. F, folios 79r, 79v, and 80r.
55. Ms. F, folio 79v.
56. See Gohau, *A History of Geology*.
57. See Capra, *The Science of Leonardo*, p. 156.
58. See ibid., p. 139.
59. See Capra, *The Science of Leonardo*, p. 155.
60. See ibid., p. 148.
61. See Gohau, *A History of Geology*, pp. 20–21.
62. See Kemp, *Leonardo da Vinci*, p. 314.
63. Codex Arundel, folio 19r.
64. Ms. F, folio 62v.
65. Codex Leicester, folio 35v.
66. Gould, "The Upwardly Mobile Fossils of Leonardo's Living Earth," p. 39.
67. Ms. E, folio 4v.
68. Codex Atlanticus, folio 350.
69. Ms. F, folio 70r.
70. Codex Leicester, folio 36r.
71. See Gohau, *A History of Geology*, p. 143.
72. See Wicander and Monroe, *Essentials of Geology*, pp. 8–10.
73. See Gohau, *A History of Geology*, p. 145.
74. See Capra, *The Science of Leonardo*, pp. 109 ff.
75. Ellenberger, *History of Geology*, vol. 1, p. 96.
76. Codex Leicester, folio 8v.
77. Codex Leicester, folio 9r.

Chapter 3. The Growth of Plants

1. Emboden, *Leonardo da Vinci on Plants and Gardens*.
2. Parts of this chapter were published in a separate book, titled *Leonardo's Botany: A Discourse on the Science of Qualities*; see Capra, *La botanica di Leonardo*.
3. Emboden, *Leonardo da Vinci on Plants and Gardens*, p. 77.
4. See ibid., p. 131.

5. Ibid., p. 139.

6. See Brown, "Verrocchio and Leonardo."

7. See Ames-Lewis, "Leonardo's Botanical Drawings."

8. Codex Atlanticus, folio 888r.

9. See Emboden, *Leonardo da Vinci on Plants and Gardens*, p. 125.

10. See ibid., pp. 127ff.

11. Ibid., p. 126.

12. Ibid., pp. 109 and 132.

13. Richter, *The Notebooks of Leonardo da Vinci*, vol. I, pp. 203ff.

14. Codex Arundel, folio 114v.

15. Ms. G, folio 8r.

16. Emboden, *Leonardo da Vinci on Plants and Gardens*, p. 101.

17. Windsor Collection, *Landscapes, Plants, and Water Studies*, folio 8v.

18. Emboden, *Leonardo da Vinci on Plants and Gardens*, p. 116.

19. See Capra, *The Science of Leonardo*, pp. 61–63.

20. Roberts, "The Drawings and Manuscripts."

21. Emboden, *Leonardo da Vinci on Plants and Gardens*, p. 141.

22. See Capra, *The Science of Leonardo*, p. 152.

23. Emboden, *Leonardo da Vinci on Plants and Gardens*, p. 82.

24. Ibid., p. 84.

25. See ibid., p. 86.

26. Ibid., p. 87.

27. See Capra, *The Science of Leonardo*, p. 141.

28. See Emboden, *Leonardo da Vinci on Plants and Gardens*, p. 88.

29. Ibid., p. 89.

30. See Capra, *The Science of Leonardo*, pp. 155–56.

31. Codex Atlanticus, folio 323r.

32. Emboden, *Leonardo da Vinci on Plants and Gardens*, p. 174.

33. See Capra, *The Science of Leonardo*, p. 131.

34. See Emboden, *Leonardo da Vinci on Plants and Gardens*, p. 24.

35. Anatomical Studies, folio 117r.

36. Emboden, *Leonardo da Vinci on Plants and Gardens*, p. 171.

37. See Capra, *The Web of Life*, pp. 21ff.

38. See ibid., pp. 18ff.

39. Emboden, *Leonardo da Vinci on Plants and Gardens*, p. 165.

40. Ms. G, folio 16v.

41. *Trattato*, chapter 822.

42. See Emboden, *Leonardo da Vinci on Plants and Gardens*, p. 170.

43. Ms. G, folio 16v.

44. Ms. G, folio 33v.

45. Ms. G, folio 17r.

46. Emboden, *Leonardo da Vinci on Plants and Gardens*, p. 173.

47. Ibid.
48. Ms. I, folio 12v.
49. See Laurenza, "La forme come matrici universali."
50. See Emboden, *Leonardo da Vinci on Plants and Gardens*, p. 110; see also Eloy, "Leonardo's Rule, Self-Similarity, and Wind-Induced Stresses in Trees."
51. Ms. G, folio 32v.
52. Ibid.
53. Emboden, *Leonardo da Vinci on Plants and Gardens*, p. 168.
54. Ibid.
55. Ms. G, folio 35v.
56. *Trattato*, chapter 832.
57. See Emboden, *Leonardo da Vinci on Plants and Gardens*, p. 173.
58. *Trattato*, chapter 838.
59. Ibid.
60. See Emboden, *Leonardo da Vinci on Plants and Gardens*, p. 172.
61. *Trattato*, chapter 842.
62. *Trattato*, chapter 829.
63. Emboden, *Leonardo da Vinci on Plants and Gardens*, p. 169.
64. *Trattato*, chapter 839.
65. See Emboden, *Leonardo da Vinci on Plants and Gardens*, p. 169.
66. Codex Atlanticus, folio 207r.
67. Emboden, *Leonardo da Vinci on Plants and Gardens*, p. 172.
68. Ms. B, folio 17v.
69. Emboden, *Leonardo da Vinci on Plants and Gardens*, p. 172.
70. Ms. G, folio 1r.
71. *Trattato*, chapter 841.
72. Emboden, *Leonardo da Vinci on Plants and Gardens*, p. 169.
73. See Capra, *The Science of Leonardo*, pp. 114–15.
74. Windsor Collection, *Anatomical Studies*, folio 196v.
75. Emboden, *Leonardo da Vinci on Plants and Gardens*, p. 171.
76. Bottazzi, "Leonardo as Physiologist."

Chapter 4. The Human Figure

1. For a recent detailed chronology of Leonardo's research in anatomy, including several new datings and interpretations, see Laurenza, *Leonardo: L'anatomia*.
2. See Capra, *The Science of Leonardo*, p. 216.
3. See Keele, *Leonardo da Vinci's Elements of the Science of Man*, pp. 64–65.
4. Capra, *The Science of Leonardo*, pp. 212ff.
5. Windsor Collection, *Anatomical Studies*, folio 153r.
6. Windsor Collection, *Anatomical Studies*, folio 154r. See Capra, *The Science of Leonardo*, p. 115.
7. Windsor Collection, *Anatomical Studies*, folio 143r.

8. See Capra, *The Science of Leonardo*, pp. 117–18.

9. Windsor Collection, *Anatomical Studies*, folio 81v.

10. See Laurenza, *Leonardo: L'anatomia*, pp. 74–77.

11. See Capra, *The Science of Leonardo*, p. 95.

12. See Zöllner and Nathan, *Leonardo da Vinci*, p. 384.

13. For a comprehensive collection of Leonardo's proportion drawings, see Zöllner and Nathan, *Leonardo da Vinci*, pp. 348ff.

14. Ackerman, "Science and Art in the Work of Leonardo."

15. See Laurenza, "La grammatica delle forme."

16. Zöllner and Nathan, *Leonardo da Vinci*, p. 348.

17. Laurenza, "La grammatica delle forme."

18. Venetian folio.

19. See Capra, *The Science of Leonardo*, pp. 163–64.

20. Laurenza, "L'uomo geometrico."

21. See ibid.

22. See Laurenza, "La grammatica delle forme."

23. See Capra, *The Science of Leonardo*, pp. 205ff.

24. *Trattato*, chapter 21.

25. Laurenza, "La grammatica delle forme"; my own translation.

26. See O'Malley and Saunders, *Leonardo da Vinci on the Human Body*, p. 456.

27. Johannes Nathan in Zöllner and Nathan, *Leonardo da Vinci*, p. 402.

28. Laurenza, *La ricerca dell'armonia*.

29. See Capra, *The Science of Leonardo*, p. 154.

30. Ibid.

31. Ibid., p. 156.

32. See Laurenza, *La ricerca dell'armonia*, pp. 73ff.

33. See Saunders and O'Malley, *The Illustrations from the Works of Andreas Vesalius*.

34. See ibid., p. 28.

35. Laurenza, *La ricerca dell'armonia*, pp. 82ff.

36. Ibid., p. 85 (my translation).

37. Ibid., p. viii.

38. See Capra, *The Science of Leonardo*, pp. 3ff.

39. Laurenza, *La ricerca dell'armonia*, p. ix.

40. Windsor Collection, *Anatomical Studies*, folios 173r and 77r.

41. Windsor Collection, *Anatomical Studies*, folio 173r.

42. Windsor Collection, *Anatomical Studies*, folio 156v.

43. Windsor Collection, *Anatomical Studies*, folio 173r.

Chapter 5. The Elements of Mechanics

1. See Capra, *The Science of Leonardo*, pp. 91–92.

2. Codex *Sul Volo*, folio 3r.

3. See, e.g. Keele, *Leonardo da Vinci's Elements of the Science of Man*, pp. 252 and 267.

4. Galluzzi, *Renaissance Engineers*, p. 226.

5. See Capra, *The Turning Point*, pp. 60ff.

6. Windsor Collection, *Anatomical Studies*, folio 114v.

7. Capra, *The Science of Leonardo*, pp. 249ff.

8. Windsor Collection, *Anatomical Studies*, folio 153r.

9. Codex Atlanticus, folio 434r.

10. Windsor Collection, *Anatomical Studies*, folio 114v.

11. Windsor Collection, *Anatomical Studies*, folios 179v and 148v.

12. Ms. H, folio 95r.

13. Codex Leicester, folio 17v.

14. See Capra, *The Science of Leonardo*, pp. 34ff.

15. See ibid., pp. 89ff.

16. Codex Atlanticus, folio 21r.

17. See Capra, *The Science of Leonardo*, p. 163.

18. See Bedini and Reti, "Horology."

19. See Capra, *The Science of Leonardo*, pp. 102 and 107.

20. See Bramly, *Leonardo*, p. 342.

21. Capra, *The Science of Leonardo*, pp. 22ff.

22. See ibid., pp. 25–26.

23. Galluzzi, *Rensaissance Engineeers*.

24. Galluzzi, *Léonard de Vinci*.

25. See, e.g., Kemp and Roberts, *Leonardo da Vinci*, pp. 218–41; see also Dibner, "Leonardo"; Dibner, "Machines and Weaponry"; Pedretti, *Leonardo*; Laurenza, Taddei, and Zanon, *Le macchine di Leonardo*.

26. See Kemp, *Leonardo da Vinci*, p. 88.

27. Clark, *Leonardo da Vinci*, p. 110.

28. See Capra, *The Science of Leonardo*, pp. 191ff.

29. Ibid., p. 91.

30. See Reti, "Elements of Machines."

31. Ibid.

32. For a detailed description of the functioning of this machine, see Dibner, "Leonardo."

33. See Moody and Clagett, *The Medieval Science of Weights*.

34. Ms. E, folio 8v.

35. Codex Atlanticus, folios 416v, 499, and 450c.

36. Codex Atlanticus, folio 481r.

37. Codex Arundel, folio 1v.

38. Clagett, "Leonardo da Vinci: Mechanics."

39. See Capra, *The Science of Leonardo*, p. 150.

40. Codex Madrid I, folio 181r; Codex Madrid II, folio 80r.

41. Codex Madrid I, folio 123v; see also Codex Forster II, folio 65v.

42. Codex Leicester, folio 6r.

43. Codex Leicester, folio 26v.

44. Macagno, "Mechanics of Fluids in the Madrid Codices."

45. Fasso, "Birth of Hydraulics During the Renaissance Period."

46. Codex Madrid I, folio 169r.

47. Codex Madrid I, folio 148v.

48. Codex Leicester, folio 6r.

49. Codex Leicester, folio 11r.

50. Codex Atlanticus, folio 589v.

51. Fasso, "Birth of Hydraulics During the Renaissance Period."

52. Bertoloni-Meli, *Thinking with Objects*, p. 60.

53. See Capra, *The Science of Leonardo*, pp. 149–50.

54. Ms. I, folio 130v.

55. Codex Madrid I, folio 128v.

56. See Capra, *The Science of Leonardo*, pp. 193ff.

57. Codex Atlanticus, folio 407r.

58. Bertoloni-Meli, *Thinking with Objects*, p. 72.

59. See Capra, *The Science of Leonardo*, p. 195.

60. Codex Atlanticus, folio 421v.

61. Codex Arundel, folio 37v.

62. Codex Arundel, folio 176.

63. Codex Atlanticus, folio 1098r.

64. Codex Atlanticus, folio 215r.

65. Ms. E, folio 42r.

66. Ms. E, folio 36r.

67. Codex Atlanticus, folio 813.

68. Ms. I, folio 99v.

69. Windsor Collection, *Anatomical Studies*, folio 115v; see also Ms. E, folio 29r.

70. Codex Atlanticus, folio 362r.

71. Quoted in Bertoloni-Meli, *Thinking with Objects*, p. 1.

72. Ms. A, folio 21v.

73. Codex Trivulzianus, folio 43r.

74. Ms. A, folio 61r; see Capra, *The Science of Leonardo*, p. 230.

75. Ms. G, folio 73r.

76. See Clagett, "Leonardo da Vinci: Mechanics."

77. Codex Leicester, folio 29v.

78. Ms. E, folio 29r.

79. See Bertoloni-Meli, *Thinking with Objects*, p. 64.

80. See Bertoloni-Meli, *Thinking with Objects*, pp. 297–98.

81. See Capra, *The Science of Leonardo*, pp. 200ff.

82. Ms. M, folio 66v.

83. See Keele, *Leonardo da Vinci's Elements of the Science of Man*, p. 37.
84. Codex Madrid I, folio 134v.
85. See Fasso, "Birth of Hydraulics During the Renaissance Period," for the details of this calculation; if we were to take into account the viscosity (friction) of the water, the calculation would be more complicated.
86. See Reti, "Leonardo on Bearings and Gears."
87. Ibid.
88. Codex Forster II, folios 86r and 87r.
89. Codex Madrid I, folio 95r.
90. Codex Leicester, folio 25r; see p. 39 above.
91. Ms. E, folio 54r.
92. Codex Atlanticus, folio 407r.
93. Ms. G, folio 1r; see also Codex Leicester, folio 29v.
94. Codex Atlanticus, folio 398v.
95. Codex Madrid I, cover.
96. See Capra, *The Turning Point*, pp. 72ff.
97. Codex Atlanticus, folio 508v.
98. Codex Arundel, folio 176r.
99. Codex Atlanticus, folio 195r.
100. See Laurenza, "Moti di 'consumazione.'"
101. Codex Atlanticus, folio 257r.
102. Ibid.
103. Codex Atlanticus, folio 1060r.
104. See Clagett, "Leonardo da Vinci: Mechanics," for extensive discussions of Leonardo's work on centers of gravity.
105. See Codex Arundel, folio 218v.
106. Codex Arundel, folio 193v.
107. Codex Atlanticus, folio 480r.
108. See Capra, *The Science of Leonardo*, pp. 193ff.
109. Ms. M, folio 59v.
110. Ms. M, folio 45r.
111. See Kline, *Mathematical Thought*, p. 338.
112. Ms. M, folio 42v.
113. Ms. M, folio 44v.
114. See Bertoloni-Meli, *Thinking with Objects*, p. 76.
115. See Bedini and Reti, "Horology."
116. Codex Madrid I, folio 147r.
117. See Capra, *The Science of Leonardo*, p. 106.
118. See Codex Atlanticus, folio 132r.
119. See Dibner, "Machines and Weaponry."
120. Codex Madrid I, folio 147r.
121. Ms. C, folio 7r.

122. See Bertolini-Meli, *Thinking with Objects*, pp. 135ff.

123. Quoted in ibid., p. 302.

124. See Capra, *The Science of Leonardo*, pp. 244ff.

125. Codex Atlanticus, folio 826r.

126. Ibid.

127. Ms. F, folio 26r.

128. Ms. A, folio 24r.

129. Codex Atlanticus, folio 1058v.; c. 1485.

130. Codex Atlanticus, folio 479r; see also Ms. F, folio 37v; Codex Atlanticus, folio 571r.

131. Ms. G, folio 75r.

132. Codex Atlanticus, folio 949v.

133. Ms. A, folio 34v.

134. Ms. A, folio 53v.

135. Codex Atlanticus, folio 973v.

136. Codex Atlanticus, folio 1014v.

137. See Capra, *The Science of Leonardo*, pp. 233–34.

138. Ms. A, folio 22v.

139. See Capra, *The Science of Leonardo*, pp. 228ff.

140. Ms. M, folios 52r and 52v; Codex Madrid I, folio 188v.

141. Codex Madrid, folio 59r.

142. Codex Arundel, folio 83v.

143. Ms. I, folio 41v.

144. Codex Leicester, folio 8r.

145. Codex Arundel, folio 82v.

146. Ms. A, folio 19r.

147. See Capra, *The Science of Leonardo*, pp. 220–21.

148. Ms. A, folio 19r.

149. See Ms. I, folio 115r.

Chapter 6. The Body in Motion

1. See Clark, *Leonardo da Vinci*, p. 38.

2. See Capra, *The Science of Leonardo*, pp. 226 and 244ff.

3. Codex Arundel, folios 151r,v.

4. See Capra, *The Science of Leonardo*, pp. 243ff.

5. See Laurenza, *La ricerca dell'armonia*, p. 37.

6. See Keele, *Leonardo da Vinci's Elements of the Science of Man*, p. 103.

7. Windsor Collection, *Anatomical Studies*, folio 147v.

8. *Trattato*, chapter 304.

9. See Keele, *Leonardo da Vinci's Elements of the Science of Man*, pp. 163ff.

10. *Trattato*, chapter 317.

11. *Trattato*, chapter 398.

12. See Keele, *Leonardo da Vinci's Elements of the Science of Man*, pp. 173ff.

13. *Trattato*, 68.

14. Ms. Ashburnham II, 20r.

15. *Trattato*, chapter 367.

16. See Capra, *The Science of Leonardo*, pp. 93ff.

17. Windsor Collection, *Anatomical Studies*, folio 154r.

18. Laurenza, *Leonardo: L'anatomia*.

19. See ibid., pp. 17ff.

20. See Capra, *The Science of Leonardo*, pp. 70ff.

21. See Keele, *Leonardo da Vinci's Elements of the Science of Man*, p. 8.

22. See Laurenza, *Leonardo: L'anatomia*, pp. 22–23.

23. See Capra, *The Science of Leonardo*, p. 33.

24. See Laurenza, *Leonardo: L'anatomia*, p. 33.

25. See Capra, *The Science of Leonardo*, pp. 81ff.

26. See Laurenza, *Leonardo: L'anatomia*, p. 136.

27. See Capra, *The Science of Leonardo*, pp. 153ff.

28. See Laurenza, *Leonardo: L'anatomia*, p. 60.

29. Windsor Collection, *Anatomical Studies*, folio 139v.

30. Laurenza, *Leonardo: L'anatomia*, p. 79; my own translation.

31. See Capra, *The Science of Leonardo*, pp. 141ff.

32. Windsor Collection, *Anatomical Studies*, folio 139v.

33. See Laurenza, *Leonardo: L'anatomia*, p. 133.

34. Windsor Collection, *Anatomical Studies*, folio 154r.

35. Windsor Collection, *Anatomical Studies*, folio 154r.

36. Windsor Collection, *Anatomical Studies*, folio 113r.

37. Ibid.

38. Codex Arundel, folio 115r.

39. Windsor Collection, *Anatomical Studies*, folio 69v.

40. Quoted by Laurenza, *Leonardo: L'anatomia*, p. 127.

41. Laurenza, *Leonardo: L'anatomia*, pp. 45 and 48; my own translation.

42. Laurenza, *Leonardo: L'anatomia*; Keele, *Leonardo da Vinci's Elements of the Science of Man*; O'Malley and Saunders, *Leonardo da Vinci on the Human Body*.

43. Keele, *Leonardo da Vinci's Elements of the Science of Man*, p. 251.

44. Ibid., p. 256.

45. Ibid., pp. 156ff.

46. Windsor Collection, *Anatomical Studies*, folio 143v.

47. Windsor Collection, *Anatomical Studies*, folio 154r.

48. See Keele, *Leonardo da Vinci's Elements of the Science of Man*, p. 266.

49. Windsor Collection, *Anatomical Studies*, folio 113r.

50. Windsor Collection, *Anatomical Studies*, folio 154r.

51. Keele, *Leonardo da Vinci's Elements of the Science of Man*, p. 267.

52. Windsor Collection, *Anatomical Studies*, folio 177r.

53. Windsor Collection, *Anatomical Studies*, folio 175v.

54. Windsor Collection, *Anatomical Studies*, folio 179v.

55. Nuland, *Leonardo da Vinci*, p. 137.

56. Windsor Collection, *Anatomical Studies*, folio 154r.

57. See Keele, *Leonardo da Vinci's Elements of the Science of Man*, p. 275.

58. Windsor Collection, *Anatomical Studies*, folio 151r.

59. Windsor Collection, *Anatomical Studies*, folio 143r.

60. Windsor Collection, *Anatomical Studies*, folio 152r.

61. O'Malley and Saunders, *Leonardo da Vinci on the Human Body*, pp. 186ff.

62. Windsor Collection, *Anatomical Studies*, folio 140v.

63. See O'Malley and Saunders, *Leonardo da Vinci on the Human Body*, pp. 132ff.

64. See ibid., p. 140, for a detailed analysis of these drawings.

65. Windsor Collection, *Anatomical Studies*, folio 135v.

66. See O'Malley and Saunders, *Leonardo da Vinci on the Human Body*, p. 188.

67. Ibid.

68. Codex Arundel, folios 151r,v.

69. Capra, *The Science of Leonardo*, pp. 243ff.

70. See ibid., pp. 249ff.

71. Windsor Collection, *Anatomical Studies*, folio 39r.

72. Windsor Collection, *Anatomical Studies*, folio 76v.

73. Ibid.

74. See O'Malley and Saunders, *Leonardo da Vinci on the Human Body*, p. 378, for identifications of the detailed anatomical features of this study.

75. Keele, *Leonardo da Vinci's Elements of the Science of Man*, p. 249.

Chapter 7. The Science of Flight

1. See Capra, *The Science of Leonardo*, p. 21.

2. Laurenza, *Leonardo: L'anatomia*, p. 78; my own translation.

3. Windsor Collection, *Anatomical Studies*, folio 89r.

4. See Capra, *The Science of Leonardo*, p. 50.

5. See O'Malley and Saunders, *Leonardo da Vinci on the Human Body*, pp. 204ff.

6. Ms. G, folio 5v.

7. See Capra, *The Science of Leonardo*, p. 34.

8. See Richards, *The Romantic Conception of Life*, pp. 8ff.

9. Ibid., p. 533.

10. See Capra, *The Web of Life*, pp. 48–49.

11. Codex Arundel, folio 156v.

12. Codex Atlanticus, folio 815r.

13. See Laurenza, *Leonardo on Flight*, pp. 17–18.

14. See Arasse, *Leonardo da Vinci*, pp. 294–95.

15. For a detailed account of Leonardo's studies of flight, see Laurenza, *Leonardo on Flight*.

16. See Capra, *The Science of Leonardo*, pp. 90ff.

17. Giacomelli, "The Aerodynamics of Leonardo da Vinci."

18. Codex Atlanticus, folio 434r.

19. Codex Trivulzianus, folio 6v.

20. Codex Atlanticus, folio 1058v.

21. See Laurenza, *Leonardo on Flight*, p. 38.

22. Ibid., p. 44.

23. Ms. B, folio 83v.

24. See Reti, "Helicopters and Whirligigs."

25. For a detailed analysis of the dynamics of this machine, see Laurenza, Taddei, and Zanon, *Le macchine di Leonardo*, pp. 55ff.

26. Ms. B, folio 74v.

27. See Kemp and Roberts, *Leonardo da Vinci*, pp. 236ff.

28. See Capra, *The Science of Leonardo*, pp. 105ff.

29. Codex *Sul Volo*, folio 10v.

30. See Moin and Kim, "Tackling Turbulence with Supercomputers."

31. See, for example, Alexander, *Nature's Flyers*.

32. See Alexander, *Nature's Flyers*, pp. 36ff.

33. Ibid., pp. 72ff.

34. Ms. E, folio 54r.

35. Codex Atlanticus, folio 493r.

36. Codex Atlanticus, folio 77r.

37. Ms. E, folio 45v.

38. Ms. G, folio 50v.

39. See Codex Arundel, 96v.

40. Ms. K, folio 3r.

41. See Marinoni, "Introduction."

42. See, for example, Burton, *Bird Flight*, p. 28.

43. Codex *Sul Volo*, folio 18r.

44. Ms. F, folio 41v.

45. Codex *Sul Volo*, folio 14v.

46. Ms. K, folio 9r.

47. Anderson, *A History of Aerodynamics*, p. 25.

48. Codex Atlanticus, 45r.

49. Codex Atlanticus, folios 825r, 843r, 1030r.

50. Codex *Sul Volo*, folio 16r.

51. Ibid.

52. Codex Atlanticus, folio 434r.

53. Codex *Sul Volo*, inside back cover.

54. Codex *Sul Volo*, folio 18v.

55. Codex *Sul Volo*, folio 9r.

56. Codex *Sul Volo*, folios 8v, 8r, 7v.

57. Ms. G, folio 64r.

58. Codex *Sul Volo*, folio 6v.
59. See Capra, *The Science of Leonardo*, pp. 119ff.
60. Ibid., p. ix.
61. Ms. E, folio 54r.
62. Ms. E, folios 43r and 47v.
63. Ms. E, folio 38r; see also Codex Atlanticus, folio 1058v, quoted on p. 253 above.
64. Anderson, *A History of Aerodynamics*, p. 20.
65. Martin Kemp in Kemp and Roberts, *Leonardo da Vinci*, p. 239.

Chapter 8. *The Mystery of Life*

1. Ms. C, folio 26v.
2. Codex Leicester, folio 34r.
3. Ms. H, folio 89v; Codex Atlanticus, folio 207v.
4. See Capra, *The Web of Life*, p. 34.
5. Ms. G, folio 32v.
6. Ms. G, folio 35v.
7. Codex Atlanticus, 728r.
8. See, for example, Grmek, *Western Medical Thought*.
9. See Keele, *Leonardo da Vinci's Elements of the Science of Man*, pp. 304–5.
10. See Keele, *Leonardo da Vinci on Movement of the Heart and Blood*, pp. 53ff.
11. See, for example, Windelband, *A History of Philosophy*, pp. 186–87.
12. For a detailed analysis of the anatomical features of these drawings, see O'Malley and Saunders, *Leonardo da Vinci on the Human Body*, pp. 216 and 218.
13. Ms. G, folio 1v.
14. Windsor Collection, *Anatomical Studies*, folio 59v.
15. Windsor Collection, *Anatomical Studies*, folio 105r.
16. Ms. G, folio 1v.
17. Windsor Collection, *Anatomical Studies*, folio 155r.
18. Windsor Collection, *Anatomical Studies*, folio 174v.
19. See Keele, *Leonardo da Vinci's Elements of the Science of Man*, pp. 310ff.
20. Windsor Collection, *Anatomical Studies*, folio 165v.
21. See Schott, "Historical Notes."
22. Windsor Collection, *Anatomical Studies*, folio 50r.
23. Windsor Collection, *Anatomical Studies*, folio 115r.
24. See Keele, *William Harvey*, p. 127.
25. See Keele, *Leonardo da Vinci on Movement of the Heart and Blood*, p. 128; Harvey apparently performed the same experiment on living dogs a century later; see Keele, *William Harvey*.
26. For a detailed discussion of Leonardo's studies of the physiology of respiration (the roles of the bones and joints of the spine and ribs in the mechanics of breathing, the muscle forces that dilate the lungs, and the anatomy of the

trachea and lung), see Keele, *Leonardo da Vinci's Elements of the Science of Man*, pp. 289ff.

27. Windsor Collection, *Anatomical Studies*, folio 155r.
28. See Keele, *William Harvey*, p. 122.
29. Windsor Collection, *Anatomical Studies*, folio 156v.
30. See Keele, *William Harvey*, p. 146.
31. See Capra, *The Science of Leonardo*, p. 254.
32. Ibid., pp. 244ff.
33. Windsor Collection, *Anatomical Studies*, folio 164r.
34. Windsor Collection, *Anatomical Studies*, folio 156v.
35. Windsor Collection, *Anatomical Studies*, folio 115r.
36. Windsor Collection, *Anatomical Studies*, folios 115r, 115v, 171r, 172v.
37. Windsor Collection, *Anatomical Studies*, folio 172v.
38. Ibid.
39. Windsor Collection, *Anatomical Studies*, 171r; for Leonardo's earlier use of this technique, see Capra, *The Science of Leonardo*, p. 163.
40. See Gharib, et al., "Leonardo's Vision of Flow Visualization."
41. Nuland, *Leonardo da Vinci*, pp. 146ff.
42. O'Malley and Saunders, *Leonardo da Vinci on the Human Body*, p. 28.
43. Keele, *Leonardo da Vinci on Movement of the Heart and Blood*, p. 73.
44. For a detailed analysis of the anatomical features of these drawings, see O'Malley and Saunders, *Leonardo da Vinci on the Human Body*, p. 304.
45. Ibid.
46. Windsor Collection, *Anatomical Studies*, folio 69r.
47. Windsor Collection, *Anatomical Studies*, folio 69v.
48. O'Malley and Saunders, *Leonardo da Vinci on the Human Body*, p. 300.
49. See Leibowitz, *The History of Coronary Heart Disease*, pp. 7–8.
50. See Keele, *Leonardo da Vinci's Elements of the Science of Man*, pp. 327ff., for a detailed analysis of Leonardo's studies of the digestive system.
51. Windsor Collection, *Anatomical Studies*, 158v.
52. Windsor Collection, *Anatomical Studies*, folio 39v.
53. Windsor Collection, *Anatomical Studies*, folio 50r.
54. Windsor Collection, *Anatomical Studies*, folio 164r.
55. Windsor Collection, *Anatomical Studies*, folio 50r.
56. Quoted in Capra, *The Web of Life*, p. 48.
57. See ibid., pp. 86ff.
58. Keele, *Leonardo da Vinci on Movement of the Heart and Blood*, p. 122.
59. See Laurenza, *Leonardo: L'anatomia*, pp. 164–65.
60. Windsor Collection, *Anatomical Studies*, folio 198v.
61. See O'Malley and Saunders, *Leonardo da Vinci on the Human Body*, pp. 482ff.
62. See Keele, *Leonardo da Vinci's Elements of the Science of Man*, p. 357.
63. Windsor Collection, *Anatomical Studies*, folio 196v.

64. Keele, *Leonardo da Vinci's Elements of the Science of Man*, p. 362.

65. Nuland, *Leonardo da Vinci*, p. 161.

66. Windsor Collection, *Anatomical Studies*, folio 197r.

67. See Needham, *A History of Embryology*, p. 80.

68. Windsor Collection, *Anatomical Studies*, folio 198r.

69. Windsor Collection, *Anatomical Studies*, folio 197r.

70. Needham, *A History of Embryology*, p. 81.

71. Windsor Collection, *Anatomical Studies*, folio 198r.

72. Windsor Collection, *Anatomical Studies*, folio 114v.

73. See Laurenza, *Leonardo: L'anatomia*, pp. 160ff.

74. Capra, ibid., pp. 249ff.

75. See Capra, *The Science of Leonardo*, pp. 124ff.

76. Arasse, *Leonardo da Vinci*, p. 394.

77. Kemp, *Leonardo da Vinci*, p. 265.

78. Keele, "The Genesis of Mona Lisa."

79. Keele, *Leonardo da Vinci's Elements of the Science of Man*, p. 35.

80. Ms. I, folio 18r.

81. Clark, *Leonardo da Vinci*, p. 250.

Coda. Leonardo's Legacy

1. *Trattato*, chapter 80.

2. Ms. I, folio 15r.

3. Windsor Collection, *Anatomical Studies*, folio 114v.

4. See Capra, *The Hidden Connections*, pp. 229ff.

5. See Capra, ibid.

6. See Capra, *The Science of Leonardo*, pp. 57ff.

7. Codex Leicester, folio 13r.

LEONARDO'S NOTEBOOKS

Facsimiles and Transcriptions

Windsor Collection, *Anatomical Studies*
Keele, Kenneth, and Carlo Pedretti. *Leonardo da Vinci: Corpus of the Anatomical Studies in the Collection of Her Majesty the Queen at Windsor Castle.* 3 vols. New York: Harcourt Brace Jovanovich, 1978–80.

Windsor Collection, *Drawings and Miscellaneous Papers*
Pedretti, Carlo. *The Drawings and Miscellaneous Papers of Leonardo da Vinci in the Collection of Her Majesty the Queen at Windsor Castle.* Vol. I: *Landscapes, Plants, and Water Studies.* New York: Harcourt Brace Jovanovich, 1982. Volume II: *Horses and Other Animals.* New York: Harcourt Brace Jovanovich, 1987.

Complete edition to comprise four volumes; Vol. III (*Figure Studies, Profiles, and Caricatures*) and Vol. IV (*Miscellaneous Papers*) not yet published in facsimile; folios in these volumes are identified in the footnotes of this book by their Royal Library (RL) catalogue numbers.

Codex Arundel
Leonardo da Vinci. *Il Codice Arundel 263 nella British Library: edizione in facsimile nel riordinamento cronologico dei suoi fascicoli; a cura di Carlo Pedretti; trascrizioni e note critiche a cura di Carlo Vecce.* Florence: Giunti, 1998.

Codex Atlanticus
Leonardo da Vinci. *Il codice atlantico della Biblioteca ambrosiana di Milano: trascrizione diplomatica e critica di Augusto Marinoni.* Florence: Giunti, 1975–80.

Codex Sul Volo
Leonardo da Vinci. *The Codex on the Flight of Birds in the Royal Library at Turin.* Edited by Augusto Marinoni. New York: Johnson Reprint, 1982.

Codices Forster I, II, III

Leonardo da Vinci. *I Codici Forster del Victoria and Albert Museum di Londra; trascrizione diplomatica e critica di Augusto Marinoni, edizione in facsimile.* 3 vols. Florence: Giunti, 1992.

Codex Leicester (formerly "Codex Hammer")

Leonardo da Vinci. *The Codex Hammer.* Translated into English and annotated by Carlo Pedretti. Florence: Giunti, 1987.

Madrid Codices I, II

Leonardo da Vinci. *The Madrid Codices.* Transcribed and translated by Ladislao Reti. New York: McGraw-Hill, 1974.

Codex Trivulzianus

Leonardo da Vinci. *Il codice di Leonardo da Vinci nella Biblioteca trivulziana di Milano; trascrizione diplomatica e critica di Anna Maria Brizio.* Florence: Giunti, 1980.

Manuscripts at Institut de France

Leonardo da Vinci. *I manoscritti dell'Institut de France, edizione in facsimile sotto gli auspici della Commissione nazionale vinciana e dell'Institut de France; trascrizione diplomatica e critica di Augusto Marinoni.* Florence: Giunti, 1986–90.

Mss. A, B, C, D, E, F, G, H, I, K, L, M; ms. A includes as a supplement *Ashburnham II*, also listed as B.N. 2038; ms. B includes as a supplement *Ashburnham I*, also listed as B.N. 2037.

Trattato della pittura (Codex Urbinas)

Leonardo da Vinci. *Libro di pittura, Codice urbinate lat. 1270 nella Biblioteca apostolica vaticana, a cura di Carlo Pedretti; trascrizione critica di Carlo Vecce.* Florence: Giunti, 1995.

Venetian Folio (The Vitruvian Man)

Leonardo da Vinci. *I disegni di Leonardo da Vinci e della sua cerchia.* Vol. IV: *Venezia, Galleria dell'Accademia, Catalogo a cura di Giovanna Nepi Sciré e Annalisa Perissa Torrini.* Florence: Giunti, 2003.

BIBLIOGRAPHY

Ackerman, James. "Science and Art in the Work of Leonardo." In *Leonardo's Legacy: An International Symposium,* edited by C. D. O'Malley, pp. 205–25. Berkeley & Los Angeles: University of California Press, 1969.

Alexander, David. *Nature's Flyers.* Baltimore: Johns Hopkins University Press, 2002.

Ames-Lewis, Francis. "Leonardo's Botanical Drawings." In *Leonardo's Science and Technology,* edited by Claire Farago, pp. 275–82. New York: Garland Publishing, 1999.

Anderson, John D., Jr. *A History of Aerodynamics: And Its Impact on Flying Machines.* Cambridge: Cambridge University Press, 1997.

Arasse, Daniel. *Leonardo da Vinci: The Rhythm of the World.* New York: Konecky & Konecky, 1998.

Bedini, Silvio A., and Ladislao Reti. "Horology." In *The Unknown Leonardo,* edited by Ladislao Reti, pp. 240–63. New York: McGraw-Hill, 1974.

Bertoloni-Meli, Domenico. *Thinking with Objects: The Transformation of Mechanics in the Seventeenth Century.* Baltimore: Johns Hopkins University Press, 2006.

Bottazzi, Filippo. "Leonardo as Physiologist." In *Leonardo da Vinci,* by Reynal & Company, pp. 373–88. Novara, Italy: De Agostini, 1956.

Bramly, Serge. *Leonardo.* New York: HarperCollins, 1991.

Brown, David Alan. "Verrocchio and Leonardo: Studies for the *Giostra*." In *Florentine Drawing at the Time of Lorenzo the Magnificent,* edited by Elizabeth Cropper, pp. 99–109. Bologna: Nuova Alfa Editoriale, 1994.

Burton, Robert. *Bird Flight.* New York: Facts on File, 1990.

Capra, Fritjof. *La botanica di Leonardo: un discorso sulla scienza delle qualità.* Sansepolero, Italy: Aboca Edizioni, 2008.

———. *The Hidden Connections.* New York: Doubleday, 2002.

———. *The Science of Leonardo.* New York: Doubleday, 2007.

———. *The Turning Point.* New York: Simon & Schuster, 1982.

———. *The Web of Life.* New York: Doubleday, 1996.

Clagett, Marshall. "Leonardo da Vinci: Mechanics." In *Leonardo's Science and Technology,* edited by Claire Farago, pp. 1-20. New York: Garland Publishing, 1999.

Clark, Kenneth. *Civilisation* (prepared for the author's television series of the same name). New York: Harper & Row, 1969.

———. *Leonardo da Vinci*. London: Penguin, 1993.

Davidson, P. A. *Turbulence*. Oxford: Oxford University Press, 2004.

Dibner, Bern. "Leonardo: Prophet of Automation." In *Leonardo's Legacy: An International Symposium*, edited by C. D. O'Malley, pp. 101–23. Berkeley & Los Angeles: University of California Press, 1969.

——— "Machines and Weaponry." In *The Unknown Leonardo*, edited by Ladislao Reti, pp. 166–89. New York: McGraw-Hill, 1974.

Ellenberger, François. *History of Geology*. 2 vols. New Delhi: Oxford & IBH, 1996.

Eloy, Christophe. "Leonardo's Rule, Self-Similarity, and Wind-Induced Stresses in Trees." *Physical Review Letters* 107 (16 December 2011). Epub 2011, 258101.

Emboden, William. *Leonardo da Vinci on Plants and Gardens*. Portland, Ore.: Dioscorides Press, 1987.

Fasso, Constantino. "Birth of Hydraulics During the Renaissance Period." In *Hydraulics and Hydraulic Research: A Historical Review*, edited by Günther Garbrecht, pp. 55–79. Boston: Balkema, 1987.

Galluzzi, Paolo, ed. *Léonard de Vinci: ingénieur et architecte*. Montreal: Musée des beaux-arts de Montréal, 1987.

———. *Renaissance Engineers: From Brunelleschi to Leonardo da Vinci*. Florence: Giunti, 1996.

Gharib, M., D. Kremers, M. M. Koochesfahani, and M. Kemp. "Leonardo's Vision of Flow Visualization." *Experiments in Fluids* 33 (2002): 219–23.

Giacomelli, Raffaele. "The Aerodynamics of Leonardo da Vinci." *Journal of the Royal Aeronautical Society* 34 (1930): 1016–38.

———. "La scienza dei venti di Leonardo da Vinci." In *Atti del Convegno di Studi Vinciani*, pp. 374–98. Florence: Olschki, 1953.

Gohau, Gabriel. *A History of Geology*. New Brunswick, N. J.: Rutgers University Press, 1991.

Gould, Stephen Jay. "The Upwardly Mobile Fossils of Leonardo's Living Earth." In *Leonardo's Mountain of Clams and the Diet of Worms*, pp. 17–44. New York: Harmony Books, 1998.

Grmek, Mirko, ed. *Western Medical Thought from Antiquity to the Middle Ages*. Cambridge, Mass.: Harvard University Press, 1998.

Keele, Kenneth, "The Genesis of Mona Lisa." *Journal of the History of Medicine and Allied Sciences* 14 (1959): 135–59.

———. *Leonardo da Vinci on Movement of the Heart and Blood*. Philadelphia: Lippincott, 1952.

———. *Leonardo da Vinci's Elements of the Science of Man*. New York: Academic Press, 1983.

———. *William Harvey*. London: Nelson, 1965.

Kemp, Martin. "Analogy and Observation in the Codex Hammer." In *Leonardo's Science and Technology*, edited by Claire Farago, pp. 345–75. New York: Garland Publishing, 1999.

———. "Body of Earth and Body of Man." In *Leonardo da Vinci: Artist, Scientist, Inventor*, edited by Martin Kemp and Jane Roberts, pp. 104–17. Exhibition catalogue. New Haven, Conn.: Yale University Press, 1989.

———. "The Crisis of Received Wisdom in Leonardo's Late Thought." In *Leonardo e l'Età della Ragione*, edited by E. Bellone and P. Rossi, pp. 27–42. Milan: Scientia, 1982.

———. *Leonardo da Vinci: The Marvellous Works of Nature and Man*. Cambridge, Mass.: Harvard University Press, 1981.

Kemp, Martin, and Jane Roberts, eds. *Leonardo da Vinci: Artist, Scientist, Inventor*. Exhibition catalogue. New Haven, Conn.: Yale University Press, 1989.

Kline, Morris. *Mathematical Thought from Ancient to Modern Times*. New York: Oxford University Press, 1972.

Laurenza, Domenico. "Le forme come matrici universali." In *La mente di Leonardo*, edited by Paolo Galluzzi, pp. 188–89. Exhibition catalogue. Florence: Giunti, 2006.

———. "La grammatica delle forme." In *La mente di Leonardo*, edited by Paolo Galluzzi, pp. 152–57. Exhibition catalogue. Florence: Giunti, 2006.

———. *Leonardo: L'anatomia*, Florence: Giunti, 2009.

———. *Leonardo on Flight*. Florence: Giunti, 2004.

———. "Moti di 'consumazione'." In *La mente di Leonardo*, edited by Paolo Galluzzi, pp. 284–85. Exhibition catalogue. Florence: Giunti, 2006.

———. *La ricerca dell'armonia: rappresentazioni anatomiche del Rinascimento*. Florence: Olschki, 2003.

———. "L'uomo geometrico." In *La mente di Leonardo*, edited by Paolo Galluzzi, p. 158. Exhibition catalogue. Florence: Giunti, 2006.

Laurenza, Domenico, Mario Taddei, and Edoardo Zanon. *Le macchine di Leonardo*. Florence: Giunti, 2005.

Leibowitz, Joshua. *The History of Coronary Heart Disease*. Berkeley & Los Angeles: University of California Press, 1970.

Livio, Mario. *The Golden Ratio*. New York: Broadway Books, 2002.

Lovelock, James. *Healing Gaia*. New York: Harmony Books, 1991.

Macagno, Enzo. *Leonardian Fluid Mechanics in the Codex Atlanticus*. Iowa City: Iowa Institute of Hydraulic Research, University of Iowa, 1986.

———. "Leonardo da Vinci: Engineer and Scientist." In *Hydraulics and Hydraulic Research: A Historical Review*, edited by Günther Garbrecht, pp. 33–53. Boston: Balkema, 1987.

———. "Mechanics of Fluids in the Madrid Codices." In *Leonardo e l'Età della Ragione*, edited by E. Belloni and P. Rossi, pp. 333–74. Milan: Scientia, 1982.

Maccagni, Carlo. "The Italian School of Hydraulics in the 16th and 17th Centuries." In *Hydraulics and Hydraulic Research: A Historical Review*, edited by Günther Garbrecht, pp. 81–88. Boston: Balkema, 1987.

Marinoni, Augusto. "Introduction." In *The Codex on the Flight of Birds in the Royal Library at Turin*, by Leonardo da Vinci. Facsimile edition. Edited by Augusto Marioni. New York: Johnson Reprint, 1982.

Merzkirch, Wolfgang. *Flow Visualization*. 2nd ed. New York: Academic Press, 1987.

Moin, Parviz, and John Kim. "Tackling Turbulence with Supercomputers." *Scientific American* 276, no. 1 (January 1997): 62–68.

Moody, Ernest, and Marshall Clagett, eds. *The Medieval Science of Weights*. Madison: University of Wisconsin Press, 1952.

Murray, Penelope, ed. *Genius: The History of an Idea*. New York: Basil Blackwell, 1989.

Needham, Joseph. *A History of Embryology*. London: Cambridge University Press, 1934.

Nuland, Sherwin B. *Leonardo da Vinci*. New York: Viking Penguin, 2000.

O'Malley, C. D., and J. B. Saunders. *Leonardo da Vinci on the Human Body*. New York: Crown, 1982; Gramercy Books, 2003.

Orr, David. *Hope Is an Imperative: The Essential David Orr*. Washington, D.C.: Island Press, 2011.

Pedretti, Carlo. *Leonardo: The Machines*. Florence: Giunti, 1999.

Pizzorusso, Ann. "Leonardo's Geology." *Leonardo* 29, no. 3 (1996): 197–200.

Reti, Ladislao. "Elements of Machines." In *The Unknown Leonardo*, edited by Ladislao Reti, 264–87. New York: McGraw-Hill, 1974.

———. "Helicopters and Whirligigs." *Raccolta Vinciana* 20 (1964): 331–38.

———. "Leonardo on Bearings and Gears." *Scientific American* 224, no. 2 (February 1971): 100–110.

Richards, Robert. *The Romantic Conception of Life*. Chicago: University of Chicago Press, 2002.

Richter, J.-P. *The Notebooks of Leonardo da Vinci Compiled and Edited from the Original Manuscripts*. 2 vols. Originally published in 1883. Reprint, New York: Dover Publications, 1970.

Roberts, Jane. "The Drawings and Manuscripts." In *Leonardo da Vinci: Artist, Scientist, Inventor*, edited by Martin Kemp and Jane Roberts, pp. 17–22. Exhibition Catalogue. New Haven, Conn.: Yale University Press, 1989.

Sabersky, Rolf H., et al. *Fluid Flow: A First Course in Fluid Mechanics*. Englewood Cliffs, N.J.: Prentice-Hall, 1999.

Saunders, J. B., and Charles D. O'Malley. *The Illustrations from the Works of Andreas Vesalius*. Cleveland: World Publishing, 1950; paperback ed. New York: Dover, 1973.

Schott, A. "Historical Notes on the Mechanism of Closure of the Atrioventricular Valves." *Medical History* 24 (1980): 163–84.

Steptoe, Andrew, ed. *Genius and the Mind*. New York: Oxford University Press, 1998.

Wheatley, Margaret. *Leadership and the New Science*. 2nd ed. San Francisco: Berrett-Koehler, 1999.

Wicander, Reed, and James Monroe. *Essentials of Geology*. New York: West Publishing, 1995.

Windelband, Wilhelm. *A History of Philosophy*. Reprint, Cresskill, N.J.: The Paper Tiger, 2001. Originally published 1901 by Macmillan.

Zammattio, Carlo. "The Mechanics of Water and Stone." In *The Unknown Leonardo*, edited by Ladislao Reti, 190–215. New York: McGraw-Hill, 1974.

Zöllner, Frank, and Johannes Nathan. *Leonardo da Vinci: The Complete Paintings and Drawings*. London: Taschen, 2003.

RESOURCES FOR LEONARDO SCHOLARSHIP

If you are interested in learning more about Leonardo, please visit http://www.bkconnection.com/Leonardo, where you can learn how to access his original manuscripts and further explore his fascinating work.

PHOTO CREDITS

and to Siggy Zerweckh for inspiring discussions of the history of aerodynamics, and for his critical reading of my chapter on "The Science of Flight."

I am also very grateful to Carlo Pedretti for his continuing encouragement and support, and to Linda Warren, former head librarian of the Elmer Belt Library, for giving me unrestricted access to its collections of the complete facsimile editions of Leonardo's manuscripts, as well as for her generous help with bibliographical research.

I am grateful to Françoise Viatte for her support and for helpful discussions of Leonardo's *Virgin of the Rocks* and his manuscripts in the Bibliothèque de l'Institut de France.

I am indebted to Satish Kumar and Inga Page for giving me the opportunity to teach a course on Leonardo's synthesis of art and science at Schumacher College in England during the spring of 2010, to Peter Adams for co-teaching the course with me, and to the course participants for many critical questions and helpful suggestions.

I am deeply grateful to my brother, Bernt Capra, for reading the entire manuscript, and for his enthusiastic support and numerous helpful suggestions. I am also very grateful to Ernest Callenbach and to my daughter, Juliette Capra, for reading portions of the manuscript and offering many critical comments; and to Borys Czernichowski for his excellent rendering and prompt delivery of the technical drawings in this book.

My special thanks go to my assistant, Trena Cleland, for her careful and sensitive editing of the first draft of the manuscript, and for skillfully managing the flow of communications into my home office while I was concentrating on my writing.

I am grateful to Steve Piersanti, Jeevan Sivasubramaniam, and the entire Berrett-Koehler team for their enthusiastic support; to Sara Galinetto and Jordi Pigem for their help in communicating with various museums and libraries to obtain reproductions of Leonardo's works and the permissions to use them as illustrations in this book; and to Christine Taylor and her team at Wilsted & Taylor for seeing the manuscript through the publishing process.

Last but not least, I wish to express my deep gratitude to my wife, Elizabeth Hawk, for many discussions about Renaissance art and culture, for suggesting and helping me design the timeline of scientific discoveries, and for her patience and unwavering support during many years of research and writing.

PHOTO CREDITS

INDEX

*Page references given in **bold** indicate illustrations.*

ABOUT THE AUTHOR

Fritjof Capra, Ph.D., physicist and systems theorist, is a founding director of the Center for Ecoliteracy in Berkeley, California, which is dedicated to promoting ecology and systems thinking in primary and secondary education. He serves on the faculty of Schumacher College, an international center for ecological studies in the United Kingdom.

Author photo: Basso Cannarsa

After receiving his Ph.D. in theoretical physics from the University of Vienna in 1966, Capra did research in particle physics at the University of Paris (1966–68), the University of California at Santa Cruz (1968–70), the Stanford Linear Accelerator Center (1970), Imperial College, University of London (1971–74), and the Lawrence Berkeley Laboratory at the University of California (1975–88).

In addition to his research in physics and systems theory, Capra has been engaged in a systematic examination of the philosophical and social implications of contemporary science for the past forty years. His books on this subject have been acclaimed internationally, and he has lectured widely to lay and professional audiences in Europe, Asia, and North and South America.

Capra is the author of several international bestsellers, including *The Tao of Physics* (1975), *The Turning Point* (1982), *The Web of Life* (1996), *The Hidden Connections* (2002), and *The Science of Leonardo* (2007). He

379

has been the focus of more than fifty television interviews, documentaries, and talk shows in Europe, the United States, Brazil, Argentina, and Japan, and has been featured in major newspapers and magazines internationally. He was the first subject of the BBC's documentary series *Beautiful Minds* (2002).

Capra holds an Honorary Doctor of Science degree from the University of Plymouth and is the recipient of many other awards, including the Gold Medal of the UK Systems Society, the Neil Postman Award for Career Achievement in Public Intellectual Activity from the Media Ecology Association, the Medal of the Presidency of the Italian Republic, the Leonardo da Vinci Medallion of Honor from the University of Advancing Technology in Tempe, Arizona, the Bioneers Award, the New Dimensions Broadcaster Award, and the American Book Award.

Fritjof Capra lives in Berkeley with his wife and daughter.

www.fritjofcapra.net

Berrett–Koehler
Publishers

Berrett-Koehler is an independent publisher dedicated to an ambitious mission: *Creating a World That Works for All*.

We believe that to truly create a better world, action is needed at all levels—individual, organizational, and societal. At the individual level, our publications help people align their lives with their values and with their aspirations for a better world. At the organizational level, our publications promote progressive leadership and management practices, socially responsible approaches to business, and humane and effective organizations. At the societal level, our publications advance social and economic justice, shared prosperity, sustainability, and new solutions to national and global issues.

A major theme of our publications is "Opening Up New Space." Berrett-Koehler titles challenge conventional thinking, introduce new ideas, and foster positive change. Their common quest is changing the underlying beliefs, mindsets, institutions, and structures that keep generating the same cycles of problems, no matter who our leaders are or what improvement programs we adopt.

We strive to practice what we preach—to operate our publishing company in line with the ideas in our books. At the core of our approach is stewardship, which we define as a deep sense of responsibility to administer the company for the benefit of all of our "stakeholder" groups: authors, customers, employees, investors, service providers, and the communities and environment around us.

We are grateful to the thousands of readers, authors, and other friends of the company who consider themselves to be part of the "BK Community." We hope that you, too, will join us in our mission.

A BK Currents Book

This book is part of our BK Currents series. BK Currents books advance social and economic justice by exploring the critical intersections between business and society. Offering a unique combination of thoughtful analysis and progressive alternatives, BK Currents books promote positive change at the national and global levels. To find out more, visit **www.bkconnection.com**.

Berrett–Koehler
Publishers

A community dedicated to creating
a world that works for all

Dear Reader,

Thank you for picking up this book and joining our worldwide community of Berrett-Koehler readers. We share ideas that bring positive change into people's lives, organizations, and society.

To welcome you, we'd like to offer you a free e-book. You can pick from among twelve of our bestselling books by entering the promotional code **BKP92E** here: http://www.bkconnection.com/welcome.

When you claim your free e-book, we'll also send you a copy of our e-newsletter, the *BK Communiqué*. Although you're free to unsubscribe, there are many benefits to sticking around. In every issue of our newsletter you'll find

- A free e-book
- Tips from famous authors
- Discounts on spotlight titles
- Hilarious insider publishing news
- A chance to win a prize for answering a riddle

Best of all, our readers tell us, "Your newsletter is the only one I actually read." So claim your gift today, and please stay in touch!

Sincerely,

Charlotte Ashlock
Steward of the BK Website

Questions? Comments? Contact me at bkcommunity@bkpub.com.

Certified

Corporation
bcorporation.net